Microplastics in Aquatic Environments: Occurrence, Distribution and Effects

Microplastics in Aquatic Environments: Occurrence, Distribution and Effects

Editors

Costanza Scopetani
Tania Martellini
Diana Campos

MDPI • Basel • Beijing • Wuhan • Barcelona • Belgrade • Manchester • Tokyo • Cluj • Tianjin

Editors
Costanza Scopetani
University of Helsinki
Finland

Tania Martellini
University of Florence
Italy

Diana Campos
University of Aveiro
Portugal

Editorial Office
MDPI
St. Alban-Anlage 66
4052 Basel, Switzerland

This is a reprint of articles from the Special Issue published online in the open access journal *Toxics* (ISSN 2305-6304) (available at: https://www.mdpi.com/journal/toxics/special_issues/microplastics_aquatic).

For citation purposes, cite each article independently as indicated on the article page online and as indicated below:

LastName, A.A.; LastName, B.B.; LastName, C.C. Article Title. *Journal Name* **Year**, *Volume Number*, Page Range.

ISBN 978-3-0365-5047-3 (Hbk)
ISBN 978-3-0365-5048-0 (PDF)

Contents

About the Editors

Costanza Scopetani

Costanza Scopetani is a postdoctoral researcher in the Faculty of Biological and Environmental Sciences at the University of Helsinki. She has a Ph.D. and a M.A. in Chemical Sciences from the University of Florence. Her academic studies began with a B.S. in Technology for the Conservation and Restoration of Cultural Heritage at the University of Florence. Her research fields include environmental science and analytical chemistry with a comprehensive background in transport, distribution and accumulation mechanisms of organic compounds and microplastics in organisms and different environmental matrices. She is also a member of the editorial board of the journals *Environmental Protection Research* and *Frontiers in Environmental Science* (IF: 4.24).

Tania Martellini

Tania Martellini is an associate professor in Analytical Chemistry in the Department of Chemistry "Ugo Schiff", University of Florence (Italy). She holds a doctorate in chemical sciences and a MSc in chemistry with an analytical-environmental background. She has been conducting research for about 20 years in the environmental field with a special focus on emerging contaminants in all environmental matrices to study the sources, distribution and environmental fate of persistent organic contaminants and microplastics. She has participated in numerous national and international projects studying both temperate and remote areas and indoor environments. In particular, in 2009 she was part of the scientific staff at the Italian base Dirigibile Italia in Ny Alesund and took part in the XXVI Italian expedition to Antarctica with the project "Flows of POPs between poLar Abiotic and Biotic compartments (POP-LAB)" at the MZS station in Terranova Bay (Antarctica). She is the author of numerous publications, and serves as a referee and editor for several scientific journals.

Diana Campos

Diana Campos is a researcher in the Centre for Environmental and Marine Studies & Department of Biology, University of Aveiro, Portugal. She holds Ph.D. in Biology (2018, major in Ecotoxicology and Environmental Biology) and an M.Sc. in Applied Biology (2012), both from the University of Aveiro, Portugal. Her early research career has focused on evaluating the effects of emergent contaminants in the aquatic ecosystem under realistic exposure scenarios, more recently focusing on micro and nano plastics. Her multidisciplinarity can be seen by her expertise in using different aquatic invertebrate species and different levels of biological organisation including subcellular responses (e.g., biochemical markers), organismal endpoints and community and ecosystem level responses (e.g., using aquatic mesocosms).

Editorial

Editorial for the Special Issue "Microplastics in Aquatic Environments: Occurrence, Distribution and Effects"

Costanza Scopetani [1,*], Tania Martellini [2,3] and Diana Campos [4]

[1] Faculty of Biological and Environmental Sciences Ecosystems and Environment Research Programme, University of Helsinki, Niemenkatu 73, FI-15140 Lahti, Finland
[2] Department of Chemistry "Ugo Schiff", University of Florence, Via della Lastruccia 3—Sesto Fiorentino, 50019 Florence, Italy; tania.martellini@unifi.it
[3] Consorzio Interuniversitario per lo Sviluppo dei Sistemi a Grande Interfase (CSGI), University of Florence, Via della Lastruccia 3—Sesto Fiorentino, 50019 Florence, Italy
[4] CESAM—Centre for Environmental and Marine Studies, Department of Biology, University of Aveiro, 3810-193 Aveiro, Portugal; diana.campos@ua.pt
* Correspondence: costanza.scopetani@helsinki.fi

Citation: Scopetani, C.; Martellini, T.; Campos, D. Editorial for the Special Issue "Microplastics in Aquatic Environments: Occurrence, Distribution and Effects". *Toxics* **2022**, *10*, 407. https://doi.org/10.3390/toxics10070407

Received: 18 July 2022
Accepted: 19 July 2022
Published: 21 July 2022

Publisher's Note: MDPI stays neutral with regard to jurisdictional claims in published maps and institutional affiliations.

The large production and widespread daily consumption of plastic materials—which began in the last century—together with the often-inadequate collection and recycling systems, have made plastics and, consequently, microplastics (MPs) ubiquitous pollutants [1].

The scientific community is increasingly concerned about microplastic pollution and its possible effects on biota and the environment. Aquatic ecosystems such as rivers, lakes, estuaries, seas, and oceans seem to act as important sinks for plastics and microplastics. Microplastic pollution is so widespread that we might assume no aquatic environment has been left untouched [2–5].

Microplastic pollution as a global concern is confirmed by the research papers collected in this Special Issue; these papers come from 28 Universities and research institutions and are spread across ten countries in three continents.

The Special Issue "Microplastics in Aquatic Environments: Occurrence, Distribution and Effects" collected and published 11 novel contributions focusing on microplastics in aquatic environments, their occurrence and distribution, and the effects they might have on the environment and biota. The selected papers comprise three reviews and eight research articles. In their review, Yang et al. (2021) [6] summarized the current literature on MPs in the marine environment, focusing on the sources and fates of MPs and their impacts on marine organisms; moreover, they highlighted the potential of bacteria in plastic degradation processes and the need to further study this subject.

Santini et al. (2022) [7] addressed the occurrence of natural and synthetic microfibers in waters, sediments, and biota in the Mediterranean Sea, emphasizing the challenges in distinguishing natural fibers from plastics ones, and the need to further study the environmental impact of both.

Lim et al. (2022) [8] conducted a meta-analysis of the characterization of plastic ingested by fish on a global scale, and found that plastic fibers are the most-ingested items (70.6%). Additionally, the authors observed that polyethylene (15.7%) and polyester (11.6%) are the most abundant polymers found in fishes' digestive organs. In terms of size, the most frequently ingested plastics were small microplastics (<1 mm).

The eight selected research papers can be grouped into three main themes: (1) the effects of microplastic exposure to aquatic biota (rotifers, mussels, fish larvae, and microalgae), encompassing 55% of the published papers in this SI [9–14]; (2) the distribution and seasonal variation of microplastics in aquatic environments [15]; and (3) the contaminants associated with microplastics in freshwater environments [16].

As Guest Editors of this Special Issue, we were pleased to receive several papers concerning the interaction between microplastics and biota; despite a large number of peer-reviewed papers published on this research topic, there are still several gaps that need to be filled [17,18]. Zhang et al. (2022) [12], for instance, investigated the toxicity of fluorescent nano- and microplastics (80 nm and 8 µm) on grass carp embryos and larvae using scanning electron microscopy (SEM) and fluorescence imaging. Their results showed that nanoplastics accumulated in the chorion and did not penetrate the embryo's chorionic membrane. The larvae were prone not only to ingesting microplastics and expelling them with their feces, but also to ingesting the expelled microplastics again while feeding on their own excrement, re-accumulating the plastic particles in their oral cavities. Furthermore, the authors showed that microplastics around 1 µm in size could accumulate in the larvae's nasal cavities.

Drago and Weithoff (2021) [9] analyzed the fitness responses of two rotifer species, *Brachionus calyciflorus* and *Brachionus fernandoi*, when exposed to polystyrene (1-, 3-, 6-µm), polyamide microplastics (5–25 µm) and silica beads (3 µm, SiO_2). The results showed that 3-µm polystyrene had a significant effect on the population growth rate of both rotifer species, whereas no effect was evidenced after exposure to polyamide microplastics and silica beads.

In another study, von Hellfeld and co-authors (2022) evaluated the toxicity of polystyrene MPs in marine mussels *Mytilus galloprovincialis* when exposed to two different polystyrene microplastic sizes (45 µm and 4.5 µm) [10]. The exposure was carried out with pristine and contaminated microplastics, with cadmium (Cd) and benzo(a)pyrene (BaP). The pristine microplastics (both tested sizes) were found in the digestive gland after 1 day of exposure, while after 3 days of depuration, 4.5 µm microplastics had accumulated within the gill filaments. In contrast to Cd, BaP body burdens increased significantly in mussels exposed to BaP-contaminated microplastics, causing histological changes in the digestive gland. These results show that polystyrene microplastics can act as a carrier of organic contaminants and pose a threat to aquatic biota.

The toxicity of microplastics on *M. galloprovincialis* was studied also by Rodrigues et al. (2022) [11]. Mussels were exposed to polyamide microplastics alone and in combination with the toxic exudate from the invasive red seaweed *Asparagopsis armata*. The study showed that microplastics accumulated mainly in the digestive gland of the organisms and that the combined exposure to microplastics and *A. armata* induced oxidative damage at the protein level in the gills and reduced the production of byssus. This study highlights the need to assess microplastics' toxicity in combination with other stress factors, such as invasive species and contaminants. In this regard, Scott et al. (2021) [16] studied the interactions between different polymer types of microplastics and per- and polyfluoroalkyl substances (PFAS) in a lacustrine and a controlled environment. The polymers were kept submerged in the lake water in the presence of associated organic/inorganic matter and biofilm for one and three months; meanwhile, in the laboratory experiment, the polymers were kept in water contaminated with PFAS but without inorganic and organic matter. The results indicated that the presence of inorganic and organic matter considerably enhances the adsorption of PFAS by polymers; this emphasizes the need to assess the risks posed by microplastic pollution under realistic environmental conditions.

All the exposure experiments described so far suggest that microplastic pollution may constitute a serious hazard to aquatic biota. For instance, according to Hadiyanto et al. (2021) [13], Styrofoam microplastics can inhibit the photosynthesis process of *Spirulina platensis*, as well as being a source of nutrients, especially carbon, for the microalgae.

Other organisms that have been found to be capable of ingesting microplastics are blackfly larvae (Simuliidae), as shown by Corami et al. (2022) [14]. Two species of blackfly larvae, *Simulium equinum* and *Simulium ornatum*, were sampled from two rivers in Italy and analyzed for microplastics (<100 µm), and natural and non-plastic synthetic fibers. The authors showed, for the first time, that blackfly larvae can ingest microplastics from their habitat and suggested that these organisms could be employed as bioindicators for microplastic pollution in rivers, as they are already bioindicators used to assess river water quality. Indeed, rivers can be heavily contaminated with microplastics, as Wicaksono et al.

(2021) pointed out in their study [15]. The authors collected water and sediment samples along the Tallo River (Indonesia) during the wet and dry seasons. Microplastic concentration was up to 3.41 ± 0.13 item/m^3 and 150 ± 36.06 item/kg for water and sediment samples, respectively. As in many other aquatic environments, the most abundant polymers found in the Tallo River were polyethylene and polypropylene [15].

The results of the contributions collected herein have helped to fill some knowledge gaps about the occurrence, distribution, and effects of microplastics on aquatic ecosystems. The outcomes clearly indicate that microplastic pollution is a serious environmental issue; the scientific community should increase its knowledge and understanding of how it could affect the environment, biota, and humans, and how it could be reduced and prevented. Nevertheless, to adopt adequate mitigation strategies and contribute to preserving biodiversity and environmental health towards zero pollution, it is pivotal that the studies consider realistic and environmentally relevant conditions.

We would like to thank all the authors for submitting their original contributions to this Special Issue. We greatly appreciate the support of all the reviewers who spent time evaluating and improving the quality of the manuscripts. Last but not least, we would like to thank the editors of *Toxics* for their kind invitation, and Mia Yan, Selena Li, and Linda Li of the *Toxics* Editorial Office for their precious support.

Funding: This research received no external funding.

Conflicts of Interest: The authors declare no conflict of interest.

References

1. Wu, P.; Huang, J.; Zheng, Y.; Yang, Y.; Zhang, Y.; He, F.; Chen, H.; Quan, G.; Yan, J.; Li, T.; et al. Environmental occurrences, fate, and impacts of microplastics. *Ecotoxicol. Environ. Saf.* **2019**, *184*, 109612. [CrossRef] [PubMed]
2. Nizzetto, L.; Bussi, G.; Futter, M.N.; Butterfield, D.; Whitehead, P.G. A theoretical assessment of microplastic transport in river catchments and their retention by soils and river sediments. *Environ. Sci. Process. Impacts* **2016**, *18*, 1050–1059. [CrossRef] [PubMed]
3. Wagner, M.; Scherer, C.; Alvarez-Muñoz, D.; Brennholt, N.; Bourrain, X.; Buchinger, S.; Fries, E.; Grosbois, C.; Klasmeier, J.; Marti, T.; et al. Microplastics in freshwater ecosystems: What we know and what we need to know. *Environ. Sci. Eur.* **2014**, *26*, 12. [CrossRef] [PubMed]
4. Scopetani, C.; Chelazzi, D.; Martellini, T.; Pellinen, J.; Ugolini, A.; Sarti, C.; Cincinelli, A. Occurrence and characterization of microplastic and mesoplastic pollution in the Migliarino San Rossore, Massaciuccoli Nature Park (Italy). *Mar. Pollut. Bull.* **2021**, *171*, 112712. [CrossRef] [PubMed]
5. Antunes, J.; Frias, J.; Sobral, P. Microplastics on the Portuguese coast. *Mar. Pollut. Bull.* **2018**, *131*, 294–302. [CrossRef] [PubMed]
6. Yang, H.; Chen, G.; Wang, J. Microplastics in the Marine Environment: Sources, Fates, Impacts and Microbial Degradation. *Toxics* **2021**, *9*, 41. [CrossRef] [PubMed]
7. Santini, S.; De Beni, E.; Martellini, T.; Sarti, C.; Randazzo, D.; Ciraolo, R.; Scopetani, C.; Cincinelli, A. Occurrence of Natural and Synthetic Micro Fibers in the Mediterranean Sea: A Review. *Toxics* **2022**, *10*, 391. [CrossRef]
8. Lim, K.P.; Lim, P.E.; Yusoff, S.; Sun, C.; Ding, J.; Loh, K.H. A Meta-Analysis of the Characterisations of Plastic Ingested by Fish Globally. *Toxics* **2022**, *10*, 186. [CrossRef] [PubMed]
9. Drago, C.; Weithoff, G. Variable Fitness Response of Two Rotifer Species Exposed to Microplastics Particles: The Role of Food Quantity and Quality. *Toxics* **2021**, *9*, 305. [CrossRef] [PubMed]
10. Von Hellfeld, R.; Zarzuelo, M.; Zaldibar, B.; Cajaraville, M.P.; Orbea, A. Accumulation, Depuration, and Biological Effects of Polystyrene Microplastic Spheres and Adsorbed Cadmium and Benzo(a)pyrene on the Mussel *Mytilus galloprovincialis*. *Toxics* **2022**, *10*, 18. [CrossRef] [PubMed]
11. Rodrigues, F.G.; Vieira, H.C.; Campos, D.; Pires, S.F.S.; Rodrigues, A.C.M.; Silva, A.L.P.; Soares, A.M.V.M.; Oliveira, J.M.M.; Bordalo, M.D. Co-Exposure with an Invasive Seaweed Exudate Increases Toxicity of Polyamide Microplastics in the Marine Mussel *Mytilus galloprovincialis*. *Toxics* **2022**, *10*, 43. [CrossRef] [PubMed]
12. Zhang, C.; Zuo, Z.; Wang, Q.; Wang, S.; Lv, L.; Zou, J. Size Effects of Microplastics on Embryos and Observation of Toxicity Kinetics in Larvae of Grass Carp (*Ctenopharyngodon idella*). *Toxics* **2022**, *10*, 76. [CrossRef] [PubMed]
13. Hadiyanto, H.; Haris, A.; Muhammad, F.; Afiati, N.; Khoironi, A. Interaction between Styrofoam and Microalgae *Spirulina platensis* in Brackish Water System. *Toxics* **2021**, *9*, 43. [CrossRef] [PubMed]
14. Corami, F.; Rosso, B.; Iannilli, V.; Ciadamidaro, S.; Bravo, B.; Barbante, C. Occurrence and Characterization of Small Microplastics (<100 μm), Additives, and Plasticizers in Larvae of Simuliidae. *Toxics* **2022**, *10*, 383. [CrossRef]
15. Wicaksono, E.A.; Werorilangi, S.; Galloway, T.S.; Tahir, A. Distribution and Seasonal Variation of Microplastics in Tallo River, Makassar, Eastern Indonesia. *Toxics* **2021**, *9*, 129. [CrossRef] [PubMed]

16. Scott, J.W.; Gunderson, K.G.; Green, L.A.; Rediske, R.R.; Steinman, A.D. Perfluoroalkylated Substances (PFAS) Associated with Microplastics in a Lake Environment. *Toxics* **2021**, *9*, 106. [CrossRef] [PubMed]
17. Scopetani, C.; Esterhuizen, M.; Cincinelli, A.; Pflugmacher, S. Microplastics Exposure Causes Negligible Effects on the Oxidative Response Enzymes Glutathione Reductase and Peroxidase in the Oligochaete Tubifex tubifex. *Toxics* **2020**, *8*, 14. [CrossRef] [PubMed]
18. Silva, C.J.; Machado, A.L.; Campos, D.; Rodrigues, A.C.; Silva, A.L.P.; Soares, A.M.; Pestana, J.L. Microplastics in freshwater sediments: Effects on benthic invertebrate communities and ecosystem functioning assessed in artificial streams. *Sci. Total Environ.* **2022**, *804*, 150118. [CrossRef] [PubMed]

Article

Occurrence and Characterization of Small Microplastics (<100 μm), Additives, and Plasticizers in Larvae of Simuliidae

Fabiana Corami [1,2,*], Beatrice Rosso [1,2], Valentina Iannilli [3], Simone Ciadamidaro [4], Barbara Bravo [5] and Carlo Barbante [1,2]

[1] Institute of Polar Sciences, CNR-ISP, Campus Scientifico—Ca' Foscari University of Venice, Via Torino 155, 30172 Venice, Italy; beatrice.rosso@unive.it (B.R.); barbante@unive.it (C.B.)

[2] Department of Environmental Sciences, Informatics and Statistics, Ca' Foscari University of Venice, Via Torino 155, 30172 Venice, Italy

[3] ENEA, SSPT-PROTER-BES, C. R. Casaccia, Via Anguillarese 301, 00123 Roma, Italy; valentina.iannilli@enea.it

[4] ENEA, SSPT-PROTER-BES, C. R. Saluggia, Via Crescentino Snc., 13040 Saluggia, Italy; simone.ciadamidaro@enea.it

[5] ThermoFisher Scientific, Str. Rivoltana, Km 4, 20090 Rodano, Italy; barbara.bravo@thermofisher.com

* Correspondence: fabiana.corami@cnr.it; Tel.: +39-0412348568

Abstract: This study is the first to investigate the ingestion of microplastics (MPs), plasticizers, additives, and particles of micro-litter < 100 μm by larvae of Simuliidae (Diptera) in rivers. Blackflies belong to a small cosmopolitan insect family whose larvae are present alongside river courses, often with a torrential regime, up to their mouths. Specimens of two species of blackfly larvae, *Simulium equinum* and *Simulium ornatum*, were collected in two rivers in Central Italy, the Mignone and the Treja. Small microplastics (SMPs, <100 μm), plasticizers, additives, and other micro-litter components, e.g., natural and non-plastic synthetic fibers (APFs) ingested by blackfly larvae were, for the first time, quantified and concurrently identified via MicroFTIR. The pretreatment allowed for simultaneous extraction of the ingested SMPs and APFs. Strong acids or strong oxidizing reagents and the application of temperatures well above the glass transition temperature of polyamide 6 and 6.6 (55–60 °C) were not employed to avoid further denaturation/degradation of polymers and underestimating the quantification. Reagent and procedural blanks did not show any SMPs or APFs. The method's yield was >90%. Differences in the abundances of the SMPs and APFs ingested by the two species under exam were statistically significant. Additives and plasticizers can be specific to a particular polymer; thus, these compounds can be proxies for the presence of plastic polymers in the environment.

Keywords: blackfly larvae; freshwaters; Simuliidae; microplastics; additives; plasticizers

Citation: Corami, F.; Rosso, B.; Iannilli, V.; Ciadamidaro, S.; Bravo, B.; Barbante, C. Occurrence and Characterization of Small Microplastics (<100 μm), Additives, and Plasticizers in Larvae of Simuliidae. *Toxics* **2022**, *10*, 383. https://doi.org/10.3390/toxics10070383

Academic Editor: Costanza Scopetani

Received: 27 June 2022
Accepted: 8 July 2022
Published: 10 July 2022

Publisher's Note: MDPI stays neutral with regard to jurisdictional claims in published maps and institutional affiliations.

1. Introduction

The ingestion of ubiquitous and persistent microplastics (MPs) in biota, i.e., in macroinvertebrates, is documented in polar environments [1–3], marine environments [4–8], and riverine environments [9–12]. Invertebrates ingest food particles according to the size of their mouthparts; the size of these particles is usually <100 μm. MPs < 100 μm (small microplastics, SMPs), as well as additives, plasticizers, and other micro-litter components <100 μm (e.g., natural and non-plastic synthetic fibers; APFs), can be mistaken for food particles, ingested, and enter the trophic web. SMPs can be primary, e.g., those released from the discharge of washing machines [13], or secondary, e.g., those derived from the fragmentation of macroplastics and large microplastic pieces. It should be underlined that the fragmentation of large MPs can release or expose additives and plasticizers employed in the plastic industry and can be polymer-use specific; these compounds are thought to be responsible for the toxicity of plastic polymers toward biota [14]. However, assessment of the additives and plasticizers in environmental matrices and biota has been overlooked. Some studies have tested the ingestion of MPs by macroinvertebrates in lab conditions, but these controlled exposure studies may lack environmental realism, and the concentration of the ingested MPs cannot correspond to those observed in nature [10]. Hence, the focus of this study is to investigate the ingestion of SMPs and APFs in two blackflies species, *Simulium equinum* (Linnaeus, 1758) and *Simulium ornatum* (Meigen, 1818 (complex)), for the first time. Specimens of these two species were collected in their habitat.

Blackflies (Diptera, Simuliidae) form a relatively small and uniform family of insects, numbering nearly 2300 known species worldwide [15]. They are passive filter feeders, filtering suspended particulate matter from the water and staying fixed to smooth surfaces in the lotic reaches of watercourses. Blackfly larvae are crucial in watercourses' ecologies, making the filtered matter available for other invertebrates, amphibians, and fishes that feed on them [16]. Blackfly larvae spend most of their time attached to the substrate in watercourses, and, in this sedentary mode, they feed. The primary feeding device, and distinguishing feature of the family, is a pair of large cephalic filtering "fans," which are complex oral structures consisting of many serially arranged rays fixed on the two fans' stems. These filtering "fans" are chitinous–mucous structures. Opened in riverine waters, they can trap fine suspended organic (e.g., detritus, bacteria, algae, animal matter) and inorganic matter with a passive and undiscriminating collection system; if it can be manipulated in the mouth and can enter the *cibarium*, any catchable particulate filtrate is taken into the gut. If compressible, even larger particles can be swallowed [17,18]. Concerning their feeding mode, blackfly larvae may ingest SMPs and APFs.

A previously developed pretreatment method (at CNR-ISP, [7]) was optimized to assess the abundance of SMPs, APFs, and other microlitter components ingested by blackfly larvae; the method allows for concurrent extraction of all the aforementioned particles and does not contribute to these particles' further degradation/denaturation. Many pretreatment methods employ strong oxidizing agents or strong acids, which can modify particle sizes and contribute to discoloration, degradation, and loss of several polymers [19], especially nylon 6 and nylon 6,6 (PA 6 and PA 6,6). Moreover, these pretreatment methods employ temperatures $\geq 60\,^{\circ}\text{C}$, which can contribute to the loss of polymers, in particular, PA 6 and PA 6,6, as the range of their glass transition temperature (Tg) is 55–60 $^{\circ}\text{C}$ [7,19,20]. Hence, these pretreatment methods can result in underestimation of the actual abundance of MPs/SMPs in the samples and samples that are not representative. SMPs and APFs will be simultaneously quantified (microscopic count) and identified via Micro-FTIR.

2. Material and Methods

2.1. Sampling Sites and Macroinvertebrate Sampling

When sampling macroinvertebrates for water quality monitoring, organisms of *Simulium equinum* and *Simulium ornatum* were collected in the summer of 2018 from theTreja River (42.18402, 12.37895), a few kilometers downstream from Mazzano Romano, near the Monte Gelato waterfalls, an attractive place for tourists during spring and summer, and the Mignone River (42.19557, 11.79347), near Tarquinia (Figure 1). Because of their characteristics, these rivers may well represent environments influenced by various pressures and impacts.

The Treja River is the third major right tributary of the Tiber River. Its source is in Monte Lagusiello near Lake Bracciano, and the river flows through a valley that gives it its name, which is characterized by tuffaceous material. Along the river's course, the natural environments are in a good state of conservation; there are alternating areas of cultivated countryside, livestock activities, and woods.

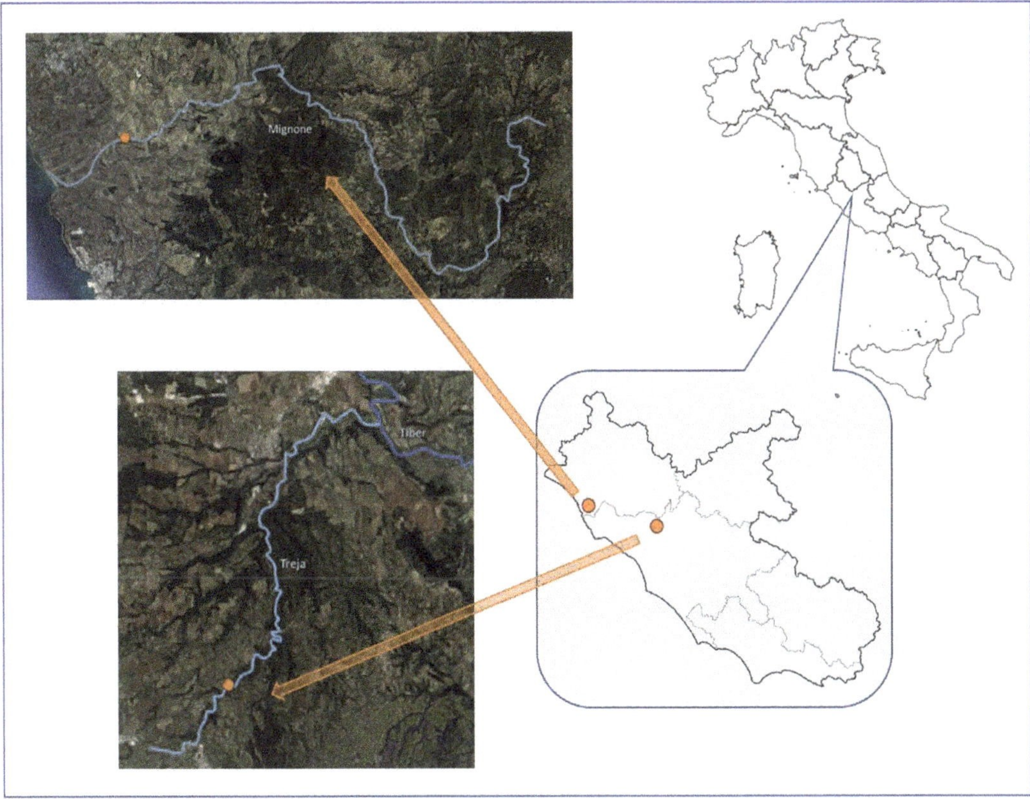

Figure 1. Sampling sites where blackfly larvae (Simuliidae) were collected; the Mignone and Treja rivers are located near Rome, in Lazio, Italy.

The Mignone River is 62 km long, originating in the Sabatini Mountains in the territory of the town of Vejano, located northwest of Lake Bracciano. In its initial part, this river is almost a stream, which has carved its bed within deep valleys, while the remaining stretch was once navigable. It reaches the Tyrrhenian Sea, north of Rome, in Tarquinia, after a course of 60 km. The river and its catchment area represent one of the most remarkable environmental areas of Lazio, due to high conservational preservation as Sites of Community Importance. However, the qualitative state of the river in the lower course is influenced by anthropogenic activities.

Moreover, they are frequently visited nature reserves, and the entire catchment areas of the Mignone and Treja are the object of historical and artistic tourism. Therefore, agriculture, WWTP (wastewater treatment plant) discharges, and various tourist activities in these two areas may be significant sources of SMPs and APFs.

At each of the sites, which are, as a general rule, monitored for water quality status, macroinvertebrates, including blackfly larvae, were collected using a hand net by placing it on the riverbed and moving the substratum in front of the net opening with the free hand or a foot. Sampling was performed in riffle mesohabitat, which is the most suitable for blackfly larvae according to their ecology [18]. In order to cover the highest diversity of the local habitat conditions where macroinvertebrates and different blackfly species could be found, all microhabitats were surveyed in the riffle mesohabitat, giving priority to the stable substratum on which blackfly larvae can anchor themselves. The finalized sample for each site was sorted in the field to separate the substratum from organisms. All blackfly larvae were sorted among the macroinvertebrates collected; they were immediately fixed in ethanol 70% (absolute, for HPLC, $\geq$99.8%, Sigma Aldrich, Merck Darmstadt, Germany) to prevent gut content excretion. Different species of blackfly larvae were identified through microscopic morphological examination at the ENEA laboratory. The two species, *Simulium equinum* (Linnaeus, 1758) and *Simulium ornatum* (Meigen, 1818 (complex)), were identified at both sampling sites.

Thirty organisms were collected for each of the two identified species of Simuliidae at each sampling site in the rivers under study. Before their identification, the organisms were carefully rinsed several times with ultrapure water (Milli-Q®, Merck Darmstadt, Germany), followed by a fresh 70% ethanol solution to remove materials on the body surface, which were, therefore, not ingested. Then, 10 organisms per species were employed for taxonomic identification and dry weight detection. The average dry weight per organism of *S. ornatum* was 0.5 mg, while for the *S. equinum*, it was 0.6 mg.

The organisms designated explicitly for the analysis of microplastics and other microlitter components (20 organisms per species at each sampling site, which is monitored for water quality) were preserved in ethanol 80% and then transferred to the laboratory of CNR-ISP (spring 2020).

2.2. Quality Assurance and Quality Control (QA/QC)

Decontamination and pretreatment procedures were performed at CNR-ISP Venezia in a plastic-free cleanroom ISO 7. This cleanroom (a controlled-atmosphere laboratory where atmospheric pressure, humidity, temperature, and particle pollution are controlled) is entirely free of plastic materials, even in the air pre-filters. The environmental contamination in the pretreatment procedures for the analysis of SMPs and APFs is efficiently minimized.

Samples were pre-treated (extraction and purification) and filtered in batches on aluminum oxide filters (ANODISC filters, Supported Anopore Inorganic Membrane, 0.2 µ, 47 mm, Whatman™; Merck, Darmstadt, Germany). The pretreatment procedures and filtration were performed under a decontaminated steel fume hood. Operators wore cotton lab coats and nitrile gloves. All glassware was previously washed with a 1% Citranox® solution (Citranox® acid detergent, Sigma Aldrich purchased from Merck Darmstadt, Germany), rinsed with ultrapure water (UW, produced by UW system, Elga Lab Water, Veolia, High Wycombe, UK), and decontaminated with acetone (suitable for HPLC, 99.9%, Sigma Aldrich, Merck Darmstadt, Germany). Then, the glassware was rinsed with a 50% (v/v) solution of methanol (suitable for HPLC, 99.9%, Sigma Aldrich, Merck Darmstadt, Germany) and ethanol (absolute, for HPLC, $\geq$99.8%, Sigma Aldrich, Merck Darmstadt, Germany), and, finally, with ethanol. The steelware was previously rinsed with UW, decontaminated with methanol, a 50% (v/v) solution of methanol and ethanol, and ethanol. Reagent (e.g., UW from Milli-Q® (Millipore, Merck, Darmstadt, Germany), ethanol, H_2O_2, etc.) and procedural blanks were performed for each batch.

After filtration, all filters were stored in decontaminated glass Petri dishes covered with aluminum foil. Before the analysis, filters were transferred from the fume hood in the cleanroom to the Micro-FTIR laboratory, carefully covered with aluminum foil to avoid any external contamination.

Certified reference materials for MPs in biota are lacking; therefore, to estimate the yield of the pretreatment procedure used in this study, a model organism that was accessible and easy to sample was chosen. The choice was *Monocorophium insidiosum* (Corophidae, Amphipoda), whose specimens were sampled in the Pordelio Channel, Venice Lagoon, in the summer of 2020; three pooled samples were then spiked with silver–grey beads of polyamide 12 (average size 90 µm; Goodfellow Cambridge Limited, Huntingdon, UK). The polymer to be employed was selected by the particle color, size, and ease of mixing it in the sample.

2.3. Extraction, Purification, and Filtration of APFs and SMPs Ingested by Blackfly Larvae

For the extraction and purification of the APFs and SMPs ingested by blackfly larvae, the method developed by Corami et al. [7] was employed with slight modifications. Due to the small size of blackfly larvae, the organisms were not dissected; hence, the APFs and SMPs were extracted from the whole organism.

Briefly, under a decontaminated fume hood in the cleanroom, organisms were put in a decontaminated Erlenmeyer flask with H_2O_2, ethanol, and UW (1:2:1 ratio) and stirred for 96 h on a multipurpose orbital shaker at room temperature. The aim of this step is not thorough digestion (i.e., strong acids or strong oxidants); rather, it is an extraction of the ingested particles by dissolving the organic matter with no further denaturation of polymers. The residual dissolved organic matter was removed through the following purification procedure: flushing ethanol and a 70% (v/v) ethanol–methanol solution alternated with the extracted slurry directly onto the aluminum oxide filter during vacuum filtration.

Filtration was performed with a decontaminated glass filtering apparatus and a vacuum pump Laboport® (VWR International, Milan, Italy) under a decontaminated fume hood in the cleanroom; aluminum oxide filters were rinsed by alternating 50 mL of a 50% (v/v) solution of ethanol with 50 mL of 70% (v/v) solution of ethanol–methanol before the filtration. The filter was rinsed several times with a 50% (v/v) ethanol solution at the end of filtration. Each filter was stored in decontaminated glass Petri dishes for at least 72 h under a fume hood in the cleanroom before the analysis via Micro-FTIR.

2.4. Quantitative and Chemical Characterization of APFs and SMPs via Micro-FTIR

A Nicolet™ iN™ 10 infrared microscope (Thermo Fisher Scientific, Madison, WI, USA), equipped with an ultra-fast motorized stage and liquid-nitrogen-cooled MCT detector (mercury cadmium telluride detector), was employed for the analysis. The settings were: transmittance mode, a spectral range of 4000–1200 cm^{-1}, 100-µm step size scanning (spatial resolution), 100–100 µm aperture, and 64 co-added scans at a spectral resolution of 4 cm^{-1} [7,13,20].

Microscopic counting was performed according to Corami et al. [7,20]. Microscopic counting has been employed for bacteria, phytoplankton, pollen, spores, and microplastics as well [21–31]. A significant advantage of microscopic counting is that there is no doubt about how many organisms, cells, or particles are present within reliable computable limits and degrees of chance. When filters are employed as a support for counting, the measurement of complete filters is very time-consuming [28,30,31]. However, analyzed filter areas, i.e., counting areas or count fields, need to represent the entire filter to avoid issues regarding representativeness and reproducibility. Since the loading of the filters cannot be known in advance, counting areas with different abundances should be considered to avoid issues regarding the accuracy of the extrapolation of microplastics, organisms, cells, or bacteria findings.

In our study, at least 14 known-sized areas (i.e., count fields) were randomly chosen with no overlapping on the surface of the filter (the different approaches to choosing representative measurement areas are in the Supplementary Information, Figure S1). Moreover, a significant number of particles (250–350 particles per count field) were analyzed using the PARTICLES WIZARD of the Omnic™ Picta™ software. The spectral background was acquired on a clean point in each count field. The IR spectrum was retrieved for each particle, and the spectral background was deduced; the resulting spectrum was then compared with several reference libraries (the list of reference libraries is in the Supplementary Information). In PARTICLE WIZARDS, particles were identified and counted when the identification match percentage was ≥65%; when operating with this software section, the optimal range of match percentage is between 65% and 75%. Moreover, particle sizes (length and width) were collected using the Imaging of PARTICLE WIZARDS.

The total number of SMPs and APFs per organism was then calculated according to Equation (1) (modified from Corami et al., 2020b [13]):

$$\frac{N_{tot}}{Specimen} = \frac{(n * F)}{n\ specimens} \tag{1}$$

where n = SMPs or APFs counted on every field, n specimens = the total number of organisms analyzed, and F = count factor, calculated as follows:

$$F = \frac{Total\ area\ of\ the\ filter}{Area\ of\ a\ count\ field\ *\ n\ count\ fields} \tag{2}$$

The weight of microplastics per specimen can be calculated according to Equation (3) (modified from Corami et al., 2020b [13]):

$$\frac{W_{tot}}{specimen} = \frac{N_{tot*V*\rho}}{n\ specimens} \tag{3}$$

where W_{tot} = total weight of SMPs or APFs, n specimens = the total number of organisms analyzed, V is the volume of each particle calculated based on its AR, and ρ is the identified polymer's density, additive, plasticizer, etc.. The aspect ratio (AR); [13,32,33] is the ratio between the maximum length (L) and the maximum width (W) of the smallest rectangle (bounding box) enclosing the particle chosen with the Imaging of PARTICLE WIZARDS, employed for the analysis. When the AR $\leq$ 1, particles are considered spherical; when the AR $\leq$ 2, particles are elongated/ellipsoidal. When the AR $\geq$ 3, particles are considered cylindrical. The volumes of SMPs and APFs can be calculated according to their geometrical shape (i.e., sphere, ellipse, and cylinder).

2.5. Statistical Analysis

The abundance and distribution of SMPs and APFs, as well as their weights, are expressed as the average number of particles per organism. Statistical analyses were performed using STATISTICA software (TIBCO, Palo Alto, CA, USA). Fisher's exact test was performed to test whether the variances of the abundance of SMPs and APFs were homogenous (F test, α = 0.05). After invalidation of the homogeneity of variances, non-parametric statistical tests were performed to assess significant differences in the abundance of ingested APFs, SMPs, and other components of the microlitter. While the Kruskal–Wallis test ($p < 0.05$) was employed for multiples comparison, the Mann–Whitney U test ($p < 0.05$) was performed for pairwise comparisons. Since particles' abundance data are count data, they follow a Poisson distribution [20,34,35]; Poisson's confidence interval was calculated accordingly.

3. Results

3.1. SMPs Ingested by Blackfly Larvae

SMPs and APFs were not detected on reagent and procedural blanks. Contamination was minimized during all steps of the pretreatment and analysis.

The complete list of polymers identified and quantified is reported in Table 1. The abundance of the SMPs ingested (n SMPs/organism) by the specimens of *S. equinum* and *S. ornatum* in the two rivers under study is shown in Figure 2, while the weight of the ingested SMPs is shown in Figure 3. The fiducial interval (FI, or confidence interval) was calculated according to Poisson's distribution.

Table 1. List of the polymers identified and quantified in the specimens of *S. equinuum* and *S. ornatum*, collected in the Treja and Mignone rivers.

HDPE	High Density Polyethylene
PA	Nylon 6
PFA	Pefluoroalcoxy Fluorocarbon
PPA	Polyphtalamide
PES	Polyester
ECTFE	Ethylene chlorotrifluoroethylene
PC/ABS	Polycarbonate/Acrylonitrile Styrene Butadiene
ARAMID	Aramid
PO	Olefin fiber
PEAA-Zn	Polyethylene acrylic acid copolymer—Zinc salt
EVOH	Ethyl vinyl alcohol
MODACRILIC	Modacrilic
PP	Polypropylene
PEA	Polyethylacrylate
PAA	Polyarylamide
EPM	Ethylene propylene rubber
PBA	Polybutylacrylate
FKM	Fluoroelastomer
PA 12	Grilamid tr 55
PTFE	Polytetrafluoroethylene
BR	Butadien rubber

Polymers with a wide range of densities were identified and quantified, e.g., from PP (density = 0.9005 g cm^{-3}) to PTFE (density = 2.2 g cm^{-3}) and FKM (density = 2.1 g cm^{-3}). The match percentage (i.e., the correlation coefficient between the measured spectrum and the reference spectrum for each polymer identified or the match %) was in the optimal range (65–75%) for all of the identified polymers. Moreover, the match percentage of several spectra identified in the analyzed samples was well above 75% of the optimal match percentage (i.e., >85%, HDPE, PO, PP, PTFE). Some spectra are shown as examples in the Supplementary Information (Figure S1). Only optimally identified SMPs (match % $\geq$ 65%) were quantified.

The highest abundance of SMPs was shown by the *S. ornatum* collected in the Mignone River (1101 ± 47 SMPs/organism) at almost five times higher than the abundance of the same species collected in the Treja River (248 ± 22 SMPs/organism). Regarding *S. equinum*, the specimens of the Mignone River showed the lowest abundance (144 ± 17 SMPs/organism) at almost 70% lower than the abundance of the same species in the Treja River (462 ± 30 SMPs/organism).

Most of the SMPs ingested by the two species in the two rivers studied were less than 52 µm in length. According to their AR (Figure 4), ellipsoidal particles were prevalent for all the polymers identified. The average length of particles in the Treja River, ingested by *S. equinum* (46 µm), was higher than that of the *S. ornatum* (39 µm); in contrast, the latter ingested larger particles in the Mignone River (52 µm and 42 µm for *S. equinum* and *S. ornatum*, respectively).

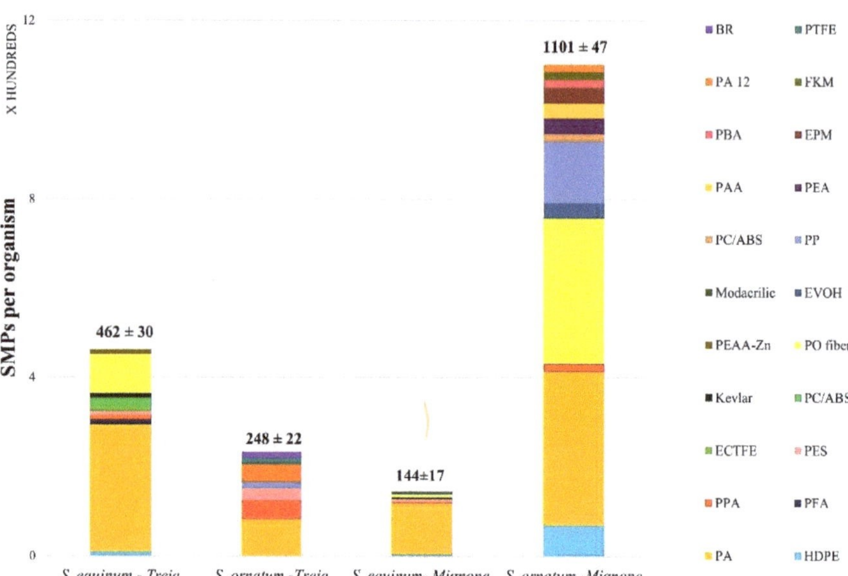

Figure 2. The average abundance of SMPs per organism in the two species of blackfly larvae under examination, *Simulium equinum* and *Simulium ornatum* (20 organisms per species for each sampling site were analyzed). The fiducial interval according to Poisson's distribution is reported for each species in the sampling sites studied. The distribution of polymers ingested is shown as well. Complete names of the polymers can be found in Table 1.

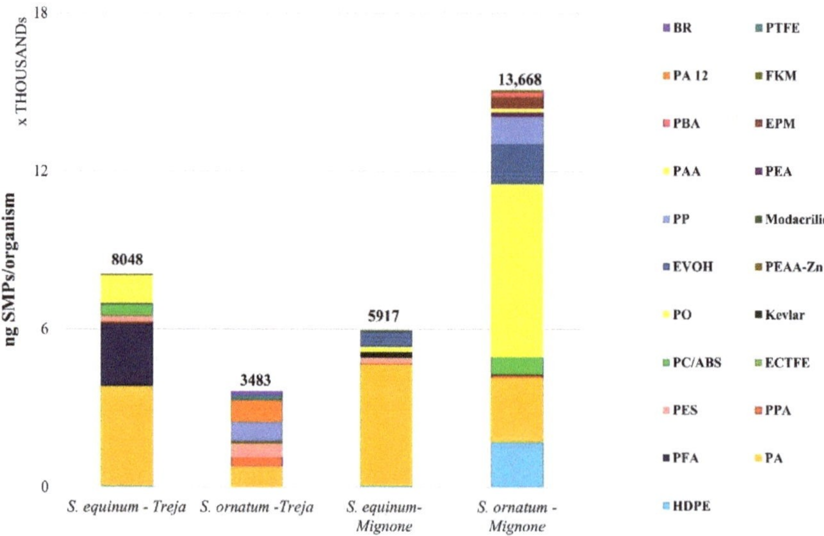

Figure 3. Weight of ingested SMPs (ng SMPs/organism) by *S. equinum* and *S. ornatum* collected in the Treja and Mignone rivers.

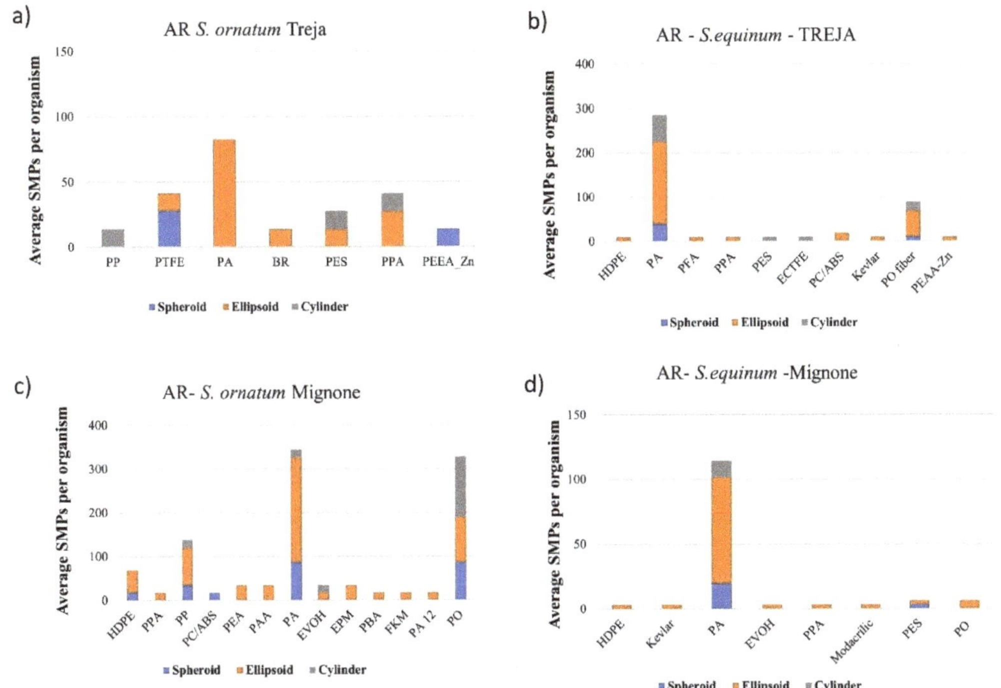

Figure 4. Aspect ratio (AR) of the polymers identified and quantified in specimens of *S. ornatum* (**a,c**) and *S. equinum* (**b,d**) under examination. The number of the spheroid, ellipsoid, and cylinder particle shapes is reported for the average abundance of each polymer identified and quantified via microscopic counting.

3.2. APFs and Other Components of Micro-Litter Ingested by Blackfly Larvae

The same pretreatment method allowed for simultaneous extraction of the SMPs and APFs, which were then filtered on the same filter. Afterward, APFs were quantified and detected concurrently with SMPs in the same analysis via MicroFTIR.

The abundance of the APFs ingested (n APFs/organism) by the two species investigated is shown in Figure 5. *S. ornatum* in the Mignone River showed the highest abundance of APFs (1565 ± 56 APFs/organism) at almost four times higher than the abundance of APFs in *S. equinuum* (442 ± 30 APFs/organism). The lowest abundance of APFs was observed in the Treja River, once again in *S. ornatum* (358 ± 27 APFs/organism), while *S. equinum* showed a comparable concentration (423 ± 29 APFs/organism) to that observed in the Mignone River. The weights of the AFPs ingested by *S. equinum* and *S. ornatum* in the two rivers are shown in Figure 6. The specimens showed approximately the same weight of APFs (ng/organism), except for *S. ornatum* in the Mignone River, which showed the highest weight of APFs (58 mg/organism). Rayon was the most represented among the APFs observed.

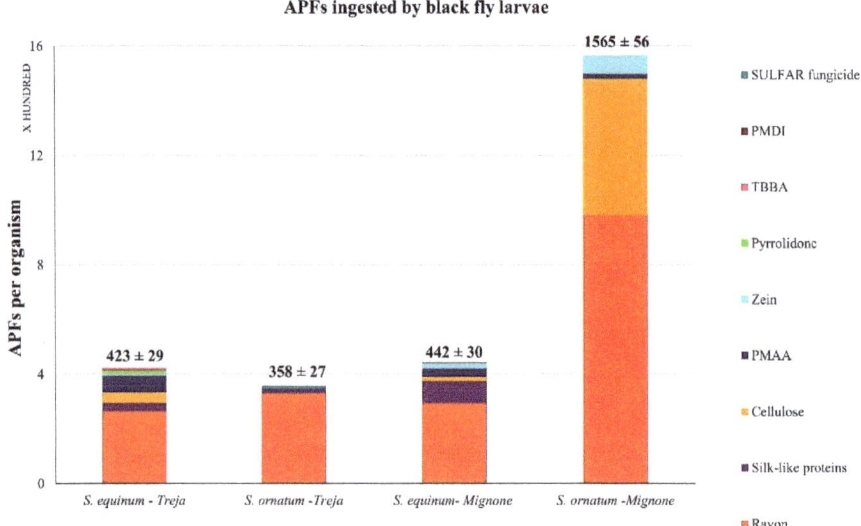

Figure 5. The average abundance of APFs per organism in the two species of blackfly larvae under exam, *Simulium equinum* and *Simulium ornatum* (20 organisms per species for each sampling site was analyzed). The distribution of ingested additives, plasticizers, and other microlitter components is also shown. Rayon is a non-plastic synthetic fiber, which is preeminent in all the specimens studied. Simuliidae can ingest larger particles if compressible; some rayon fragments in *S. ornatum* in the Mignone River were >150 μm in length. The fiducial interval according to Poisson's distribution is reported for each species in the sampling sites studied.

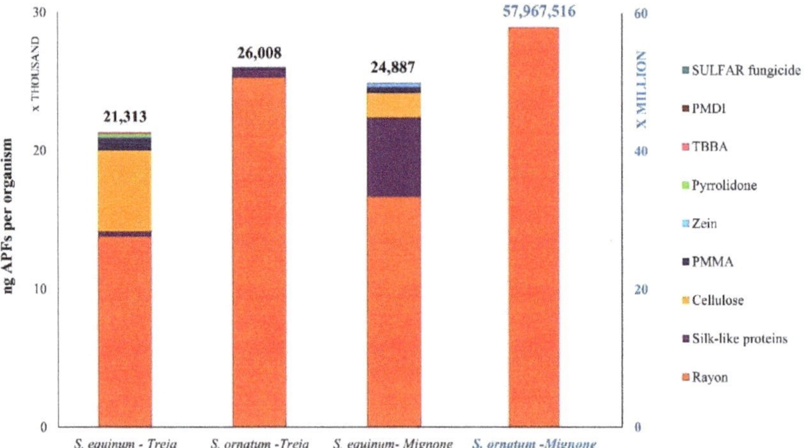

Figure 6. Weight of ingested APFs (ng APFs/organism) by *S. equinum* and *S. ornatum* collected in the Treja and Mignone rivers.

As noted for the AR of SMPs, the ellipsoidal shape was prevalent for APFs (Figure 7). The average sizes of the APFs ingested by *S. ornatum* (length 70 μm, width 35 μm in the Mignone River; length 69 μm, width 32 μm in the Treja River) were higher than those ingested by *S. equinum* (length 55 μm, width 29 μm in the Mignone River; length 55 μm, width 28 μm in the Treja River). It should be noted that the high abundance and amount of rayon observed in the *S. ornatum* in the Mignone river is due to the presence of fragments higher than 150 µm in length.

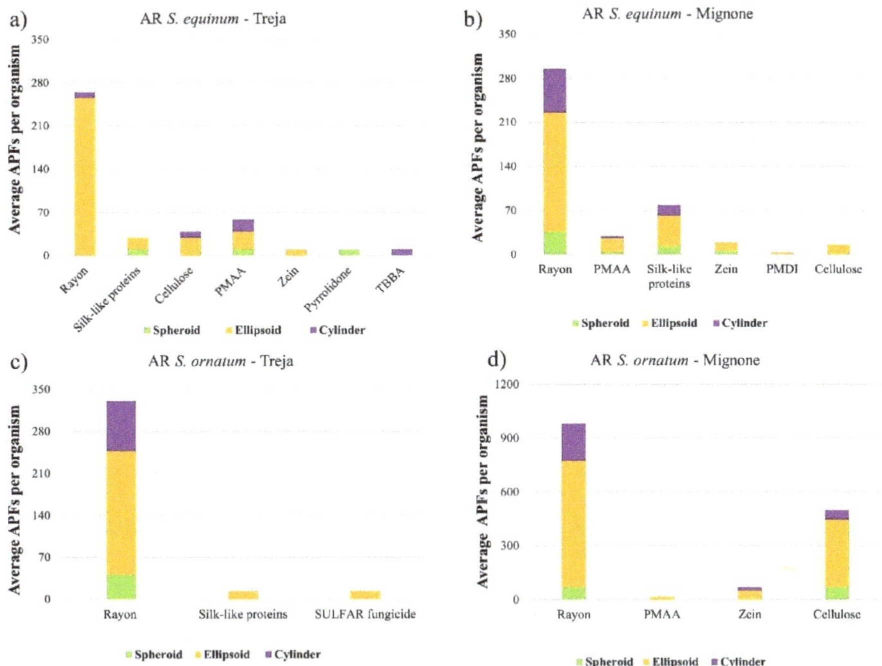

Figure 7. Aspect ratio (AR) of the APFs identified and quantified in specimens of *S. equinum* (**a,c**) and *S. ornatum* (**b,d**) under examination. The number of the spheroid, ellipsoid, and cylinder particle shapes is reported for the average abundance of each particle identified and quantified via microscopic counting.

4. Discussion

4.1. SMPs Ingested by Blackfly Larvae

The variances of polymer distributions for *S. equinum* and *S. ornatum* in the two rivers were different (F test, $\alpha = 0.05$); according to the non-parametric Kruskall–Wallis test, the observed differences in the abundances and polymer distributions for both species in the two rivers were statistically significant ($p < 0.05$). Statistical analysis (Mann–Whitney U test, $p < 0.05$) showed that the differences in the SMPs' observed abundances for the same species in the two rivers under study were significantly different, just as the SMPs' abundances of the two species studied in the same river were also consistently dissimilar.

Specimens of *S. ornatum* showed a wider variety of polymers ingested than the organisms of *S. equinum* in the studied rivers. Several different factors (e.g., environmental, chemical, biological, etc.) could affect the ingestion of SMPs by blackfly larvae. The observed differences might be related to the type of polymer, the sources and pathways that the specific polymer followed before entering the riverine water, and where the blackfly larvae of the two species are located in these rivers.

The most abundant polymer was PA; this was followed by PO (maximum value for *S. ornatum* in the Mignone River, 327 ± 14 SMPs/organism, 2464 ng/organism), which has many usages in fabrics and textiles and may have diffuse sources. PA's abundance in the Treja River was 285 ± 19 SMPs per organism (3806 ng/organism) of *S. equinum* and 83 ± 7 SMPs per organism (786 ng/organism) of *S. ornatum*, while in the Mignone River, PA's abundances were 115 ± 14 SMPs/organism (4605 ng/organism) and

344 ± 15 SMPs/organism (2464 ng/organism), respectively. Another polymer present in all of the organisms in both rivers is PPA, primarily employed in electronics and electrical equipment. As assumed for PAA and EPM ingested by *S. ornatum* in the Mignone River, sources could also be diffuse for PPA. PES was ingested by the two species in the Treja River and only by *S. equinum* in the Mignone River. It should be noted that the shapes of the PES particles ingested by *S. equinum* were quite different, i.e., ellipsoidal and cylindric in the Treja River, while ellipsoidal and spheroidal in the Mignone River; this might support the notion that the pathways to the two rivers and the larval preference for the sizes of ingested particles may be somewhat different.

Regarding fluorinated polymers, the two species in the Treja River ingested a variety of them, i.e., PFA, ECTFE, and PTFE, while the *S. ornatum* in the Mignone River ingested only FKM. As a group of polymers, fluorinated polymers are employed for several purposes, from insulation to piping, waterborne coating systems, cookware, fabric and carpet protection, and the mechanical and automotive industries, to name a few. The presence and pathways of these polymers are a function of their widespread and extensive use; the ingestion by blackfly larvae may have been affected by the fragments' shape.

The ingestion of BR in *S. ornatum* in the Treja River should be highlighted; 70% of this polymer is employed in the manufacturing of tires. Tire wear particles can enter the environment through atmospheric transport, WWTP effluents, and road runoff, and then accumulate in sediments and surface waters [36] where biota can ingest them.

Some other studies have dealt with the presence of MPs by riverine insects [10–12,37–43]. Some of these have dealt with the ingestion of MPs by riverine insects [10–12,40–43]; however, the insects studied were not Simulidae, and some studies were mainly exposure experiments to few native polymers. Caddisfly cases (Trichoptera) from the same area of the Mignone River were investigated for the presence of plastics [39]. Nevertheless, the fragments studied had sizes ($\sim$1 mm) well above those observed for the ingested SMPs by the two species under examination here, and they were analyzed only by a visual exam (microscopical examination); thus, the polymers were not properly identified.

The polymers identified and quantified in this study were neither virgin nor native; they were discharged into the environment, and they reached the rivers through, e.g., atmospheric transport, rains, winds, and soil runoff. They were finally ingested by *S. equinum* and *S.ornatum* in the two rivers where specimens were collected.

Furthermore, a wide variety of polymers were identified and quantified thanks to the pretreatment method, which allowed for the recovery of low-density polymers, e.g., PE and PA, and high-density polymers, such as PTFE and FKM. It should be highlighted that the experimental conditions used for pretreatment did not affect particle size [19] and made it possible to identify PA and other polymers unambiguously [7,19], which allows for a more adequate and representative quantification of what is ingested by the organisms.

4.2. APFs Ingested by Blackfly Larvae

According to the Mann–Whitney U test, significant differences were observed for the same species in the same rivers in the two rivers studied and for the two species in the same river ($p < 0.05$). The variances of the APFs' distributions for *S. equinum* and *S. ornatum* in the two rivers were different (F test, $\alpha = 0.05$). According to the non-parametric Kruskall–Wallis test, the differences observed in abundance and distribution of APFs for both species in the two rivers were highly statistically significant ($p < 0.01$).

Non-plastic synthetic fibers, i.e., rayon, and natural fibers, such as cellulose, are often identified in several organisms [5,21,44–47]. Several lotic insects produce silk-like proteins or silk, e.g., caddisflies, aquatic moths, and dipterans [48,49]. Rayon and silk-like proteins were predominant in both species in the Treja and Mignone rivers. While the blackfly larvae produce silk-like proteins, it should be noted that washing machine discharges can contain rayon fragments, which are then released into the environment [50] after flowing through wastewater treatment plants. Another potential source of rayon in the environment is the decomposition of cigarette butts unwisely abandoned by tourists in the woods at the most visited places near the Treja and Mignone rivers.

However, additives and plasticizers are often overlooked. These compounds are added to polymers to impart specific features and can be released into the environments when plastic objects and macroplastics are broken into smaller fragments [51,52]; thus, they can be employed as proxies of the presence of polymers. Moreover, additives and plasticizers can exert toxic effects on biota [14]; therefore, the quantification and the identification of these compounds are relevant for an in-depth knowledge of plastic pollution and the potential hazards for biota in the whole trophic web.

Additives are, e.g., PMAA (polymethylacrylamide) employed as a flocculant in wastewater treatment and coatings such as those found in specimens for both rivers (i.e., PEAA-Zinc); TBBA (tetrabromobisphenol A), employed as a flame retardant and present in sewage sludge; and PMDI (methylene diphenyl diisocyanate), which is employed for polyurethane manufacturing.

Cellulose ingested by the organisms might not be human-made but rather part of the food they usually eat. The other compounds ingested by the organisms have the most diverse usages. While Sulfar® is a fungicide used for vine cultivation, pyrrolidone is employed in pharmaceutics and as an additive for inkjet cartridges; these compounds are generally contained in plastic packaging, and their residues may have remained on plastic fragments that were subsequently ingested. Zein is a component of biopolymers.

Due to their sizes (<50 μm in length), most of these compounds may reach the two rivers alongside water leaving the treatment plants in the area (for instance, near the sampling site at the Treja River, there is a wastewater treatment plant at Mazzano Romano). It is worth noting that polyurethane was not found in the specimens collected, but the specific additive PMDI was identified. Hence, additives and plasticizers may be significant proxies of plastic polymers.

5. Conclusions

This study is the first to show that blackfly larvae (Simuliidae), members of a cosmopolitan insect family employed to test the quality of river waters via several status assessment methods, can ingest SMPs and APFs in their own habitat. Moreover, this is the first study to show that additives and plasticizers can be ingested by biota. The quantification and identification of additives and plasticizers will be relevant to assessing the MPs' pollution and the potential threat they may pose to biota.

The pretreatment method allowed for retrieval of the ingested SMPs and APFs simultaneously and efficiently because the yield is >90%. Moreover, the pretreatment method employed did not further denaturate the polymers that could be optimally identified, as shown by the identification of PA; this polymer can be easily overlooked due to the temperatures and aggressive reagents employed, resulting in an underestimation of the actual MP abundance.

Statistically significant differences were observed intra-species in the abundance of SMPs and APFs at both the Treja and Mignone sites under examination, which are used to survey river water quality. Further, relevant statistical differences were observed inter-species in each river under investigation. Based on these preliminary results, it is somewhat difficult to address differences related to the feeding behavior of the larvae of these two species in the two rivers studied; these differences may be related to several environmental, ecological, biological, and chemical factors. However, the results of this study can be relevant to further thorough studies of the various links among the factors mentioned above.

Investigating what has been ingested by the larvae of *S. ornatum* and *S. equinum* may account for the environmental impacts, hazards, and threats that pollutants such as SMPs and APFs may pose to biota and the good environmental quality status of river waters. Since Simulidae are commonly used in biomonitoring to assess riverine waters' ecological conditions (European Water Framework Directive 2000/60/EC), these preliminary data could aid further in-depth investigations of blackfly larvae and their potential role as bioindicators of microplastic pollution.

Supplementary Materials: The following supporting information can be downloaded at: https://www.mdpi.com/article/10.3390/toxics10070383/s1, Figure S1: Polymer spectra collected, as an example, some of the spectra identified with match percentages greater than 85% are shown.

Author Contributions: F.C.: Conceptualization; methodology; validation; formal analysis; supervision; resources; data curation; writing—original draft; writing—review and editing. B.R.: methodology, validation; investigation, writing—original draft; writing—review and editing. V.I.: conceptualization; validation; resources; writing—review and editing. S.C.: resources; writing—original draft; writing—review and editing; B.B.: resources, writing—review and editing. C.B.: writing—review and editing. All authors have read and agreed to the published version of the manuscript.

Funding: This research received no external funding.

Institutional Review Board Statement: Not applicable.

Informed Consent Statement: Not applicable.

Data Availability Statement: Not applicable.

Acknowledgments: The authors thank Elga Lab Water, High Wycombe UK, which supplied the pure water system used in this study. The authors would like to thank two anonymous English-speaking reviewers for carefully editing the proper English language, grammar, punctuation, spelling, and style. The authors would also like to thank the two anonymous reviewers, the anonymous academic editor, and the guest editor, whose insightful comments contributed to the improvement of the text.

Conflicts of Interest: The authors declare no conflict of interest.

References

1. Iannilli, V.; Pasquali, V.; Setini, A.; Corami, F. First evidence of microplastics ingestion in benthic amphipods from Svalbard. *Environ. Res.* **2019**, *179*, 108811. [CrossRef] [PubMed]
2. Sfriso, A.A.; Tomio, Y.; Rosso, B.; Gambaro, A.; Sfriso, A.; Corami, F.; Rastelli, E.; Corinaldesi, C.; Mistri, M.; Munari, C. Microplastic accumulation in benthic invertebrates in Terra Nova Bay (Ross Sea, Antarctica). *Environ. Int.* **2020**, *137*, 105587. [CrossRef]
3. Fang, C.; Zheng, R.; Hong, F.; Jiang, Y.; Chen, J.; Lin, H.; Lin, L.; Lei, R.; Bailey, C.; Bo, J. Microplastics in three typical benthic species from the Arctic: Occurrence, characteristics, sources, and environmental implications. *Environ. Res.* **2021**, *192*, 110326. [CrossRef] [PubMed]
4. Alfaro-Núñez, A.; Astorga, D.; Cáceres-Farías, L.; Bastidas, L.; Villegas, C.S.; Macay, K.C.; Christensen, J.H. Microplastic pollution in seawater and marine organisms across the Tropical Eastern Pacific and Galápagos. *Sci. Rep.* **2021**, *11*, 6424. [CrossRef]
5. Cole, M.; Lindeque, P.; Fileman, E.; Halsband, C.; Goodhead, R.; Moger, J.; Galloway, T.S. Microplastic ingestion by zooplankton. *Env. Sci. Technol.* **2013**, *47*, 6646–6655. [CrossRef] [PubMed]
6. Cho, Y.; Shim, W.J.; Jang, M.; Han, G.M.; Hong, S.H. Nationwide monitoring of microplastics in bivalves from the coastal environment of Korea. *Environ. Poll.* **2021**, *270*, 116175. [CrossRef] [PubMed]
7. Corami, F.; Rosso, B.; Roman, M.; Picone, M.; Gambaro, A.; Barbante, C. Evidence of small microplastics (<100 μm) ingestion by Pacific oysters (Crassostrea gigas): A novel method of extraction, purification, and analysis using Micro-FTIR. *Mar. Pollut. Bull.* **2020**, *160*, 111606. [CrossRef] [PubMed]
8. Jiang, Y.; Yang, F.; Hassan Kazmi, S.S.U.; Zhao, Y.; Chen, M.; Wang, J. A review of microplastic pollution in seawater, sediments and organisms of the Chinese coastal and marginal seas. *Chemosphere* **2022**, *286*, 131677. [CrossRef]
9. Hurley, R.R.; Woodward, J.C.; Rothwell, J.J. Ingestion of microplastics by freshwater Tubifex worms. *Environ. Sci. Technol.* **2017**, *51*, 12844–12851. [CrossRef]
10. Windsor, F.M.; Tilley, R.M.; Tyler, C.R.; Ormerod, S.J. Microplastic ingestion by riverine macroinvertebrates. *Sci. Total Environ.* **2019**, *646*, 68–74. [CrossRef]
11. Silva, C.J.M.; Silva, A.L.P.; Gravato, C.; Pestana, J.L.T. Ingestion of small-sized and irregularly shaped polyethylene microplastics affect *Chironomus riparius* life-history traits. *Sci. Total Environ.* **2019**, *672*, 862–868. [CrossRef] [PubMed]
12. Akindele, E.O.; Ehlers, S.M.; Koop, J.H.E. Freshwater insects of different feeding guilds ingest microplastics in two Gulf of Guinea tributaries in Nigeria. *Environ. Sci. Pollut. Res.* **2020**, *27*, 33373–33379. [CrossRef] [PubMed]
13. Corami, F.; Rosso, B.; Bravo, B.; Gambaro, A.; Barbante, C. A novel method for purification, quantitative analysis, and characterization of microplastic fibers using Micro-FTIR. *Chemosphere* **2020**, *238*, 124564. [CrossRef]
14. Beiras, R.; Verdejo, E.; Campoy-López, P.; Vidal-Liñán, L. Aquatic toxicity of chemically defined microplastics can be explained by functional additives. *J. Hazard. Mater.* **2021**, *406*, 124338. [CrossRef] [PubMed]

15. Adler, P.H. World Blackflies (Diptera: Simuliidae): A Comprehensive Revision of the Taxonomic and Geographical Inventory. 2021. Available online: https://biomia.sites.clemson.edu/pdfs/blackflyinventory.pdf (accessed on 7 July 2022).

16. Ciadamidaro, S.; Mancini, L.; Rivosecchi, L. Black flies (*Diptera, Simuliidae*) as ecological indicators of stream ecosystem health in an urbanizing area (Rome, Italy). *Ann. Ist. Super Sanità* **2016**, *52*, 269–276. [CrossRef] [PubMed]

17. Hart, D.D.; Merz, R.A.; Genovese, S.J.; Clark, B.D. Feeding postures of suspension-feeding larval black flies: The conflicting demands of drag and food acquisition. *Oecologia* **1991**, *85*, 457–463. [CrossRef] [PubMed]

18. Currie, D.C.; Craig, D.A. Feeding strategies of larval Black Flies. In *Black Flies. Ecology, Population Management and Annotated World*; Kim, K.C., Merritt, R.W., Eds.; The Pennsylvania University Press: Philadelphia, PA, USA, 1986; pp. 155–170.

19. Al-Azzawi, M.S.M.; Kefer, S.; Weißer, J.; Reichel, J.; Schwaller, C.; Glas, K.; Knoop, O.; Drewes, J.E. Validation of sample preparation methods for microplastic analysis in wastewater matrices—reproducibility and standardization. *Water* **2020**, *12*, 2445. [CrossRef]

20. Corami, F.; Rosso, B.; Morabito, E.; Rensi, V.; Gambaro, A.; Barbante, C. Small microplastics (<100 µm), plasticizers and additives in seawater and sediments: Oleo-extraction, purification, quantification, and polymer characterization using Micro-FTIR. *Sci. Total Environ.* **2021**, *797*, 148937. [CrossRef]

21. Andersen, R.R. *Algal Culturing Technique*; Elsevier Press: Amsterdam, The Netherlands, 2005; pp. 1–589.

22. Brierley, B.; Carvalho, L.; Davies, S.; Krokowski, J. *Guidance on the Quantitative Analysis of Phytoplankton in Freshwater Samples*; NERC Open Research Archive: Edgewater, MD, USA, 2007; pp. 1–24.

23. Comtois, P.; Alcazar, P.; Neron, D. Pollen counts statistics and its relevance to precision. *Aerobiologia* **1999**, *15*, 19–28. [CrossRef]

24. Gough, H.L.; Stahl, D.A. Optimization of direct cell counting in sediment. *J. Microbiol. Methods* **2003**, *52*, 39–46. [CrossRef]

25. Mazziotti, C.; Fiocca, A.; Vadrucci, M.R. Phytoplankton in transitional waters: Sedimentation and Counting methods. *TWB J.* **2013**, *2*, 90–99.

26. Lisle, J.T.; Hamilton, M.A.; Willse, A.R.; McFeters, G.A. Comparison of Fluorescence Microscopy and Solid-Phase Cytometry Methods for Counting Bacteria in Water. *Appl. Environ. Microbiol.* **2004**, *70*, 5343–5348. [CrossRef] [PubMed]

27. Muthukrishnan, T.; Govender, A.; Dobretsov, S.; Abed, R.M.M. Evaluating the Reliability of Counting Bacteria Using Epifluorescence Microscopy. *J. Mar. Sci. Eng.* **2017**, *5*, 4. [CrossRef]

28. Huppertsberg, S.; Knepper, T.P. Instrumental analysis of microplastics—benefits and challenges. *Anal. Bioanal. Chem.* **2018**, *410*, 6343–6352. [CrossRef] [PubMed]

29. Oßmann, B.E.; Sarau, G.; Holtmannspötter, H.; Pischetsrieder, M.; Christiansen, S.H.; Dicke, W. Small-sized microplastics and pigmented particles in bottled mineral water. *Water Res.* **2018**, *141*, 307–316. [CrossRef] [PubMed]

30. Tong, H.; Jiang, Q.; Hu, X.; Zhong, X. Occurrence and identification of microplastics in tap water from China. *Chemosphere* **2020**, *252*, 126493. [CrossRef]

31. Imhof, H.K.; Laforsch, C.; Wiesheu, A.C.; Schmid, J.; Anger, P.M.; Niessner, R.; Ivleva, N.P. Pigments and plastic in limnetic ecosystems: A qualitative and quantitative study on microparticles of different size classes. *Water Res.* **2016**, *98*, 64–74. [CrossRef]

32. Bharti, S.K.; Kumar, D.; Anand, S.; Barman, S.C.; Kumar, N. Characterization and morphological analysis of individual aerosol of PM10 in urban area of Lucknow, India. *Micron* **2017**, *103*, 90–98. [CrossRef]

33. Hamacher-Barth, E.; Jansson, K.; Leck, C. A method for sizing submicrometer particles in air collected on Formvar films and imaged by scanning electron microscope. *Atmos. Meas. Tech.* **2013**, *6*, 3459–3475. [CrossRef]

34. Filella, M. Questions of size and numbers in environmental research on microplastics: Methodological and conceptual aspects. *Environ. Chem.* **2015**, *12*, 527–538. [CrossRef]

35. Courtene-Jones, W.; Quinn, B.; Gary, S.F.; Mogg, A.O.M.; Narayanaswamy, B.E. Microplastic pollution identified in deep-sea water and ingested by benthic invertebrates in the Rockall Trough, North Atlantic Ocean. *Environ. Poll.* **2017**, *231*, 271–280. [CrossRef] [PubMed]

36. Leads, R.R.; Weinstein, J.E. Occurrence of tire wear particles and other microplastics within the tributaries of the Charleston Harbor Estuary, South Carolina, USA. *Mar. Poll. Bull.* **2019**, *145*, 569–582. [CrossRef] [PubMed]

37. Ribeiro-Brasil, D.R.G.; Schlemmer, B.L.; Veloso, G.K.O.; Pio de Matos, T.; Silva de Lima, E.; Dias-Silva, K. The impacts of plastics on aquatic insects. *Sci. Total Environ.* **2022**, *813*, 152436. [CrossRef] [PubMed]

38. Ehlers, S.; Manz, W.; Koop, J. Microplastics of different characteristics are incorporated into the larval cases of the freshwater caddisfly *Lepidostoma basale. Aquat. Biol.* **2019**, *28*, 67–77. [CrossRef]

39. Gallitelli, L.; Cesarini, G.; Cera, A.; Sighicelli, M.; Lecce, F.; Menegoni, P.; Scalici, M. Transport and Deposition of Microplastics and Mesoplastics along the River Course: A Case Study of a Small River in Central Italy. *Hydrology* **2020**, *7*, 90. [CrossRef]

40. Khosrovyan, A.; Gabrielyan, B.; Kahru, A. Ingestion and effects of virgin polyamide microplastics on *Chironomus riparius* adult larvae and adult zebrafish *Danio rerio. Chemosphere* **2020**, *259*, 127456. [CrossRef]

41. Immerschitt, I.; Martens, A. Ejection, ingestion and fragmentation of mesoplastic fibres to microplastics by Anax imperator larvae (Odonata: Aeshnidae). *Odonatologica* **2021**, *49*, 57–66.

42. Stanković, J.; Milošević, D.; Savić-Zdraković, D.; Yalçın, G.; Yildiz, D.; Beklioğlu, M.; Jovanović, B. Exposure to a microplastic mixture is altering the life traits and is causing deformities in the non-biting midge *Chironomus riparius* meigen (1804). *Environ. Pollut.* **2020**, *262*, 114248. [CrossRef]

43. Ziajahromi, S.; Kumar, A.; Neale, P.A.; Leusch, F.D.L. Effects of polyethylene microplastics on the acute toxicity of a synthetic pyrethroid to midge larvae (*Chironomus tepperi*) in synthetic and river water. *Sci. Total Environ.* **2019**, *671*, 971–975. [CrossRef]

44. Chinfak, N.; Sompongchaiyakul, P.; Charoenpong, C.; Shi, H.; Yeemin, T.; Zhang, J. Abundance, composition, and fate of microplastics in water, sediment, and shellfish in the Tapi-Phumduang River system and Bandon Bay, Thailand. *Sci. Total Environ.* **2021**, *781*, 146700. [CrossRef]
45. Pegado, T.D.S.E.S.; Schmid, K.; Winemiller, K.O.; Chelazzi, D.; Cincinelli, A.; Dei, T.G.L. First evidence of microplastic ingestion by fishes from the Amazon River estuary. *Mar. Poll. Bull.* **2018**, *133*, 814–821. [CrossRef] [PubMed]
46. Iannilli, V.; Corami, F.; Grasso, P.; Lecce, F.; Buttinelli, M.; Setini, A. Plastic abundance and seasonal variation on the shorelines of three volcanic lakes in Central Italy: Can amphipods help detect contamination? *Environ. Sci. Poll. Res.* **2020**, *27*, 14711–14722. [CrossRef] [PubMed]
47. Savoca, S.; Matanović, K.; d'Angelo, G.; Vetri, V.; Anselmo, S.; Bottari, T.; Mancuso, M.; Kužir, S.; Spanò, N.; Capillo, G.; et al. Ingestion of plastic and non-plastic microfibers by farmed gilthead sea bream (*Sparus aurata*) and common carp (*Cyprinus carpio*) at different life stages. *Sci. Total Environ.* **2021**, *782*, 146851. [CrossRef]
48. Bétard, F. Insects as zoogeomorphic agents: An extended review. *Earth Surf. Processes Landf.* **2021**, *46*, 89–109. [CrossRef]
49. Hamann, L.; Blanke, A. Suspension feeders: Diversity, principles of particle separation and biomimetic potential. *J. R. Soc. Interface* **2022**, *19*, 20210741. [CrossRef]
50. Hamidian, A.H.; Ozumchelouei, E.J.; Feizi, F.; Wu, C.; Zhang, Y.; Yang, M. A review on the characteristics of microplastics in wastewater treatment plants: A source for toxic chemicals. *J. Clean. Prod.* **2021**, *295*, 126480. [CrossRef]
51. Wong, J.K.H.; Lee, K.K.; Tang, K.H.D.; Yap, P.-S. Microplastics in the freshwater and terrestrial environments: Prevalence, fates, impacts and sustainable solutions. *Sci. Total Environ.* **2020**, *719*, 137512. [CrossRef]
52. Huang, W.; Song, B.; Liang, J.; Niu, Q.; Zeng, G.; Shen, M.; Deng, J.; Luo, Y.; Wen, X.; Zhang, Y. Microplastics and associated contaminants in the aquatic environment: A review on their ecotoxicological effects, trophic transfer, and potential impacts to human health. *J. Hazard. Mater.* **2020**, *405*, 124187. [CrossRef]

Article

Size Effects of Microplastics on Embryos and Observation of Toxicity Kinetics in Larvae of Grass Carp (*Ctenopharyngodon idella*)

Chaonan Zhang [1,2,†], Zhiheng Zuo [1,†], Qiujie Wang [1], Shaodan Wang [1], Liqun Lv [3] and Jixing Zou [1,2,*]

[1] Joint Laboratory of Guangdong Province and Hong Kong Region on Marine Bioresource Conservation and Exploitation, College of Marine Sciences, South China Agricultural University, Guangzhou 510642, China; zhangchaonan@stu.scau.edu.cn (C.Z.); zuozhiheng@stu.scau.edu.cn (Z.Z.); wongchaukit@stu.scau.edu.cn (Q.W.); 20201067007@stu.scau.edu.cn (S.W.)

[2] Guangdong Laboratory for Lingnan Modern Agriculture, South China Agricultural University, Guangzhou 510642, China

[3] National Pathogen Collection Center for Aquatic Animals, Shanghai Ocean University, Shanghai 201306, China; lqlv@shou.edu.cn

* Correspondence: zoujixing@scau.edu.cn

† These authors contributed equally to this article.

Abstract: Microplastics have caused great concern in recent years. However, few studies have compared the toxicity of different sizes of microplastics in fishes, especially commercial fishes, which are more related to human health. In the present study, we revealed the effects of varying sizes of microplastics on grass carp embryos and larvae using scanning electron microscopy (SEM) and fluorescence imaging. Embryos were exposed to 80 nm and 8 μm microplastics at concentrations of 5, 15, and 45 mg/L. Toxicity kinetics of various sizes of fluorescent microplastics were analyzed through microscopic observation in the larvae. Results found that nanoplastics could not penetrate the embryo's chorionic membrane, instead they conglutinated or aggregated on the chorion. Our results are the first to explore the defense mechanisms of commercial fish embryos against microplastics. Larvae were prone to ingesting their own excrement, resulting in microplastic flocculants winding around their mouth. For the first time, it was found that excreted microplastics could be reconsumed by fish and reaccumulated in the oral cavity. Microplastics of a certain size (1 μm) could be accumulated in the nasal cavity. We speculate that the presence of a special groove structure in the nasal cavity of grass carp larvae may manage to seize the microplastics with a particular size. As far as we know, this is the first report of microplastics being found in the nasal passages of fish. Fluorescence images clearly recorded the toxicity kinetics of microplastics in herbivorous fish.

Keywords: microplastic; grass carp; size; accumulation; re-consumption

Citation: Zhang, C.; Zuo, Z.; Wang, Q.; Wang, S.; Lv, L.; Zou, J. Size Effects of Microplastics on Embryos and Observation of Toxicity Kinetics in Larvae of Grass Carp (*Ctenopharyngodon idella*). *Toxics* **2022**, *10*, 76. https://doi.org/10.3390/toxics10020076

Academic Editors: Costanza Scopetani, Tania Martellini and Diana Campos

Received: 22 December 2021
Accepted: 2 February 2022
Published: 7 February 2022

1. Introduction

The last five years have witnessed a rapid surge of published articles on microplastic pollution, which testifies to the great concern this pollutant has posed in recent years [1,2]. Although first raised as an issue by Thompson et al., 2004 [3], microplastics were first discovered in North America in the 1970s in the form of small spheres in plankton off the coast of New England [4]. Subsequently, other researchers also found that these tiny particles were not only in the aquatic environment [5–7], but also in soil [8,9], organisms [10–12], and even in the atmosphere [13,14]. According to the US National Oceanic and Atmospheric Administration (NOAA) in 2008, plastics smaller than 5 mm in size were identified as microplastics (MPs) [15]. With the development of cognition and technology, smaller microplastics were classified into nanoplastics (NPs). Although not clearly defined, particles within 100 nm in scale were commonly referred to as nanoplastics [16–18]. The 21st century has been called the age of plastics [19], largely because plastics are indispensable in contemporary life. Unfortunately, used plastics are not recycled or managed well, resulting in an increasing amount of waste getting discarded into the environment every year [20,21]. After physical, chemical, and biological degradation, plastics turn into microplastics or nanoplastics, which have become a threat to the ecological environment and human health [22,23]. People are now horrified by their huge numbers and extremely worried about the potential threat microplastics pose when they enter living organisms, because it means the plastics could threaten our health through the food chain, and even through drinking and simply breathing [24–26].

Many researchers have focused on the impact of microplastics on aquatic organisms, especially on algae [27–29] and shellfish [30,31], whereas relatively few studies have been conducted on fish [32,33]. In addition to the type, shape, concentration, and color of microplastics, particle size is one of the key factors influencing microplastics toxicological effects [34–36]. In general, the smaller particle size, the more toxic they are to organisms [16–18]. Specifically, on the one hand, microplastics with larger specific surface areas can adsorb more pollutants, resulting in enhanced toxicity. On the other hand, the smaller size of the microplastics, the longer they are retained in the body, increasing the risk of potential damage. For example, Ivleva et al. (2017) found rapid accumulation of <15 µm microplastics and concluded that smaller particles were of more concern than the larger ones [37]. Both 0.05 and 10 µm microplastics increased oxidative stress in marine copepod, but smaller microplastics raised more reactive oxygen species (ROS) [38]. The growth and reproduction of copepod showed a size-dependent decline after exposure to microplastics for 16 d [39]. These studies speculated that the effects of microplastics with different sizes on organisms are different, and toxicity usually increases with decreasing size. However, few studies compared the toxicity of varying sizes of microplastics in fish, especially commercial fish. Commercial fish refers to fish that can be bought in the market and cooked in the kitchen, and are more directly related to human health.

Compared to adult fishes, larvae are more sensitive to environmental stress [40,41]. Especially in its early stages, the pigment on the fish body surface is not fully formed, but the fish can feed and swim freely, making them ideal specimens to study dynamic distribution processes of microplastics in the body [42]. Fish eggs with lipophilic chorionic membranes could be potential surfaces for increased microplastic deposition and accumulation. Both periods (the larval and eggs) are critical for fish populations because of their high sensitivity to pollutants [43,44]. Batel et al. (2018) found that smaller and heavier microplastics (1–5 mm) accumulated in high numbers on the surface of zebrafish egg chorions [45]. Zhang et al. (2020) speculated that weak physical forces and/or electrostatic interactions operated between the chorion membrane and microplastics [46]. Fluorescence images of accumulation and egestion of microplastics in filter feeding tadpoles (*Xenopus tropicalis*) were concentration dependent [47]. The impacts of microplastics on embryo and larval fish can be directly reflected by fluorescence micrograph and SEM images. Our research group have focused on the differences of toxicity kinetics of microplastics in larvae with three feeding types and found that the effects of microplastics on fish were species-specific [42]. The results showed that the ingestion of microplastics in hybrid snakehead (carnivores) was lower than that in bighead carp (filter feeders) and mrigal (omnivores), while mrigal larvae were less effective to remove microplastics than bighead carp larvae. There is little research available on herbivorous fish [48], since this species is fewer, and samples are hard to obtain. However, grass carp (*Ctenopharyngodon idella*), as the typical representative of herbivorous fish, is a commercial fish with the largest amount of aquaculture in China [49,50].

In the present study, grass carp embryo and larvae were the model organisms, and different sizes of polystyrene microspheres were the exposure xenobiotics. Embryos at 12 h post fertilization (hpf) were exposed to 0.08 and 8 µm microplastics at various concentrations. In order to facilitate observation, green and red fluorescent microplastics were selected to visually reflect the dynamic distribution processes of microplastics in larvae. Toxicity kinetics of microplastics were analyzed through microscopic observation. This is the first study to investigate the accumulation, distribution, and egestion of microplastics in grass carp larvae. Therefore, our results aimed at bridging the gap on effects of microplastics in herbivorous fish.

2. Materials and Methods

2.1. Microplastics and Fish

We used microspheres with mean diameters of 0.08 and 8 μm (Dae Technology Co., Ltd., Tianjin, China) for the embryo toxicity assay, and fluorescent microspheres for larval exposure and elimination experiments. Green fluorescent polystyrene microspheres (excitation wavelength: 488 nm; emission wavelength: 518 nm) with mean diameters of 0.5 and 5 μm were purchased from Dae Technology Co., Ltd. (Tianjin, China). Orange fluorescent polystyrene microspheres (excitation wavelength: 540 nm; emission wavelength: 580 nm) with mean diameter of 1 μm were bought from the same company. Red fluorescent polystyrene microspheres (excitation wavelength: 620 nm; emission wavelength: 680 nm) with mean diameter of 5 μm were bought from Tianjin BaseLine ChromTech Research Centre (Tianjin, China). SEM figures of all kinds of microspheres are shown in Supplementary Figure S1.

The embryos of grass carp obtained from a stock farm in Qingyuan city, Guangdong Province, China, were packed in oxygenated bags and transferred to the lab immediately. They were then acclimatized in a 100 L glass tank prior to the exposure test. The dechlorinated circulating water conditions were as follows: water temperature 25.4 ± 1.3 °C, pH 7.0 ± 0.3, dissolved oxygen 6.5 ± 0.6 mg/L, and 14 h light/10 h dark photoperiod. The animals used in the present study were cultured and sacrificed following the terms of use of animals approved by the Animal Care and Use Committee of South China Agricultural University (identification code: 20210236; date of approval: 27 May 2021).

2.2. Embryo Toxicity Assay

The experimental embryos of grass carp were all in organogenesis stage (12 hpf). Microspheres with two sizes (0.08 and 8 μm) and at three concentrations (5, 15, and 45 mg/L) were used for the embryo toxicity assay. Each of the 15 embryos were assigned to glass Petri dishes with a diameter of 5 cm containing 5 mL test solution at random. There were two control groups that did not contain microplastics. The experiment was repeated three times. A total of 360 individuals and 24 glass Petri dishes were used. Embryo mortality was observed and recorded every two hours. The embryos were considered dead when they turned white.

2.3. SEM Analysis of Embryo

After 2, 4, 6, and 8 h exposure, embryos were collected and analyzed as described by [42,51], with slight modifications. The two sample preparation methods are as follows: (a) critical point drying: embryos were fixed in 4% paraformaldehyde for more than 24 h, rinsed thrice with 0.1 M phosphate-buffered saline (PBS, pH 7.4) for 15 min, and postfixed with 1% osmium tetroxide for 1.5 h at room temperature. Dehydration was carried out sequentially with ethanol concentrations of 30%, 50%, 70%, and 90% once for 10 min, followed by 100% ethanol twice for 10 min. After dehydration, samples were replaced with isopentyl acetate twice for 15 min, then dried in critical point desiccators (EP CPD300, Leica, Germany) overnight and stored at room temperature for SEM analysis; (b) freeze drying method: embryos were fixed in 3% glutaraldehyde for more than 24 h, and rinsed six times with 0.1 M PBS for 20 min. The dehydration procedure was similar to method (a), followed by replacement with tert-butanol twice for 20 min. After dehydration, embryos were dried in a vacuum freeze dryer (ES2030, Hitachi, Japan) and stored at room temperature for SEM analysis.

Before observation, samples were sputter-coated with an electrically conductive gold-palladium alloy in vacuum via a High Vacuum Sputter Coater (Leica EM ACE600, Germany). SEM images were taken with a Zeiss EVO MA 15 scanning electron microscope

(Carl Zeiss AG, Jena, Germany) and FEI Verios 460 scanning electron microscope (Thermo Fisher Scientific, Waltham, MA, USA).

2.4. Exposure and Elimination Experiment of Larvae

The experimental larvae were hatched from normal fertilized eggs in clean water. We chose larvae that hatched after 24 h for the exposure and elimination experiment. They were exposed to 10 mg/L microplastics with diameters of 0.5 and 5 μm (green fluorescent microplastics) and 1 and 5 μm (red fluorescent microplastics), respectively, for four days. During the experiment, five samples from each group were taken out every 12 h and rinsed with clean water, and photographed under the fluorescence microscope (Nikon C-HGFI) equipped with a Nikon SMZ18 camera.

For the elimination experiment, the remanent larvae were transferred into 200 mL glass beakers containing clean water for four days. Each of the three samples were chosen every 12 h, rinsed with clean water, and photographed as described before.

3. Results

3.1. Effects of Microplastics on Embryos

There were no significant differences in the survival rates of grass carp embryos among all groups after 8 h exposure (Supplementary Figure S2). Even in a very high concentration of microplastics (45 mg/L), embryos could still hatch normally. There was no difference in morphology or fetal heart rate either.

3.2. Effects of Microplastics on Chorion Membranes

In order to maintain the stereoscopic morphology of the embryo, we used two sample preparation methods for SEM analysis. Unfortunately, the size of the fertilized eggs of grass carp was about 4 mm, and chorion membranes were shriveled or deformed to varying degrees after drying (Supplementary Figure S3) due to the technical difficulty.

High-definition enlarged images showed that the membrane surface was uneven, and there were many irregular protuberances (Figure 1). 80 nm microplastics were conglutinated or aggregated on the embryo chorion (Figure 2). The pore structures were observed in some embryos (Figure 3), but whether they were caused by microplastics was unclear. In critical point drying, the pores on the membrane surface appeared to be torn open to show a fibrous structure (Figure 3C,D). In addition, rod-shaped bacteria appeared and attached to some of the membrane surface (Supplementary Figure S4).

3.3. Uptake and Accumulation of Green Fluorescent Microspheres in Grass Carp Larvae

Grass carp larvae (about 9 mm in length) were observed to the microplastics exposure experiment for four days. During the first 24 h of exposure, green autofluorescence was observed in the thoracic cavity of the larvae, both in the control (Supplementary Figure S5a,b) and exposed groups (Figure 4a,b). After three days of exposure, autofluorescence in the larvae faded, leaving remnant fluorescence in the yolk sac. Photographs of the control group under fluorescent lenses are shown in Supplementary Figure S5.

In the exposed group, 5 μm microplastics gradually accumulated in the intestines of the larval grass carps from 36 h to 60 h (Figure 4c–e). However, from 72 h to 96 h, there was no fluorescent signal in the intestines, and all the microplastics accumulated in the oral cavity (Figure 4f–h). Under a brightfield microscope, obvious flocculation could be observed around the oral cavity (Figure 4F–H).

In the exposed 72–96 h of 0.5 μm microplastics, the fluorescent particles in some of the intestinal tracts were not removed (Supplementary Figure S6f–h), while most of the microplastics accumulated in the oral cavity. The accumulation of 0.5 μm microplastics during 36–60 h was similar with that of 5 μm microplastics (Supplementary Figure S6c–e).

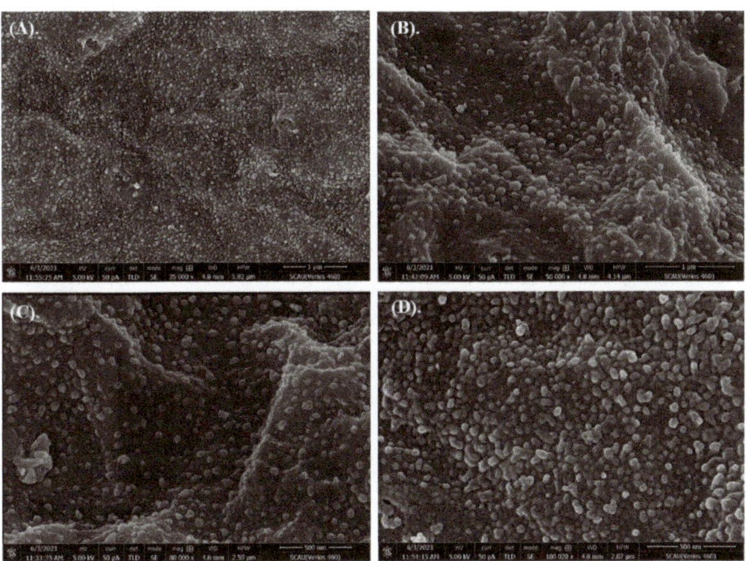

Figure 1. High-definition enlarged images of chorion membranes of grass carp. (**A**–**D**) show different parts of chorion membranes.

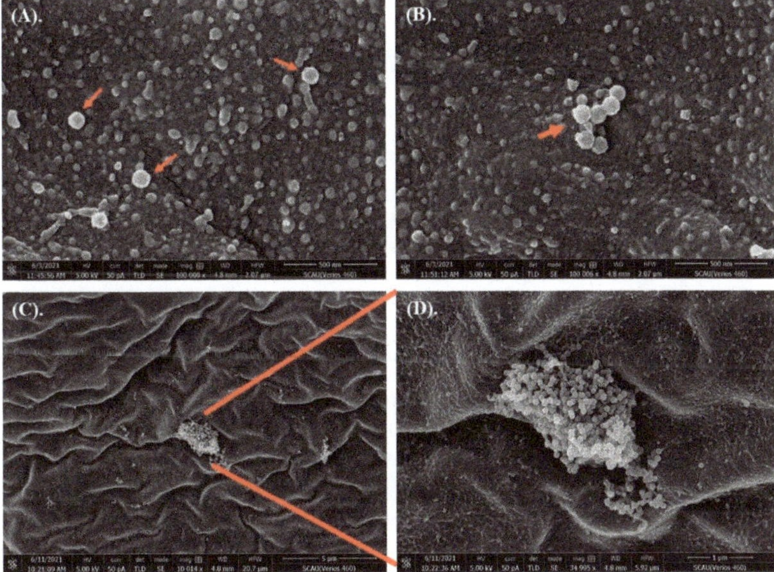

Figure 2. SEM images of the out-membrane surface of grass carp embryo after exposed to 80 nm microplastics. (**A**–**C**) show different status of microplastics on membranes. (**D**) is a larger version of (**C**).

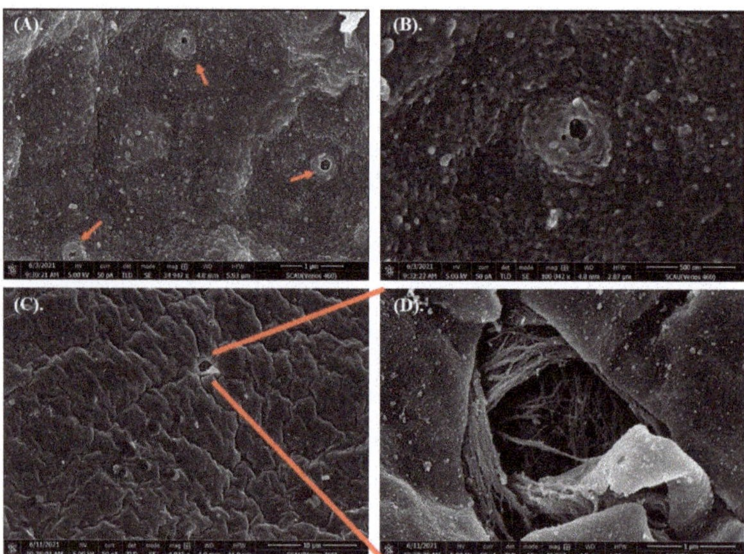

Figure 3. The pore structures of the out-membrane surface of grass carp embryo after exposed to microplastics. (**A**–**C**) show different pore structures. (**D**) is a larger version of (**C**).

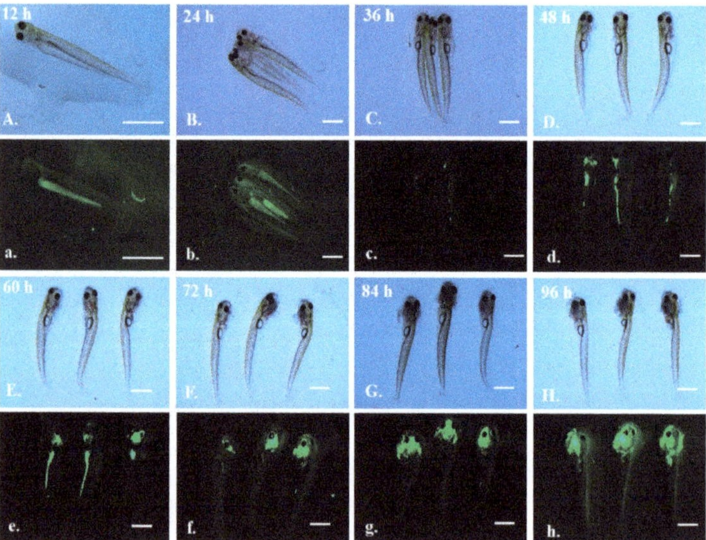

Figure 4. The larvae of grass carp after exposure to 5 μm green fluorescent microplastics. Photographs were taken under a brightfield microscope (capital letters **A**–**H**) and green fluorescent microscope (lowercase letters **a**–**h**). Observation time was labeled in the figure. Scale bar = 2 mm.

3.4. Uptake and Accumulation of Red Fluorescent Microspheres in Grass Carp Larvae

There was no red fluorescence in grass carp larvae of the control group (Supplementary Figure S7). However, grass carp larvae after exposure to 5 μm red fluorescent microplastics showed red autofluorescence in the thoracic cavity at 12–24 h (Figure 5a,b). After 36 h of exposure, red fluorescence appeared in a strip shape, indicating that the 5 µm red fluorescent microplastic had entered the intestines of the larval grass carp (Figure 5c–h). Autofluorescence in the thorax of grass carp was band-shaped. Unlike 5 µm green fluorescent microplastics, 5 µm red fluorescent microplastics accumulated in the intestines during exposure.

3.5. Elimination of Green Fluorescent Microspheres in Grass Carp Larvae

The elimination test also lasted for four days. No fluorescent microplastics were found in the intestines of grass carps in the control group (Supplementary Figure S8). As shown in Supplementary Figure S9, after 4 days of exposure to 5 µm green fluorescent microplastics, floccules and fluorescent substances around the oral cavity of the larval grass carps did not disappear during four days of the elimination test, while the larvae could swim normally. The cleaning situation was similar for larvae exposed to 0.5 µm green fluorescent microplastics (Supplementary Figure S10). It is worth noting that grass carps in the control group did not have flocculent entanglement near their mouths.

3.6. Elimination of Red Fluorescent Microspheres in Grass Carp Larvae

No fluorescent microplastics were found in the intestines of grass carps in the control group (Supplementary Figure S11). We observed that 5 µm red fluorescent microplastics accumulated in the intestines of grass carps during exposure. Over the elimination course of 48 h, microplastics were gradually removed from the intestines (Supplementary Figure S12a–d). During the 60–96 h of elimination, red fluorescence mainly concentrated in the oral cavity of grass carps, and floccules also appeared at this time (Supplementary Figure S12e–h).

The accumulation sites of 1 µm fluorescent microplastics were different from those of 5 µm fluorescent microplastics. At 24 h after exposure, red fluorescent signals appeared at the nose of the larval grass carp (Figure 6a,b). After 36 h of exposure, 1 µm microplastics gradually entered the intestines, but the red fluorescent signal in the nose was still not eliminated (Figure 6c–h). Notably, after 96 h, microplastics seemed to be more concentrated around the oral cavity (Figure 6h). Under a brightfield microscope, obvious floccules could be observed (Figure 6H).

Orange fluorescent microplastics with 1 µm size in the grass carp intestines were removed from the body at the early stage of the elimination experiment (within 12 h). However, the fluorescence in the nose always existed (Supplementary Figure S13). The close-up is shown in Figure 7. From the images of the larvae, we could not determine whether the fluorescence was in the nasal region. Compared with the appearance of adult grass carp (Supplementary Figure S14), we found that the nasal cavity of grass carp was very obvious.

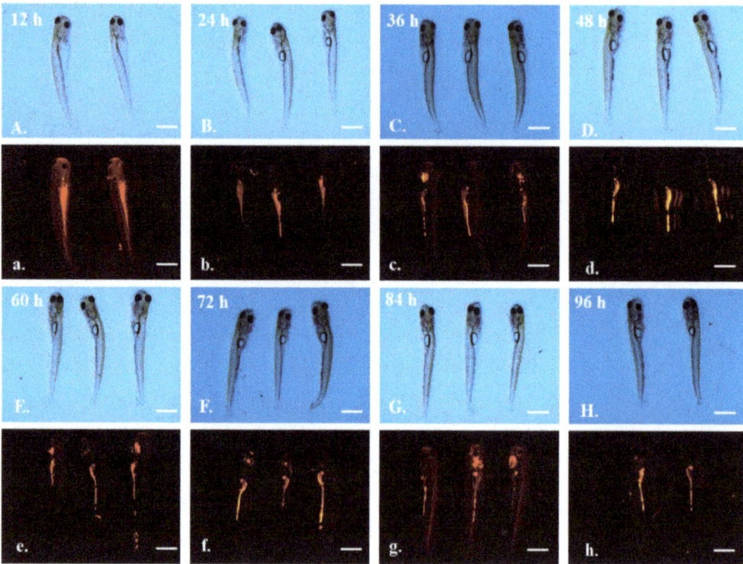

Figure 5. The larvae of grass carp after exposure to 5 μm red fluorescent microplastics. Photographs were taken under a brightfield microscope (capital letters **A–H**) and red fluorescent microscope (lowercase letters **a–h**). Observation time was labeled in the figure. Scale bar = 2 mm.

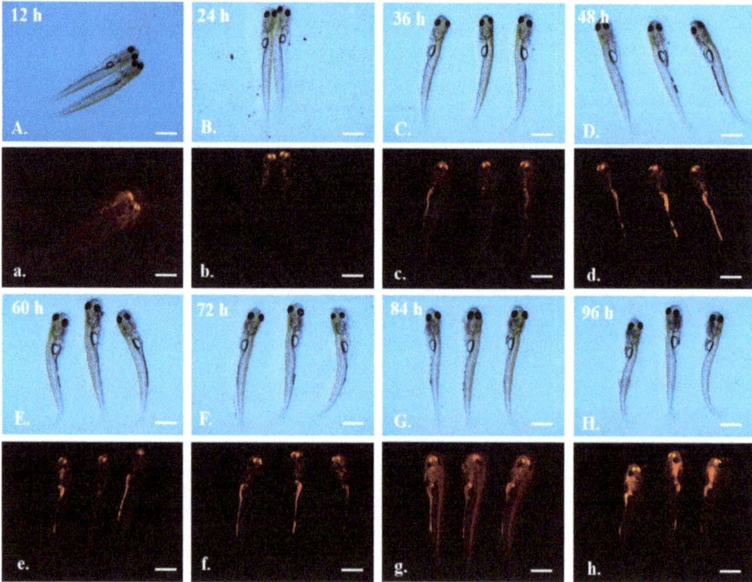

Figure 6. The larvae of grass carp after exposure to 1 μm orange fluorescent microplastics. Photographs were taken under a brightfield microscope (capital letters **A–H**) and red fluorescent microscope (lowercase letters **a–h**). Observation time was labeled in the figure. Scale bar = 2 mm.

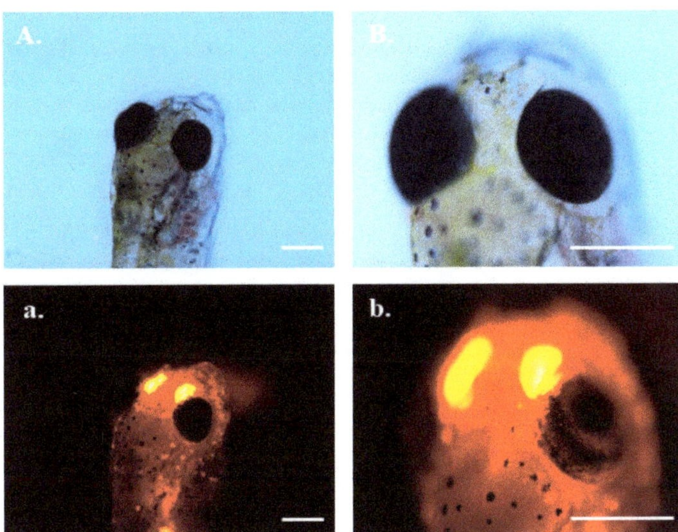

Figure 7. The larvae of grass carp after exposure to 1 μm red fluorescent microplastics. Photographs were taken under a brightfield microscope (capital letters **A,B**) and red fluorescent microscope (lowercase letters **a,b**). **B/b** is a larger version of **A/a**. Scale bar = 0.5 mm.

4. Discussion

4.1. Effects of Microplastics on Embryos

We studied the effects of microplastics of different sizes and varying concentrations on grass carp embryos. Results showed that embryos at 12 hpf were not affected by microplastics with nano size or high concentrations. SEM photos showed that microplastics centered and aggregated on the embryo chorion, but couldn't penetrate into the interior. Fertilization and development of fish eggs are in vitro. Nutrients needed for the development of the embryo come from the yolk, and there is little need to obtain nutrients or excrete waste from outside the embryo. During the development of the embryo, the dense chorionic membrane structure is helpful for protection, since the fish eggs have to face various environmental stresses. However, the function of irregular protuberances on the membrane surface (Figure 1) was unclear, and adverse effects caused by the tiny particles on chorion was unmeasurable. Our results were similar with [46], in which they also found that microplastics could be adsorbed on the outer membrane surface making the membrane layer irregular in zebrafish embryos after being exposed to 10 μm microplastics at 10 mg/L for 48 h. They deduced that there were weak physical forces and/or electrostatic interactions between the chorion membrane and microplastics. Another report showed that silver nanoparticles with an average diameter of 11.6 nm were passively diffused into zebrafish embryos through chorion pore canals [52]. However, most research results supported the conclusion that no overt embryotoxicity occurred when nanoparticles aggregated on the chorion of embryos [53].

Fish eggs can be divided into adhesive, pelagic, demersal, and floating eggs according to their specific gravity and viscosity. The zygotes of zebrafish, a model organism commonly used in the laboratory, are demersal eggs, which are characterized with a larger density than water and a smaller yolk gap [46]. However, the zygotes of grass carps used in this experiment are floating eggs, which are characterized by water absorption and expansion, large perivitelline space, and suspension in the water layer [54]. The differences in the surface chorionic membrane of various types of fish eggs might lead to the discrepancy in conglutination of microplastics, which have not been studied thoroughly. This could be of significant concern, and it is important to address the effects with individual differences.

4.2. Effects of Microplastics of Different Sizes on Fish

The effects of 5 μm microplastics with green and red fluorescence exposure results were not the same, which suggested the importance in the selection of microplastic materials. This is likely because different materials would obtain different experimental results. Even when different groups of researchers use microplastics of the same size as the material, cross-sectional comparisons should be treated with caution. Fluorescent dye-labeled microplastics bring convenience to observation, but also create a certain confusion. Catarino et al. (2019) found that manufactured fluorescent microplastics leached their fluorophores, and fluorophores possibly accumulated in the zebrafish gut, rather than the microplastics themselves [55]. By carefully comparing our experimental results with those of Catarino et al. (2019), we confirmed that what entered the grass carp guts were fluorescent microplastics, rather than fluorophores. The biggest difference was whether they were distributed in bands or strips in the body. However, although it was confirmed that they were the same particle size of 5 μm, the difference of toxicity kinetics in red and green fluorescent microplastics during the exposure experiment could not be accounted for. Commercial microplastic pellets, especially those with fluorescence, need to be carefully selected and considered.

The green fluorescent microplastics sized 0.5 and 5 μm showed no size-dependent effects. They both accumulated mainly in the digestive and oral tracts of grass carp larvae via oral ingestion regardless of exposure and depuration time. In general, small particles led to prolonged retention time and high bioavailability. A number of past results indicated that uptake of microplastics in organisms significantly depended on particle size. For example, Lu et al. (2016) found both 5 and 20 μm microplastics in the intestines and gills of adult zebrafish, while only the smaller-sized microplastics in the liver [56]. In addition, although no significant differences between histopathological changes were observed in the tissues for fish exposure to the 70 nm and 5 μm microplastics, larger-sized microplastics induced increased activities of superoxide dismutase (SOD) and catalase (CAT). Yang et al. (2020) found that 70 nm microplastics could enter the epidermis more easily than 5 μm microplastics in goldfish larvae, leading to muscle mesenchymal cell damage and nerve fiber atrophy [57]. The size-dependence effects of 0.05, 0.5 and 6 μm microplastics on rotifers were observed, such as reduction of growth rate, lifespan, and fecundity [39]. The size range of microplastics causing differences of biological effects is species-specific, which may be closely related to the organism's own tissue structure. Future research should focus on the interaction of microplastic size and the research object.

Interestingly, 1 μm orange fluorescent microplastics could accumulate in the nasal cavity of grass carp larvae, and could not be removed once they entered. We suspect that there is a special groove structure about 1 μm in the nasal cavity of grass carp larvae which manage to seize the microplastics with the particular size. As far as we know, this is the first report of microplastics being found in the nasal passages of fish. Recently, a study reported the accumulation of 23 nm microplastics in the brain of juvenile grass carp, which could cause multiple adverse effects, including impaired growth/development, behavioral changes, and anti-predatory defensive response associated with oxidative stress [58]. Another study found that microplastics were accumulated in gills close to blood vessels, indicating the respiratory system as one of the main egestion ways for microplastics in fish [59,60]. Microplastics with a diameter of 25 and 50 nm also accumulated in the eye, which could either be from outer epidermal or internal biodistribution through the intestinal epidermis [61]. The tissue specificity of microplastic accumulation in organisms and the resulting potential harm need to be studied further.

4.3. Excretion and Re-Consumption of Microplastics

In the 96 h of exposure, 5 μm red fluorescent microplastics accumulated in the digestive tract of grass carp larvae, and fluorescence intensity decreased during the elimination experiment. However, the green fluorescent microplastics, whether 0.5 or 5 μm in size, were excreted after 72 h exposure. The gut residence time of microplastics ingested by the fish seemed to be related to the fluorescent dye, independent of the size. But the retention time in rotifers likely correlated with the size of the microplastic [39]. The residence time of microplastics in organisms may depend on the gut space of organisms and the type, shape, size and concentration of the materials. The slow excretion of plastics might damage or block the digestive tract, thus affecting food consumption and the energy acquirement for vital functions. Moreover, longer retention times might prolong the negative effects. Most laboratory toxicology experiments use regular, smooth microspheres as experimental materials, which may have different residence times for experimental materials and field samples (such as fibers or fragments). The residence time of microplastics in fish and their effects are, however, still beyond our knowledge.

There was still strong fluorescent during depuration period, indicating that grass carp larvae could re-accumulate feces containing microplastics in the oral cavity. For the first time, it was found that excreted microplastics could be reconsumed by fish and reaccumulated in the oral cavity. We suspect that the mechanism of why the re-accumulated microplastics remained in the oral cavity is related to the mouth structure and fecal properties of grass carp larvae. The process of consuming-excreting-reconsuming microplastics may increase the potential for bioaccumulation. Such a process of reconsuming was not observed in the previous toxicity kinetics of carnivorous, omnivorous, and filter-feeding larvae [42]. Although most commercial freshwater fishes in the larval stage are planktivorous, the processes of uptake, accumulation, and elimination of microplastics are species-specific. Studies have shown that feces excreted by organisms after microplastics exposure carried microplastics, and changed the sedimentation rate, which was one of the major pathways for vertical translocation. Cole et al. (2016) hypothesized a mechanism in which floating plastics were transported out of surface water through a combination of microplastics and fecal pellets [62]. They found that the sinking rate of fecal pellets incorporated within microplastics decreased by 2.25-fold because of the reduction in density. However, another study pointed out that excreted polyethylene microplastics coated by intestinal liquids resulted in aggregation and sinking [36]. More studies are needed to further explain the deposition and transportation mechanisms of microplastics.

5. Conclusions

This study aimed to reveal the effects of varying microplastic particle sizes on grass carp embryos and larvae from the perspective of SEM and fluorescence imaging. The results showed that nanoplastics could not penetrate the chorionic membrane of the embryos, but could conglutinate and aggregate on the chorion. A high concentration of microplastics exposure did not affect the development of embryos during organ formation. Toxicity kinetics from green and red fluorescence microplastics with the same particle size (5 μm) exposure were unexpectedly different. Feces containing microplastics reaccumulated into the oral cavity. Green fluorescent microplastics of 0.5 and 5 μm showed no size-dependent effects. Microplastics of 1 μm accumulated in the nasal cavity. Further studies should pay more attention to the choice of microplastics as the materials and the fish as the model organisms.

Supplementary Materials: The following supporting information can be downloaded at: https:// www.mdpi.com/article/10.3390/toxics10020076/s1, Figure S1: SEM figures of all kinds of polystyrene microspheres: 5 μm green fluorescent microplastics (A); 0.5 μm green fluorescent microplastics (B); 5 μm red fluorescent microplastics (C); 1 μm orange fluorescent microplastics (D). Figure S2: The survival rates of grass carp embryos among all groups when exposure to 8 μm (A) and 80 nm (B) microspheres. Figure S3: Chorion membranes of grass carp after drying. The size of microplastics and exposure time are shown in the figure. Figure S4: Rod-shaped bacteria were attached to some of the membrane surface. Figure S5: The control group in brightfield microscope (capital letters) and green fluorescent microscope (lowercase letters). Observation time was labeled in the figure. Scale bar = 2 mm. Figure S6: The larvae of grass carp after exposure to 0.5 μm green fluorescent microplastics. Photographs were taken under a brightfield microscope (capital letters) and green fluorescent microscope (lowercase letters). Observation time was labeled in the figure. Scale bar = 2 mm. Figure S7: The control group in brightfield microscope (capital letters) and red fluorescent microscope (lowercase letters). Observation time was labeled in the figure. Scale bar = 2 mm. Figure S8: The control group during elimination test in brightfield microscope (capital letters) and green fluorescent microscope (lowercase letters). Observation time was labeled in the figure. Scale bar = 2 mm. Figure S9: The larvae of grass carp after depurating from 5 μm green fluorescent microplastics. Photographs were taken under a brightfield microscope (capital letters) and green fluorescent microscope (lowercase letters). Observation time was labeled in the figure. Scale bar = 2 mm. Figure S10: The larvae of grass carp after depurating from 0.5 μm green fluorescent microplastics. Photographs were taken under a brightfield microscope (capital letters) and green fluorescent microscope (lowercase letters). Observation time was labeled in the figure. Scale bar = 2 mm. Figure S11: The control group during elimination test in brightfield microscope (capital letters) and red fluorescent microscope (lowercase letters). Observation time was labeled in the figure. Scale bar = 2 mm. Figure S12: The larvae of grass carp after depurating from 5 μm red fluorescent microplastics. Photographs were taken under a brightfield microscope (capital letters) and red fluorescent microscope (lowercase letters). Observation time was labeled in the figure. Scale bar = 2 mm. Figure S13: The larvae of grass carp after depurating from 1 μm red fluorescent microplastics. Photographs were taken under a brightfield microscope (capital letters) and red fluorescent microscope (lowercase letters). Observation time was labeled in the figure. Scale bar = 2 mm. Figure S14: The appearance of adult grass carp. The area marked in the red box is the nasal cavity.

Author Contributions: Conceptualization, C.Z. and Z.Z.; methodology, Q.W.; software, S.W.; validation, C.Z., L.L. and J.Z.; formal analysis, Z.Z.; investigation, Z.Z.; resources, J.Z.; data curation, L.L.; writing—original draft preparation, C.Z. and Z.Z.; writing—review and editing, J.Z.; visualization, L.L.; supervision, J.Z.; project administration, C.Z.; funding acquisition, J.Z. All authors have read and agreed to the published version of the manuscript.

Funding: This study was funded by China Agriculture Research System of MOF and MARA (CARS-45-50).

Institutional Review Board Statement: All experiments were approved by the Animal Care and Use Committee of South China Agricultural University (identification code: 20210422; date of approval: 22 April 2021).

Informed Consent Statement: Informed consent was obtained from all subjects involved in the study.

Data Availability Statement: The data presented in this study are available in Section 4 and supplementary material.

Conflicts of Interest: The authors report that they have no conflict of interest.

References

1. Wang, C.; Zhao, J.; Xing, B. Environmental source, fate, and toxicity of microplastics. *J. Hazard. Mater.* **2021**, *407*, 124357. [CrossRef]
2. Rahman, A.; Sarkar, A.; Yadav, O.P.; Achari, G.; Slobodnik, J. Potential human health risks due to environmental exposure to nano-and microplastics and knowledge gaps: A scoping review. *Sci. Total Environ.* **2021**, *757*, 143872. [CrossRef]
3. Thompson, R.C.; Olsen, Y.; Mitchell, R.P.; Davis, A.; Rowland, S.J.; John, A.W.; McGonigle, D.; Russell, A.E. Lost at sea: Where is all the plastic? *Science* **2004**, *304*, 838. [CrossRef]
4. Carpenter, E.J.; Anderson, S.J.; Harvey, G.R.; Miklas, H.P.; Peck, B.B. Polystyrene spherules in coastal waters. *Science* **1972**, *17*, 749–750. [CrossRef]
5. Auta, H.S.; Emenike, C.U.; Fauziah, S.H. Distribution and importance of microplastics in the marine environment: A review of the sources, fate, effects, and potential solutions. *Environ. Int.* **2017**, *102*, 165–176. [CrossRef]
6. Guo, X.; Wang, J. The chemical behaviors of microplastics in marine environment: A review. *Mar. Pollut. Bull.* **2019**, *142*, 1–14. [CrossRef]
7. Wong, J.K.H.; Lee, K.K.; Tang, K.H.D.; Yap, P.S. Microplastics in the freshwater and terrestrial environments: Prevalence, fates, impacts and sustainable solutions. *Sci. Total Environ.* **2020**, *719*, 137512. [CrossRef]
8. He, D.; Luo, Y.; Lu, S.; Liu, M.; Song, Y.; Lei, L. Microplastics in soils: Analytical methods, pollution characteristics and ecological risks. *TrAC Trends Anal. Chem.* **2018**, *109*, 163–172. [CrossRef]
9. Li, J.; Song, Y.; Cai, Y. Focus topics on microplastics in soil: Analytical methods, occurrence, transport, and ecological risks. *Environ. Pollut.* **2020**, *257*, 113570. [CrossRef] [PubMed]
10. Li, J.; Yang, D.; Li, L.; Jabeen, K.; Shi, H. Microplastics in commercial bivalves from China. *Environ. Pollut.* **2015**, *207*, 190–195. [CrossRef] [PubMed]

11. Zhang, C.; Wang, S.; Pan, Z.; Sun, D.; Xie, S.; Zhou, A.; Wang, J.; Zou, J. Occurrence and distribution of microplastics in commercial fishes from estuarine areas of Guangdong, South China. *Chemosphere* **2020**, *260*, 127656. [CrossRef] [PubMed]

12. Zhou, A.; Zhang, Y.; Xie, S.; Chen, Y.; Li, X.; Wang, J.; Zou, J. Microplastics and their potential effects on the aquaculture systems: A critical review. *Rev. Aquac.* **2021**, *13*, 719–733. [CrossRef]

13. Chen, G.; Feng, Q.; Wang, J. Mini-review of microplastics in the atmosphere and their risks to humans. *Sci. Total Environ.* **2020**, *703*, 135504. [CrossRef]

14. Wright, S.L.; Ulke, J.; Font, A.; Chan KL, A.; Kelly, F.J. Atmospheric microplastic deposition in an urban environment and an evaluation of transport. *Environ. Int.* **2020**, *136*, 105411. [CrossRef] [PubMed]

15. Arthur, C.; Baker, J.; Bamford, H. *Proceedings of the International Research Workshop on the Occurrence, Effects, and Fate of Microplastic Marine Debris*; University of Washington Tacoma: Tacoma, DC, USA, 2009.

16. Gigault, J.; Halle, A.T.; Baudrimont, M.; Pascal, P.; Gauffre, F.; Phi, T.; El Hadri, H.; Grassl, B.; Reynaud, S. Current opinion: What is a nanoplastic? *Environ. Pollut.* **2018**, *235*, 1030–1034. [CrossRef]

17. Gonçalves, J.M.; Bebianno, M.J. Nanoplastics impact on marine biota: A review. *Environ. Pollut.* **2021**, *273*, 116426. [CrossRef]

18. Shen, M.; Zhang, Y.; Zhu, Y.; Song, B.; Zeng, G.; Hu, D.; Wen, X.; Ren, X. Recent advances in toxicological research of nanoplastics in the environment: A review. *Environ. Pollut.* **2019**, *252*, 511–521. [CrossRef]

19. Cózar, A.; Echevarría, F.; González-Gordillo, J.I.; Irigoien, X.; Úbeda, B.; Hernández-León, S.; Palma, Á.T.; Navarro, S.; García-de-Lomas, J.; Ruiz, A.; et al. Plastic debris in the open ocean. *Proc. Natl. Acad. Sci. USA* **2014**, *111*, 10239–10244. [CrossRef]

20. Geyer, R.; Jambeck, J.R.; Law, K.L. Production, use, and fate of all plastics ever made. *Sci. Adv.* **2017**, *3*, e1700782. [CrossRef] [PubMed]

21. de Souza Machado, A.A.; Kloas, W.; Zarfl, C.; Hempel, S.; Rillig, M.C. Microplastics as an emerging threat to terrestrial ecosystems. *Glob. Change Biol.* **2018**, *24*, 1405–1416. [CrossRef]

22. Karbalaei, S.; Hanachi, P.; Walker, T.R.; Cole, M. Occurrence, sources, human health impacts and mitigation of microplastic pollution. *Environ. Sci. Pollut. Res.* **2018**, *25*, 36046–36063. [CrossRef]

23. Huang, D.; Tao, J.; Cheng, M.; Deng, R.; Chen, S.; Yin, L.; Li, R. Microplastics and nanoplastics in the environment: Macroscopic transport and effects on creatures. *J. Hazard. Mater.* **2021**, *407*, 124399. [CrossRef]

24. Wright, S.L.; Kelly, F.J. Plastic and Human Health: A Micro Issue? *Environ. Sci. Technol.* **2017**, *51*, 6634–6647. [CrossRef]

25. Carbery, M.; O'Connor, W.; Palanisami, T. Trophic transfer of microplastics and mixed contaminants in the marine food web and implications for human health. *Environ. Int.* **2018**, *115*, 400–409. [CrossRef] [PubMed]

26. Prata, J.C.; Da, C.J.; Lopes, I.; Duarte, A.C.; Rocha-Santos, T. Environmental exposure to microplastics: An overview on possible human health effects. *Sci. Total Environ.* **2020**, *702*, 134455. [CrossRef] [PubMed]

27. Mao, Y.; Ai, H.; Chen, Y.; Zhang, Z.; Zeng, P.; Kang, L.; Li, W.; Gu, W.; He, Q.; Li, H. Phytoplankton response to polystyrene microplastics: Perspective from an entire growth period. *Chemosphere* **2018**, *208*, 59–68. [CrossRef]

28. Wu, Y.; Guo, P.; Zhang, X.; Zhang, Y.; Xie, S.; Deng, J. Effect of microplastics exposure on the photosynthesis system of freshwater algae. *J. Hazard. Mater.* **2019**, *374*, 219–227. [CrossRef] [PubMed]

29. Zhao, T.; Tan, L.; Huang, W.; Wang, J. The interactions between micro polyvinyl chloride (mPVC) and marine dinoflagellate *Karenia mikimotoi*: The inhibition of growth, chlorophyll and photosynthetic efficiency. *Environ. Pollut.* **2019**, *247*, 883–889. [CrossRef] [PubMed]

30. Graham, P.; Palazzo, L.; Andrea De Lucia, G.; Telfer, T.C.; Baroli, M.; Carboni, S. Microplastics uptake and egestion dynamics in Pacific oysters, *Magallana gigas* (Thunberg, 1793), under controlled conditions. *Environ. Pollut.* **2019**, *252*, 742–748. [CrossRef] [PubMed]

31. Ding, J.; Li, J.; Sun, C.; Jiang, F.; He, C.; Zhang, M.; Ju, P.; Ding, N.X. An examination of the occurrence and potential risks of microplastics across various shellfish. *Sci. Total Env.* **2020**, *739*, 139887. [CrossRef]

32. Garrido Gamarro, E.; Ryder, J.; Elvevoll, E.O.; Olsen, R.L. Microplastics in Fish and Shellfish—A Threat to Seafood Safety? *J. Aquat. Food Prod. Technol.* **2020**, *29*, 417–425. [CrossRef]

33. Wang, W.; Ge, J.; Yu, X. Bioavailability and toxicity of microplastics to fish species: A review. *Ecotoxicol. Environ. Saf.* **2020**, *189*, 109913. [CrossRef] [PubMed]

34. Pelka, K.E.; Henn, K.; Keck, A.; Sapel, B.; Braunbeck, T. Size does matter—Determination of the critical molecular size for the uptake of chemicals across the chorion of zebrafish (*Danio rerio*) embryos. *Aquat. Toxicol.* **2017**, *185*, 1–10. [CrossRef] [PubMed]

35. Ding, J.; Huang, Y.; Liu, S.; Zhang, S.; Zou, H.; Wang, Z.; Zhu, W.; Geng, J. Toxicological effects of nano- and micro-polystyrene plastics on red tilapia: Are larger plastic particles more harmless? *J. Hazard. Mater.* **2020**, *396*, 122693. [CrossRef] [PubMed]

36. Hoang, T.C.; Felix-Kim, M. Microplastic consumption and excretion by fathead minnows (*Pimephales promelas*): Influence of particles size and body shape of fish. *Sci. Total Environ.* **2020**, *704*, 135433. [CrossRef] [PubMed]

37. Ivleva, N.P.; Wiesheu, A.C.; Niessner, R. Microplastic in Aquatic Ecosystems. *Angew. Chem. -Int. Ed.* **2017**, *56*, 1720–1739. [CrossRef]

38. Choi, J.S.; Hong, S.H.; Park, J. Evaluation of microplastic toxicity in accordance with different sizes and exposure times in the marine copepod *Tigriopus Japonicus*. *Mar. Environ. Res.* **2020**, *153*, 104838. [CrossRef]

39. Jeong, C.; Won, E.; Kang, H.; Lee, M.; Hwang, D.; Hwang, U.; Zhou, B.; Souissi, S.; Lee, S.; Lee, J. Microplastic Size-Dependent Toxicity, Oxidative Stress Induction, and p-JNK and p-p38 Activation in the Monogonont Rotifer (*Brachionus koreanus*). *Environ. Sci. Technol.* **2016**, *50*, 8849–8857. [CrossRef]

40. Amado, L.L.; Monserrat, J. Oxidative stress generation by microcystins in aquatic animals: Why and how. *Environ. Int.* **2010**, *36*, 226–235. [CrossRef]

41. Sun, H.; Lü, K.; Minter, E.J.A.; Chen, Y.; Yang, Z.; Montagnes, D.J.S. Combined effects of ammonia and microcystin on survival, growth, antioxidant responses, and lipid peroxidation of bighead carp *Hypophthalmythys nobilis* larvae. *J. Hazard. Mater.* **2012**, *221*, 213–219. [CrossRef]

42. Zhang, C.; Wang, J.; Zhou, A.; Ye, Q.; Feng, Y.; Wang, Z.; Wang, S.; Xu, G.; Zou, J. Species-specific effect of microplastics on fish embryos and observation of toxicity kinetics in larvae. *J. Hazard. Mater.* **2021**, *403*, 123948. [CrossRef] [PubMed]

43. Steer, M.; Cole, M.; Thompson, R.C.; Lindeque, P.K. Microplastic ingestion in fish larvae in the western English Channel. *Environ. Pollut.* **2017**, *226*, 250–259. [CrossRef] [PubMed]

44. Pannetier, P.; Morin, B.; Le Bihanic, F.; Dubreil, L.; Clérandeau, C.; Chouvellon, F.; Van Arkel, K.; Danion, M.; Cachot, J. Environmental samples of microplastics induce significant toxic effects in fish larvae. *Environ. Int.* **2020**, *134*, 105047. [CrossRef]

45. Batel, A.; Borchert, F.; Reinwald, H.; Erdinger, L.; Braunbeck, T. Microplastic accumulation patterns and transfer of benzo[a]pyrene to adult zebrafish (*Danio rerio*) gills and zebrafish embryos. *Environ. Pollut.* **2018**, *235*, 918–930. [CrossRef]

46. Zhang, R.; Wang, M.; Chen, X.; Yang, C.; Wu, L. Combined toxicity of microplastics and cadmium on the zebrafish embryos (*Danio rerio*). *Sci. Total Environ.* **2020**, *743*, 140638. [CrossRef] [PubMed]

47. Hu, L.; Su, L.; Xue, Y.; Mu, J.; Zhu, J.; Xu, J.; Shi, H. Uptake, accumulation and elimination of polystyrene microspheres in tadpoles of *Xenopus tropicalis*. *Chemosphere* **2016**, *164*, 611–617. [CrossRef]

48. Liu, Y.; Jia, X.; Zhu, H.; Zhang, Q.; He, Y.; Shen, Y.; Xu, X.; Li, J. The effects of exposure to microplastics on grass carp (*Ctenopharyngodon idella*) at the physiological, biochemical, and transcriptomic levels. *Chemosphere* **2022**, *286*, 131831. [CrossRef] [PubMed]

49. Yu, E.; Xie, J.; Wang, G.; Yu, D.; Gong, W.; Li, Z.; Wang, H.; Xia, Y.; Wei, N. Gene Expression Profiling of Grass Carp (*Ctenopharyngodon idellus*) and Crisp Grass Carp. *Int. J. Genom.* **2014**, *639687*. [CrossRef] [PubMed]

50. Tang, M.; Lu, Y.; Xiong, Z.; Chen, M.; Qin, Y. The Grass Carp Genomic Visualization Database (GCGVD): An informational platform for genome biology of grass carp. *Int. J. Biol. Sci.* **2019**, *15*, 2119–2127. [CrossRef] [PubMed]

51. Hashizume, H.; Itoh, S.; Tanaka, K.; Ushiki, T. Direct Observation of t-Butyl Alcohol Frozen and Sublimated Samples Using Low-Vacuum Scanning Electron Microscopy. *Arch. Histol. Cytol.* **1998**, *61*, 93–98. [CrossRef]

52. Lee, K.J.; Nallathamby, P.D.; Browning, L.M.; Osgood, C.J.; Xu, X.N. In Vivo Imaging of Transport and Biocompatibility of Single Silver Nanoparticles in Early Development of Zebrafish Embryos. *ACS Nano* **2007**, *1*, 133–143. [CrossRef]

53. Fent, K.; Weisbrod, C.J.; Wirth-Heller, A.; Pieles, U. Assessment of uptake and toxicity of fluorescent silica nanoparticles in zebrafish (*Danio rerio*) early life stages. *Aquat. Toxicol.* **2010**, *100*, 218–228. [CrossRef]

54. Wang, Y.; Chen, F.; He, J.; Chen, J.; Xue, G.; Zhao, Y.; Peng, Y.; Xie, P. Comparative ultrastructure and proteomics of two economic species (common carp and grass carp) egg envelope. *Aquaculture* **2021**, *546*, 737276–737284. [CrossRef]

55. Catarino, A.I.; Frutos, A.; Henry, T.B. Use of fluorescent-labelled nanoplastics (NPs) to demonstrate NP absorption is inconclusive without adequate controls. *Sci. Total Environ.* **2019**, *670*, 915–920. [CrossRef] [PubMed]

56. Lu, Y.; Zhang, Y.; Deng, Y.; Jiang, W.; Zhao, Y.; Geng, J.; Ding, L.; Ren, H. Uptake and Accumulation of Polystyrene Microplastics in Zebrafish (*Danio rerio*) and Toxic Effects in Liver. *Environ. Sci. Technol.* **2016**, *50*, 4054–4060. [CrossRef] [PubMed]

57. Yang, H.; Xiong, H.; Mi, K.; Xue, W.; Wei, W.; Zhang, Y. Toxicity comparison of nano-sized and micron-sized microplastics to Goldfish *Carassius auratus* Larvae. *J. Hazard. Mater.* **2020**, *388*, 122058. [CrossRef] [PubMed]

58. Guimarães AT, B.; Estrela, F.N.; Rodrigues AS, D.L.; Chagas, T.Q.; Pereira, P.S.; Silva, F.G.; Malafaia, G. Nanopolystyrene particles at environmentally relevant concentrations causes behavioral and biochemical changes in juvenile grass carp (*Ctenopharyngodon idella*). *J. Hazard. Mater.* **2021**, *403*, 123864. [CrossRef]

59. Yin, L.; Chen, B.; Xia, B.; Shi, X.; Qu, K. Polystyrene microplastics alter the behavior, energy reserve and nutritional composition of marine jacopever (*Sebastes schlegelii*). *J. Hazard. Mater.* **2018**, *360*, 97–105. [CrossRef]

60. Parenti, C.C.; Ghilardi, A.; Della Torre, C.; Magni, S.; Del Giacco, L.; Binelli, A. Evaluation of the infiltration of polystyrene nanobeads in zebrafish embryo tissues after short-term exposure and the related biochemical and behavioural effects. *Environ. Pollut.* **2019**, *254*, 112947. [CrossRef] [PubMed]

61. van Pomeren, M.; Brun, N.R.; Peijnenburg, W.J.G.M.; Vijver, M.G. Exploring uptake and biodistribution of polystyrene (nano)particles in zebrafish embryos at different developmental stages. *Aquat. Toxicol.* **2017**, *190*, 40–45. [CrossRef] [PubMed]

62. Cole, M.; Lindeque, P.K.; Fileman, E.; Clark, J.; Lewis, C.; Halsband, C.; Galloway, T.S. Microplastics Alter the Properties and Sinking Rates of Zooplankton Faecal Pellets. *Environ. Sci. Technol.* **2016**, *50*, 3239–3246. [CrossRef] [PubMed]

Article

Co-Exposure with an Invasive Seaweed Exudate Increases Toxicity of Polyamide Microplastics in the Marine Mussel *Mytilus galloprovincialis*

Filipa G. Rodrigues [1], Hugo C. Vieira [2], Diana Campos [2,*], Sílvia F. S. Pires [2], Andreia C. M. Rodrigues [2], Ana L. P. Silva [2], Amadeu M. V. M. Soares [2], Jacinta M. M. Oliveira [2] and Maria D. Bordalo [2]

[1] Department of Biology, University of Aveiro, 3810-193 Aveiro, Portugal; filipagrodrigues@ua.pt
[2] CESAM—Centre for Environmental and Marine Studies, Department of Biology, University of Aveiro, 3810-193 Aveiro, Portugal; hugovieira@ua.pt (H.C.V.); silviapires1@ua.pt (S.F.S.P.); rodrigues.a@ua.pt (A.C.M.R.); ana.luisa.silva@ua.pt (A.L.P.S.); asoares@ua.pt (A.M.V.M.S.); jacintaoliveira@ua.pt (J.M.M.O.); maria.bordalo@ua.pt (M.D.B.)
* Correspondence: diana.campos@ua.pt

Abstract: Plastic pollution and invasive species are recognised as pervasive threats to marine biodiversity. However, despite the extensive on-going research on microplastics' effects in the biota, knowledge on their combination with additional stressors is still limited. This study investigates the effects of polyamide microplastics (PA-MPs, 1 mg/L), alone and in combination with the toxic exudate from the invasive red seaweed *Asparagopsis armata* (2%), after a 96 h exposure, in the mussel *Mytilus galloprovincialis*. Biochemical responses associated with oxidative stress and damage, neurotoxicity, and energy metabolism were evaluated in different tissues (gills, digestive gland, and muscle). Byssus production and PA-MP accumulation were also assessed. Results demonstrated that PA-MPs accumulated the most in the digestive gland of mussels under PA-MP and exudate co-exposure. Furthermore, the combination of stressors also resulted in oxidative damage at the protein level in the gills as well as in a significant reduction in byssus production. Metabolic capacity increased in both PA-MP treatments, consequently affecting the energy balance in mussels under combined stress. Overall, results show a potential increase of PA-MPs toxicity in the presence of *A. armata* exudate, highlighting the importance of assessing the impact of microplastics in realistic scenarios, specifically in combination with co-occurring stressors, such as invasive species.

Keywords: invasive macroalgae; bivalves; marine debris; oxidative stress; energy balance; byssus production

Citation: Rodrigues, F.G.; Vieira, H.C.; Campos, D.; Pires, S.F.S.; Rodrigues, A.C.M.; Silva, A.L.P.; Soares, A.M.V.M.; Oliveira, J.M.M.; Bordalo, M.D. Co-Exposure with an Invasive Seaweed Exudate Increases Toxicity of Polyamide Microplastics in the Marine Mussel *Mytilus galloprovincialis*. *Toxics* **2022**, *10*, 43. https://doi.org/10.3390/toxics10020043

Academic Editor: Víctor Manuel León

Received: 15 December 2021
Accepted: 14 January 2022
Published: 18 January 2022

Publisher's Note: MDPI stays neutral with regard to jurisdictional claims in published maps and institutional affiliations.

1. Introduction

Marine environments represent an important life support system and one of the most complex ecosystems [1]. Nevertheless, biodiversity and marine resources are increasingly endangered due to pollution and other anthropogenic issues associated with the fast pace of human population growth and the development of the economy. The introduction of non-native marine species, overfishing, global climate change, and habitat destruction and modification are key pressure points, especially in coastal areas [2].

Global plastic production has increased dramatically in recent years, reaching almost 370 million tonnes in 2019 [3], raising growing scientific and societal concerns. In particular, microplastics (MPs: <5 mm in size) are an emerging environmental issue that accounts for the major percentage of plastic litter, having been detected in many environmental matrices [4]. These polymers are introduced in marine ecosystems through multiple pathways, such as direct disposal, airborne dispersal, terrestrial runoff, and riverine flow [5,6]. MP levels are expected to range between <0.0001 and 1.89 mg/L in the marine environment [7]. However, as these particles undergo continuous fragmentation, and considering that most surveys do not detect particles <300 μm, the concentrations found in the environment are probably underestimated [8]. Several studies have observed that MPs are widely available to the marine food web [9], as they are very similar in size to various organisms in the planktonic and benthic communities [9]. The intake of MPs can occur via gills or through direct consumption (i.e., particle ingestion) or indirectly (i.e., via trophic chains) [6,10]. Therefore, the bioavailability of MPs to marine biota is the primary environmental risk associated with this pollutant [9,11]. In this regard, filter-feeding marine organisms, such as bivalves, are probably among the most impacted groups, since they can involuntarily ingest these synthetic materials along with the natural food items while feeding by constantly filtrating substantial volumes of seawater [12]. Once ingested, small-sized MPs can be taken up into the cells by endocytosis and are accumulated or translocated to different tissues in the organisms [13–15]. MP intake may, therefore, lead to histological alterations, inflammatory reactions, and ecotoxicological responses at cellular, molecular, and biochemical levels, as they are responsible for detrimental modulations of biological functions, such as reproduction, growth, survival, and feeding [9,16].

There are different types of plastic polymers and one of the most common groups includes polyamides (PA) [17], which are important engineering plastics often used in domestic and automotive industries [18] due to their high durability and resistance. Furthermore, these particles may be released from fishing gear and aquaculture facilities [6,19], and are frequently detected in coastal waters, including biotic [20,21], water [22], and sediment compartments [23]. PA particles can be found from the intertidal to the subtidal environments [24], as they have a density similar to seawater, allowing them to remain suspended in the water-column [10], remaining available as a "food item" for filter-feeding marine organisms.

The proliferation of invasive species has also been a major cause of concern in marine ecosystems, posing a threat to biodiversity and potentially leading to severe alterations in the functioning and structure of the ecosystem. In particular, marine macroalgae constitute the main component of introduced biota, with a current global estimate varying from 163 to over 300 species [25]. The northeast Atlantic and the Mediterranean coasts support the largest number of macroalgae introductions [26], with the main human-mediated vectors responsible for their transport being maritime traffic (e.g., hull fouling, ballast waters), aquaculture, and aquarium trade [27]. Once non-native macroalgae spread beyond their natural distribution through human activities and become successfully established, they are defined as invasive [28], competing with native species, and potentially leading to their displacement. Invasive species may also modify habitats and their structure, promoting biodiversity loss, and creating cascading effects or changes in the food chain [29], which may cause significant ecological and economic damages [30]. *Asparagopsis armata* Harvey, 1855 is a red seaweed native to Southern Australia and New Zealand [31], first described in the Atlantic and Mediterranean coasts in the 1920s [32], as it is widely distributed from the British Isles to Senegal [33,34], including the Azores Islands and mainland Portugal [35,36]. It is globally known for strong invasive behaviour due to its type of life cycle (leading to fast and vast propagation mainly due to its free-living stage) and lack of predators in the invaded habitat [37]. Exudation of secondary metabolites, including halogenated compounds such as haloforms, haloketones, and haloacids, constitutes a chemical defence mechanism that is a key aspect for *A. armata* invasiveness by becoming unpalatable for predators [38,39]. Thus, this seaweed has been considered an important source of bioactive metabolites with antibacterial and antifungal properties [40], and some were also found to have mutagenic and cytotoxic effects [41]. This red macroalga is mainly found from the low intertidal to the shallow subtidal zone [42], often attached to the substrate or drifting, and tend to concentrate in rock pools during low tide [43]. In this type of environment, such chemical compounds, once exuded into the water, may be potentially toxic and pose a threat to native biota [43]. Some previous studies have already devoted attention to the impact of *A. armata* exudate on the surrounding biota. For instance, exposure to *A. armata* halogenated metabolites caused physiological impairment on the crustacean *Palaemon elegans*, the gastropod *Gibbula umbilicalis*, and the mussel *Mytilus galloprovincialis* [43–45]. Low exudate concentrations were also found to reduce feeding activity of *G. umbilicalis* and *M. galloprovincialis* as well as the byssal production and strength of *M. galloprovincialis* [44,45]. Moreover, a tendency of an increasingly toxic action of the exudate was observed in *M. galloprovincialis* under a warming temperature scenario [46].

Mussels are abundant, widespread bivalves, and key players within marine trophic chains, being frequently selected as sentinel organisms and used in ecotoxicological studies for monitoring coastal environments as representative of low-trophic level organisms [47]. The mussel *M. galloprovincialis* is considered an ecologically important organism in coastal waters and is frequently used as a bioindicator of MP pollution in marine environments [4]. The sedentary and suspension filter-feeding behaviours of this mussel species translates in a great capacity to uptake and accumulate many contaminants, consequently providing a specific response that reflects the effects of different perturbations [48]. Furthermore, this species represents an important link between benthic and pelagic ecosystems [4] and forms dense monolayered and multi-layered beds attached to the hard substrate along intertidal rocky shores providing habitat structures and shelter to various organisms, increasing habitat complexity and enhancing the biodiversity [49]. *M. galloprovincialis* also has a high socio-economic value, representing an important food resource globally consumed by human populations due to its nutritional relevance, hence representing one of the most harvested and produced species, particularly in Portugal [50].

Considerable investigations have been carried out on the effect of different MPs in the mussel, *M. galloprovincialis* [4,12,14,48,51], but none studied the consequence of this exposure in co-occurrence with the exuded compounds from an invasive seaweed. The presence of different stressors in the environment may lead to complex interactions and scenarios that need to be taken into account when evaluating their impact in order to identify realistic scenarios of exposure. Furthermore, despite being a commonly found polymer in coastal waters [20–22], there is a knowledge gap of the effect of PA-MPs in marine organisms. In this sense, the present study aimed to evaluate the consequences of PA-MP exposure in the mussel *M. galloprovincialis* and assess the influence of *A. armata* exudate on the impacts caused by this polymer. Physiological responses, including byssal thread production, oxidative damage, antioxidant defences, enzymatic activity for cholinergic neurotransmission, energy production, and metabolism, were measured.

2. Materials and Methods

2.1. Asparagopsis armata Sampling and Exudate Production

The red macroalga *A. armata* (gametophyte phase) was collected by hand through free diving in the subtidal zone at the Terceira Island in Azores (Portugal) (38°38′59.2″ N, 27°13′16.4″ W). After collecting, the macroalgae were kept in aerated seawater tanks until the next day and packed in sealed containers to be transported to the laboratory in Aveiro (Portugal). Upon arrival, *A. armata* was immediately cleared from any perceptible associated fauna and debris. Afterwards, they were allocated to a tank with artificial seawater (marine RedSea® Salt premium grade) in a 1:10 proportion (salinity: 35 ± 1, pH: 8.0 ± 0.1, temperature: 20.0 ± 0.5 °C) in the dark and with no aeration for 24 h to produce the exudate, adapted from [45]. Algae were then removed from the tank and the resulting media (considered as the stock solution, representing 100% of exudate) was preserved at −20 °C. When needed, the exudate was slowly defrosted in the dark at 4 °C, and used at a 2% concentration, chosen according to previous sublethal toxicity test results [45].

2.2. Mytilus galloprovincialis Sampling and Acclimation

In December 2020, adult specimens of *M. galloprovincialis* (4.2 ± 0.1 cm shell length) were harvested by hand, on the intertidal rocky shore of the Barra of Aveiro in Portugal (40°38′38.8″ N, 8°44′44.6″ W), during low tide. Mussels were measured with a pachymeter in the field and then transported to the laboratory, where the shell surface was gently scraped to remove algae, encrusting organisms, and debris. Afterwards, *M. galloprovincialis* individuals were allowed to depurate and acclimate during seven days in glass aquariums that contained aerated artificial seawater (salinity: 30.0 ± 0.5; temperature: 19.0 ± 0.5 °C; pH: 8.0 ± 0.1; dissolved oxygen: 8.0 ± 0.5 mg/L; oxygen saturation: >80%, measured with WTW portable meters, Weilheim, Germany) in a recirculating aquatic system (a flow-through system ensured continuous seawater renewal), with a 14 h light:10 h dark photoperiod.

2.3. Microplastic Preparation

Polyamide microplastics (PA-MP, mean size: 30–50 μm, irregularly shaped, density: 1.14 g/cm^3; CAS 32131-17-2, Figure S1) were generously provided by a company that chose to remain anonymous. A stock solution (100 mg PA-MP/L) was prepared in artificial seawater (salinity: 30; RedSea® Salt premium grade mixed with reverse osmosis water) previously filtered (0.45 μm pore size). This PA-MP solution was allowed to equilibrate for 96 h at 50 rpm at room temperature in the dark. A solution containing only artificial seawater to be used in the treatments without PA-MPs was prepared and left to shake in the same conditions. The final concentration was achieved by adding 5 mL of the stock solution to the test vials containing 495 mL of seawater, resulting in a final concentration of 1 mg/L, which fits within realistic environmental MP concentrations [7]. In the treatments without PA-MP, 5 mL of the aged artificial seawater were also added.

2.4. Experimental Setup

After acclimation, 48 mussels were exposed for 96 h to the following treatments: (i) control (artificial seawater only); (ii) *A. armata* exudate (2% concentration); (iii) PA-MPs (1 mg/L); and (iv) *A. armata* exudate (2%) and PA-MPs (1 mg/L), simultaneously. The 96 h exposure was selected in accordance with American Society for Testing and Materials E729-96 [52]. For each treatment (control; exudate exposure; PA-MP exposure; and exudate and PA-MP exposure), 12 replicates were used with 1 mussel placed individually in 1 L glass flasks containing 500 mL of aerated test medium (static exposure). Seven replicates were used for the biomarkers' analysis, and the remaining five replicates were used for PA-MP quantification. The physical–chemical test parameters were maintained at salinity—30.5 ± 0.3, temperature—18.0 ± 0.3 °C, pH—8.0 ± 0.2, dissolved oxygen—8.0 ± 0.5 mg/L, oxygen saturation—>83%, and a 14 h light:10 h dark photoperiod was used. After the 96 h of exposure, the soft tissues of each mussel were removed using a scalpel and tweezers. Tissue samples (gills, muscles, and digestive gland) for the biomarkers analysis were individually stored and weighed in microcentrifuge tubes, frozen in liquid nitrogen and subsequently stored at –80 °C prior to further analysis. Samples for the PA-MP quantification (gills and digestive gland) were kept in small glass flasks (for the microplastic quantification) and preserved at −20 °C.

2.5. Digestion of Mussel Tissues and Microplastic Quantification

The digestion and filtration procedures were adapted from the method developed by Prata et al. [53].

A 10% potassium hydroxide (KOH) (w/v ≥ 85%, Fisher Scientific, Loughborough, UK, CAS 1310-58-3) solution (100 g of KOH pellets dissolved in 1000 mL Milli-Q ultra-pure water) was freshly prepared and used to digest the mussels' tissues. Ten mL of the KOH solution were added to each glass flask containing the samples, covered with aluminium foil, and incubated at 50 °C for 48 h. After the incubation period was over, the filtration of the samples followed.

The samples were heated to boiling just before being filtered to improve the solubility of fats and soaps and, consequently, the filtration rates. Then, samples were vacuumed filtered onto glass microfiber filters (47 mm Ø; 1.2 μm pore size, Prat Dumas, Couze-St-Front, France), washed with 50 mL of boiling Milli-Q ultra-pure water, followed by the addition of 10 mL of acetone (99.5+%, Fisher Scientific, Loughborough, UK, CAS 67-64-1). Samples were then incubated for 10 min and washed with ultra-pure water.

To assure quality control during testing, the glassware was acid-washed and rinsed with Milli-Q ultra-pure water; procedural blanks (1 per every 10 samples) were prepared with the KOH solution and received the same treatment as the other samples; for digestion, tissue samples were prepared and handled under a laminar flow chamber.

After drying, each glass fibre filter of each sample (including blanks) was observed under a stereomicroscope (Zeiss, Stemi 2000, Jena, Germany), and the number of PA-MP particles was visually counted. All fibres were discarded from the analysis. In case of any doubt, PA-MPs were confirmed by applying the method of hot needle [54]. The number of PA-MPs is presented as the number of counted particles/g tissue/organism.

2.6. Biomarker Analysis

2.6.1. Sample Preparation

Samples of *M. galloprovincialis* tissues (gills, muscles, and digestive glands) were individually homogenised on ice through sonication (10% pulse mode, 250 Sonifier, Branson Ultrasonics, Danbury, CT, USA) using 1500 µL 0.1 M K-phosphate buffer, pH—7.4. Muscle samples to be analysed for energy metabolism were homogenised using the same procedure in 1500 µL ultra-pure water.

After homogenisation, one aliquot from each gill, digestive, and muscle replicate was stored with 4% butylated hydroxytoluene (BHT) in methanol to evaluate the lipid peroxidation (LPO). Aliquots for protein carbonylation (PC) determination were also stored. The remaining homogenate of gills and digestive samples was centrifuged for 15 min at 10,000 g (4 °C), and the obtained post-mitochondrial supernatant (PMS) was divided into microtubes and kept in −80 °C for posterior analysis of catalase (CAT), glutathione S-transferase (GST), and acetylcholinesterase (AChE) activities, and total glutathione (tGSH) content. The PMS from the muscle homogenate was used for determining AChE activity in this tissue.

Aliquots of muscle homogenates were also stored for the analysis of lactate dehydrogenase (LDH) activity, proteins, lipids, and sugars contents, and electron transport system (ETS) activity.

Biomarkers determinations were done in micro-assays set up in 96-well flat bottom plates and read spectrophotometrically (Microplate reader MultiSkan Spectrum (Thermo Fisher Scientific, Waltham, MA, USA).

2.6.2. Oxidative Stress and Neurophysiological Biomarkers

The protein concentration of PMS was determined according to the Bradford method [55], using bovine-globulin as a standard. The Ellman's method [56], adapted to the microplate [57], was applied to measure acetylcholinesterase (AChE) activity, using acetylthiocholine as substrate and following the absorbance increase at 412 nm. Catalase (CAT) activity was measured in the PMS by following the decomposition of the substrate hydrogen peroxide (H_2O_2) at 240 nm [58]. Glutathione-S-transferase (GST) activity was measured in PMS after the conjugation of reduced glutathione (GSH) with 1-chloro-2,4-dinitrobenzene (CDNB) at 340 nm [59]. The total glutathione (tGSH) content was determined in the PMS fraction using the recycling reaction of GSH with 5,50-dithiobis-(2-nitrobenzoic acid) (DTNB) in the presence of glutathione reductase (GR) excess at 412 nm [60–62]. To determine endogenous lipid peroxidation (LPO) thiobarbituric acid-reactive substances (TBARS) were measured at 535 nm [63]. Protein carbonylation (PC) was quantified at 450 nm based in the reaction of 2,4-dinitrophenylhydrazine (DNPH) with carbonyl groups, according to the DNPH alkaline method [64]. Lactate dehydrogenase (LDH) activity was determined by following the NADH oxidation caused by pyruvate consumption, as it leads to the decrease of absorbance at 340 nm [65], adapted to the microplate [66].

2.6.3. Cellular Energy Allocation (CEA)

CEA value is obtained from the ratio between Ea, the energy available (the sum of proteins, lipids, and sugar contents), and Ec, which is aerobic energy production (estimation of ETS activity). The CEA and ETS activity were determined based on the methods described by De Coen and Janssen [67], slightly modified for the microplate [68].

Total lipid content in muscle tissue was determined by adding chloroform, methanol, and ultra-pure water in a 2:2:1 proportion. In the organic phase of each sample, sulfuric acid (H_2SO_4) was added, followed by an incubation period of 15 min at 200 °C, and the absorbance was measured at 375 nm using tripalmitin as a lipid standard. To determine the carbohydrate and protein contents, 15% thiobarbituric acid (TCA) was added to 300 µL of homogenate and incubated for 10 min at −20 °C. Carbohydrate quantification was performed in the supernatant by adding 5% phenol and H_2SO_4 to the samples, and the absorbance was read at 492 nm, using glucose as a standard. For total protein content quantification, the remaining pellet was resuspended with 1 M NaOH (incubated for 30 min at 60 °C) and then neutralized with 1.67 HCl. Total protein content quantification followed the Bradford's method [55], using bovine serum albumin as a standard and measuring absorbance at 520 nm. Proteins, lipids, and sugar fractions were converted into energetic equivalent values using the corresponding energy of combustion: 24,000 mJ/g, 39,500 mJ/g, and 17,500 mJ/g, respectively [69].

Electron transport system (ETS) activity was evaluated using the INT (Iodonitrotetrazolium chloride) reduction assay by measuring the rate of INT reduction in the presence of the non-ionic detergent Triton X-100, at 490 nm. The stoichiometric relationship in which for 2 µmol of formazan formed, 1 µmol of oxygen is consumed was applied to calculate the cellular oxygen consumption rate. The final Ec value was converted into an energy equivalent using the specific oxyenthalpic equivalent for an average lipid, protein, and carbohydrate mixture of 480 kJ/mol O_2 [69].

2.7. Byssal Thread Production

The quantity of produced byssal threads was assessed as a physiological biomarker. Once the 96 h exposure period for the different treatments (0% exudate; 2% exudate; PA-MPs; and 2% exudate and PA-MPs) ended, the number of functional byssus produced by each *M. galloprovincialis* individual was counted, according to Coelho et al. [45]. For this evaluation, all 12 replicates were used.

2.8. Statistical Analysis

The statistical analysis of data and graphical representations of results was performed using IBM SPSS Statistics 27 and GraphPad Prism 9 for Windows. Data normality and homoscedasticity were assessed on the residuals, using the Shapiro–Wilk Test ($p > 0.05$) and the Levene's Test ($p > 0.05$), respectively. For variables not showing a normal distribution or homoscedasticity, data were square root (CAT, GST, LPO, AChE, tGSH, AChE, LDH, and AChE in the muscle) or log-transformed (lipid content, ETS activity, Ea, and PA-MP quantification in the digestive gland).

Parametric t-tests were performed to evaluate differences in the number of PA-MP particles per tissue between treatments exposed to PA-MP. One-way analysis of variance (ANOVA) with a post hoc Dunnet's test was used to investigate treatment-dependent effects on byssus production. Effects on biochemical responses among *A. armata* exudate, PA-MPs and their interactions after exposure were evaluated through two-way ANOVA, using *A. armata* exudate and PA-MPs as factors (IBM SPSS Software, Armonk, NY, USA). The *post hoc* Šidák's test was used to perform multiple comparisons and identify significant differences between treatments (GraphPad Software, CA, USA). Data were presented as mean value (mean) ± standard error of mean value (SEM).

3. Results

3.1. Polyamide Microplastics Quantification

PA-MP particles were found mainly in the digestive gland and, at a lesser amount, in the gills (Table 1). Despite the observed increase in the number of particles between the PA-MP treatment and the combined exposure, this difference was not significant ($p > 0.05$).

Table 1. Number of polyamide microplastics (PA-MPs) per gram of tissue (gills and digestive gland) in *Mytilus galloprovincialis* exposed to PA-MPs and PA-MPs together with *A. armata* exudate. All values are presented mean ± SEM. ww = wet weight.

Tissue	Number of Particles per Gram Tissue (ww)	
	PA-MP	**PA-MP + Exudate**
Gills	6.97 ± 3.08	11.95 ± 4.83
Digestive gland	35.04 ± 16.09	62.25 ± 25.98

3.2. Oxidative Stress and Neurophysiological Biomarkers

In the gills, a significant effect of PA-MPs factor was observed for CAT activity of exposed mussels (Table S1); however, despite the observed tendency to decrease CAT activity, the *post hoc* test could not discriminate significant differences among the several treatments (Figure 1a). Considering the GST activity (Figure 1b), no significant changes in the presence of *A. armata* exudate, PA-MPs, or even by the interaction between *A. armata* exudate and PA-MPs were observed (Table S1). On the other hand, significant effects were observed in the levels of tGSH in the presence of PA-MPs and in mussels exposed to both stressors, reflected by the significant interaction between *A. armata* exudate and PA-MPs (Table S1). Specifically, there were significant differences within the 2% exudate concentration ($p < 0.05$); i.e., the tGSH levels exhibited a decrease in the mussels exposed to exudate in the presence of PA-MPs, when compared to the single exposure of *A. armata* exudate (Figure 1c).

Regarding the oxidative damage in the mussel gills, no changes in PC levels were observed in mussels exposed to PA-MPs and *A. armata* exudate; however, the interaction between these factors significantly affected PC levels (Table S1). Furthermore, PC levels demonstrated a significant difference in mussels exposed to the 2% exudate concentration ($p < 0.05$), with increased values in the exposure to *A. armata* exudate in the presence of PA-MPs, when compared to the 2% exudate treatment (Figure 1d). A significant difference within the 1 mg PA-MP/L ($p < 0.05$) was also verified, whereas the exposure to PA-MPs in the presence of 2% exudate exhibited superior PC levels when compared with the exposure to PA-MPs without exudate (Figure 1d). On the other hand, LPO was not significantly affected by *A. armata* exudate, PA-MPs, or their interaction (Table S1, Figure 1e). Regarding neurotoxicity, none of the experimental treatments resulted in significant effects ($p > 0.05$) in the AChE activity (Table S1, Figure 1f).

In the digestive gland, no significant effects ($p > 0.05$) of *A. armata* exudate exposure or PA-MPs were observed in CAT activity; however, the interaction of these two factors resulted in a significant alteration ($p < 0.05$) in CAT activity (Table S2). Despite that, the *post hoc* tests did not detect statistical differences among treatments (Figure 2a). Considering the GST activity and tGSH levels, no significant effects ($p > 0.05$) of *A. armata* exudate exposure, PA-MPs, and their interaction were observed (Table S2, Figure 2b,c)

Considering the oxidative damage in the mussels' digestive gland, no significant alterations ($p > 0.05$) in PC levels were observed when organisms were exposed to *A. armata* exudate, and no interaction of *A. armata* exudate and PA-MPs was observed either (Table S2). However, the PC levels were significantly affected in mussels exposed to PA-MPs ($p < 0.05$, Table S2). The post hoc test revealed significant differences within the 2% exudate concentration ($p > 0.05$) in the levels of PC. A significant increase of PC levels was verified in mussels exposed to 2% exudate in the presence of PA-MPs, when compared to the single exposure of *A. armata* exudate without PA-MPs (Figure 2d). As observed in gills, LPO levels did not exhibit alterations in the digestive gland in none of the treatments ($p > 0.05$, Table S2, Figure 2e). Finally, the exposure to *A. armata* exudate and PA-MPs did not interfere with the activity of AChE, and there was no interaction between the two tested stressors ($p > 0.05$, Table S2).

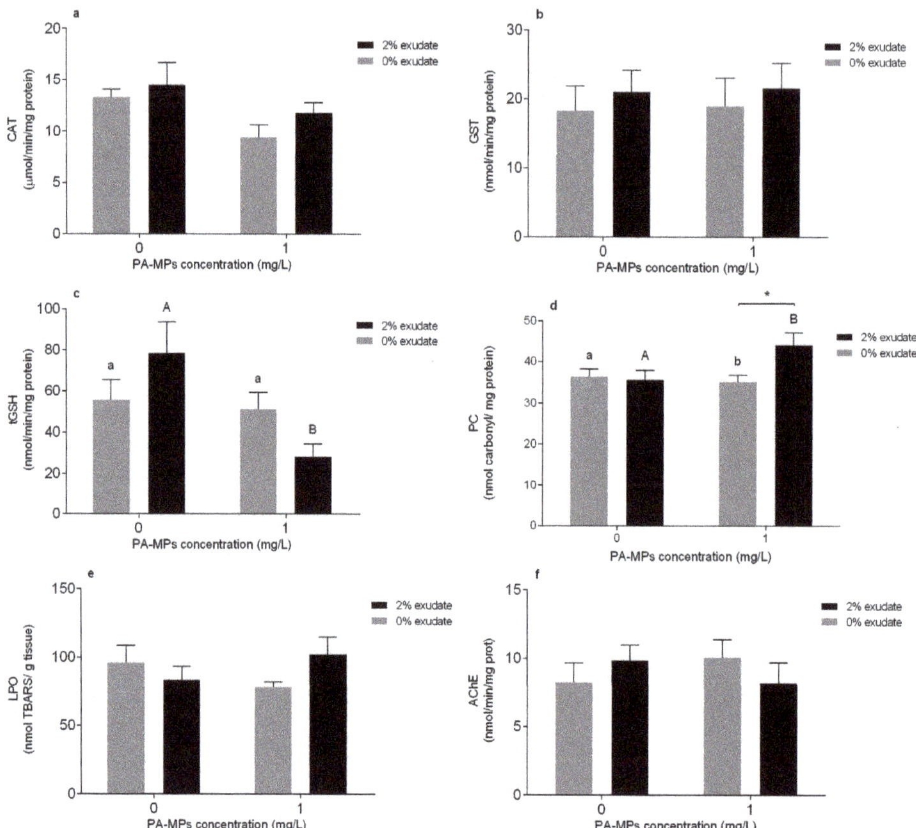

Figure 1. Oxidative stress-related biomarkers of *Mytilus galloprovincialis* gills after 96 h of exposure to *A. armata* exudate (0% and 2%) at different polyamide microplastic (PA-MPs) concentrations (0 and 1 mg/L). (**a**) Catalase activity (CAT), (**b**) glutathione-*S*-transferase activity (GST), (**c**) total glutathione contents (tGSH), (**d**) protein carbonylation levels (PC), (**e**) lipid peroxidation (LPO), and (**f**) acetylcholinesterase activity (AChE). All values are presented as mean ± SEM. * denotes a significant difference between the 0% and 2% *A. armata* exudate in the same PA-MPs concentration. The upper-case letters indicate differences in the 0% exudate treatments and the different lower-case letters represent differences in the 2% exudate treatments at the different PA-MPs concentrations.

In the muscle, LPO (Figure 3a) and PC (Figure 3b) did not undergo significant alterations ($p > 0.05$) when exposed to exudate, PA-MPs, or their interaction (Table S3). On the other hand, the AChE activity was significantly affected in mussels exposed to *A. armata* exudate ($p < 0.05$) but was not influenced ($p > 0.05$) by the presence of PA-MPs or by the interaction between factors (Table S3). Despite that, no statistical differences among treatments were observed (Figure 3c).

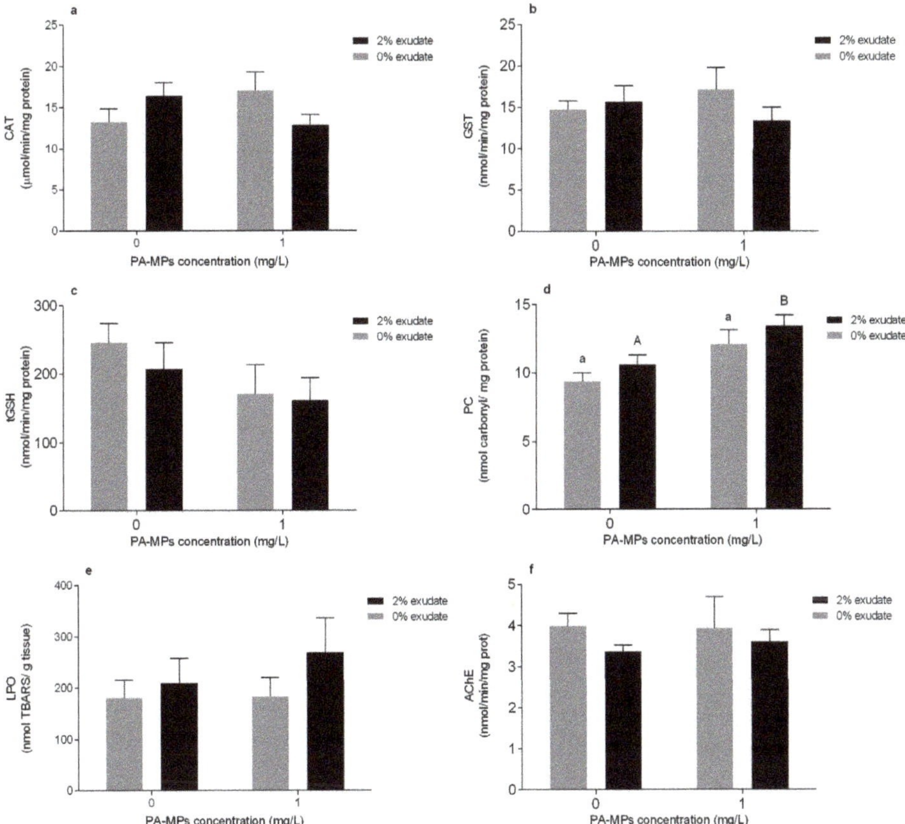

Figure 2. Oxidative stress-related biomarkers of *Mytilus galloprovincialis* digestive gland after 96 h of exposure to *A. armata* exudate (0% and 2%) at different polyamide microplastic (PA-MPs) concentrations (0 and 1 mg/L). (**a**) Catalase activity (CAT), (**b**) glutathione-*S*-transferase activity (GST), (**c**) total glutathione contents (tGSH), (**d**) protein carbonylation levels (PC), (**e**) lipid peroxidation (LPO), and (**f**) acetylcholinesterase activity (AChE). All values are presented as mean ± SEM. The upper-case letters indicate differences in the 0% exudate treatments, and the different lower-case letters represent differences in the 2% exudate treatments at the different PA-MPs concentrations.

3.3. Energy Metabolism Biomarkers

Considering the energy metabolism in the muscle tissue, the activity of LDH (Figure 4a), lipid levels (Figure 4b), and protein content (Figure 4c) were not affected by the presence of *A. armata* exudate or PA-MPs, and there was no interaction between factors ($p > 0.05$, Table S3). In addition, the single exposure to the exudate and the PA-MPs had no significant effect ($p > 0.05$). On the other hand, the interaction between *A. armata* exudate and PA-MPs demonstrated a significant impact on the sugar content ($p < 0.05$, Table S3). There was an increase in sugar content in individuals exposed to PA-MPs in the presence of 2% exudate compared to the single exposure to PA-MPs (Figure 4d, $p > 0.05$). There was also a significant increase of sugar levels in mussels exposed to exudate in the presence of PA-MPs, when compared to the treatment with only *A. armata* exudate (Figure 4d, $p < 0.05$).

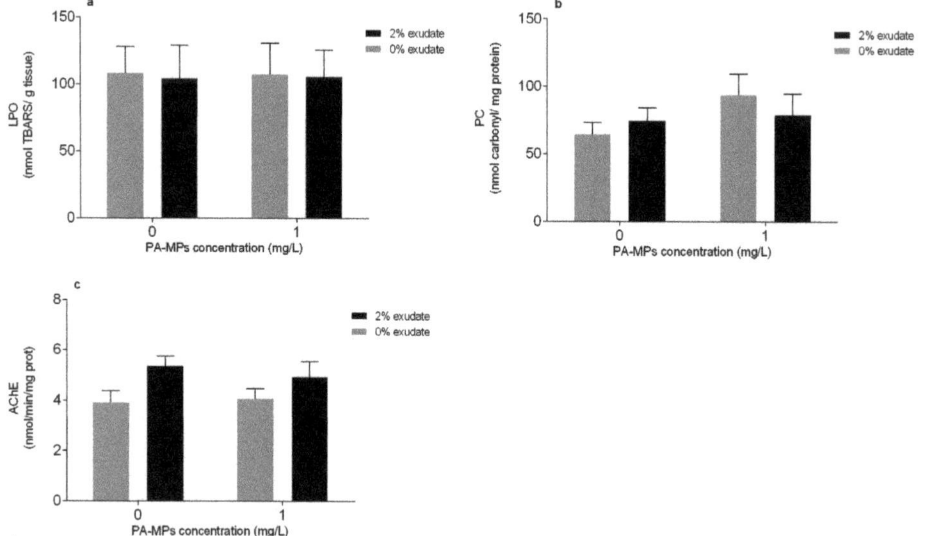

Figure 3. Oxidative stress-related biomarkers of *Mytilus galloprovincialis* muscles after 96 h of exposure to *A. armata* exudate (0% and 2%) at different polyamide microplastic (PA-MPs) concentrations (0 and 1 mg/L). (**a**) Lipid peroxidation (LPO), (**b**) protein carbonylation levels (PC), and (**c**) acetylcholinesterase activity (AChE). All values are presented as mean ± SEM.

Regarding the aerobic metabolic capacity, ETS activity (E_c) was impacted in individuals exposed to PA-MPs ($p < 0.05$) but was not affected by the presence of *A. armata* exudate or the interaction of factors ($p > 0.05$, Table S3).These alterations were not reflected in the overall energy available (E_a) in the presence of *A. armata* exudate ($p > 0.05$), PA-MPs ($p > 0.05$) and there was also no interaction ($p > 0.05$).

CEA was affected in mussels exposed to the PA-MPs treatment ($p < 0.05$), and there were no modifications in individuals exposed to exudate or both factors ($p > 0.05$, Table S3). There was a significant difference in the 2% exudate concentration ($p < 0.05$), i.e., a decrease in CEA was verified in organisms exposed to *A. armata* exudate in the presence of PA-MPs, when compared to 2% exudate in the absence of PA-MPs.

3.4. Byssal Thread Production

The number of produced byssal threads was not significantly affected in mussels exposed to *A. armata* exudate ($p > 0.05$). However, a significant decline in the number of byssus was observed in mussels exposed to both PA-MP treatments (with and without the exudate) when compared to control ($p < 0.05$, Figure 5).

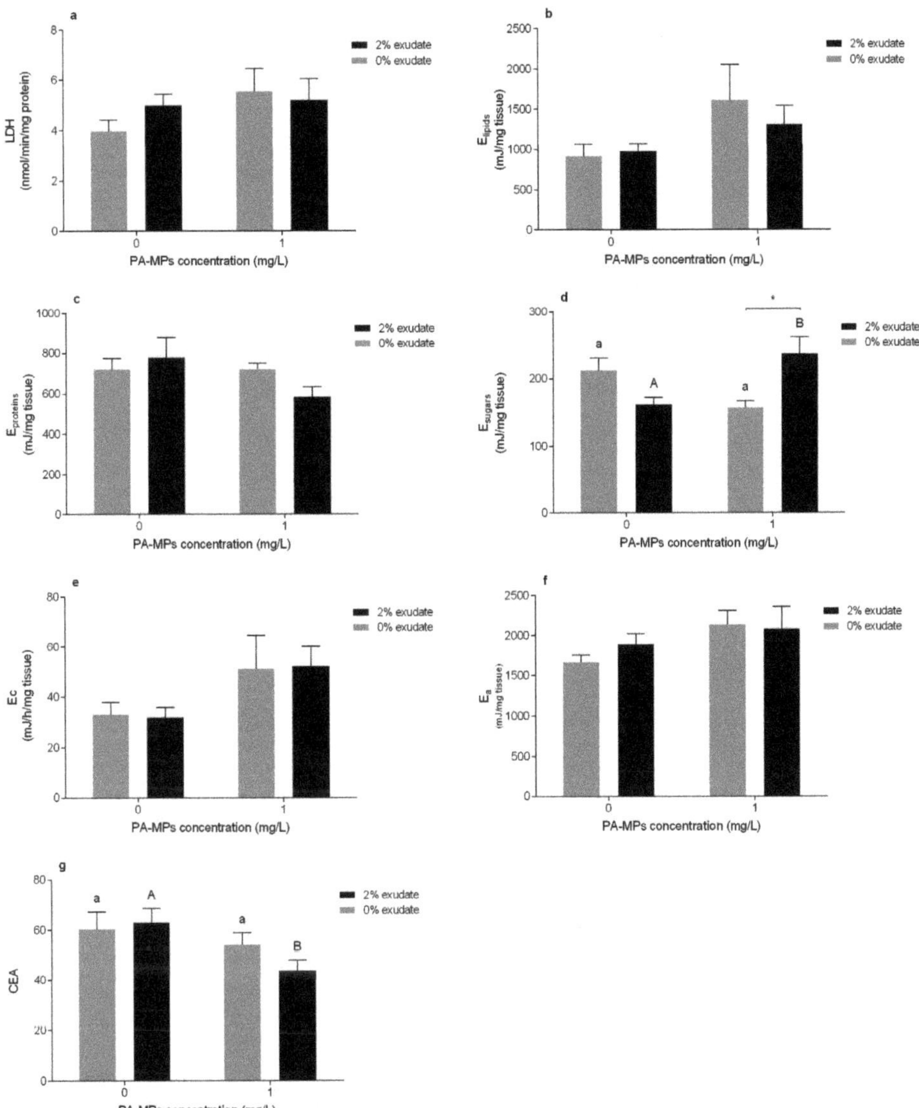

Figure 4. Energy metabolism biomarkers of *Mytilus galloprovincialis* muscles after 96 h of exposure to *A. armata* exudate (0% and 2%) at different polyamide microplastic concentrations (0 and 1 mg/L). (**a**) Lactate dehydrogenase (LDH), (**b**) lipid contents (E_{lipids}), (**c**) protein contents ($E_{proteins}$), (**d**) sugar content (E_{sugars}), (**e**) electron transport system, (**f**) energy available (Ea), and (**g**) cellular energy allocation (CEA). All values are presented as mean ± SEM. * denotes a significant difference between the 0% and 2% *A. armata* exudate in the same PA-MPs concentration. The upper-case letters indicate differences in the 0% exudate treatments and the different lower-case letters represent differences in the 2% exudate treatments at the different PA-MPs concentrations.

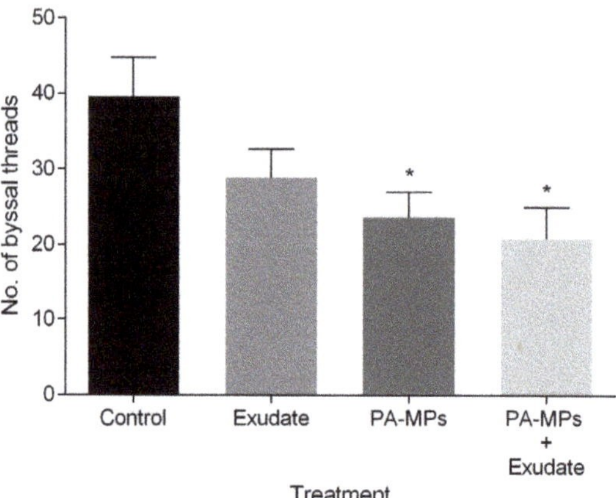

Figure 5. Number of produced byssal threads by *Mytilus galloprovincialis* during the 96 h exposure to different treatments: (i) control (0%; 0 mg/L); (ii) *A. armata* exudate (2%); (iii) PA-MPs (1 mg/L); and (iv) *A. armata* exudate (2%) and PA-MPs (1 mg/L). All values are presented as mean ± SEM. * denotes a significant difference compared with the control treatment.

4. Discussion

4.1. Microplastics in the Tissues

PA-MPs were taken up by *M. galloprovincialis*, as they are mostly found in the digestive gland, which is in line with previous studies exposing bivalves to treatments containing MPs [13,14,70–72]. A smaller amount of PA-MPs was detected in the gills. Histological analyses also revealed the presence of few particles retained in the gills epithelium of *M. galloprovincialis* exposed to polystyrene (PS) [73] and to polyethylene (PE) [51], and also of the freshwater bivalve *Corbicula fluminea* [74].

The highest number of PA-MP particles was found in the digestive gland under the presence of *A. armata* exudate. This may be explained either by the fact that the exudate presence increased the uptake of PA-MP or the exudate compounds could have compromised the mussels' ability to excrete these particles. As *A. armata* exudate was shown previously to decrease the clearance rate capacity of exposed mussels [45], the second hypothesis seems to be more plausible. The mechanism underlying this process requires further investigation. In contrast, previous studies investigating the MP effects of co-exposure with other contaminants (e.g. benzo(a)pyrene, fluoranthene) in mussels did not find differences in MP accumulation between organisms treated with MPs alone or in combination [51,73].

4.2. Oxidative Stress and Neurophysiological Biomarkers

Toxicity of MPs and *A. armata* exudate is in part mediated by increased reactive oxygen species (ROS) production, which induces antioxidant defences in the exposed organisms to prevent oxidative damage. Such responses are expected following PA-MP exposure, as this polymer may accumulate in the organisms' tissues resulting in physical damage, inflammatory responses [13,14], and the consequent activation of defence mechanisms. In addition, *Asparagopsis* seaweeds are a source of halogenated compounds that are inextricably linked to ROS production [75]. Catalase (CAT) is at the first line of defence in the elimination of ROS [76], along with other enzymatic defences, such as superoxide dismutase (SOD). GST has an important role in the phase II of biotransformation and non-enzymatic tGSH acts in the neutralization of ROS [77].

In the bivalves, gills have both a respiratory and feeding role and are the first tissue in contact with the stressor [78]. CAT activity in the gills declined in organisms exposed to PA-MPs. H_2O_2 is the main precursor of hydroxyl radical in marine organisms [72], and its formation is favoured by ROS production (mainly superoxide anion). CAT may prevent cell damage due to MPs-induced oxidative stress, as this enzyme is involved in the removal of H_2O_2 by converting the hydrogen peroxide into H_2O and O_2 and acting as a defence mechanism towards exogenous sources of H_2O_2 [77]. CAT inhibition was also observed after a 7-day exposure to PS MPs [73]. The authors hypothesised that this enzyme has a biphasic response in the neutralisation of the hydrogen peroxide production, with an activation within the first days of exposure followed by a decrease in activity [73]. Although our study assessed CAT activity after a 96 h exposure, a similar response may also explain the CAT inhibition after this period. Thus, the depletion of CAT activity observed in the PA-MPs treatment may be related with its involvement in the decomposition of hydrogen peroxide. Reduced CAT activity was also demonstrated by Abidli et al. [48] in *M. galloprovincialis* females exposed to PE at 100 and 1000 µg/L. GST activity was not altered in mussels exposed to any of the treatments. Webb et al. [79] also observed no changes in the GST activity in the mussel *Perna canaliculus* gills exposed to 0.5 g PE/L. Furthermore, results suggest a participation of tGSH as second line of defence following the depletion of CAT activity, with mussels from the combined exposure of *A. armata* exudate and PA-MPs presenting the lowest tGSH levels. tGSH is one of the most abundant scavengers in marine organisms that neutralises ROS and acts as a cofactor of various antioxidant enzymes dependent on glutathione [77], and therefore has an important role in the protection against ROS. The decrease in tGSH levels suggests an active involvement in combating excess reactive oxygen species (ROS) by increasing the consumption of total glutathione to counteract a potential increment of oxidative stress caused by the PA-MPs and the macroalga exuded secondary metabolites. Nevertheless, this decline may also reduce the competence for ROS neutralisation, which increases the oxidative damage potential [80]. In fact, although no lipid peroxidation occurred, oxidative damage at the protein level (PC) was observed in mussels exposed to both stressors combined. The imbalance between the generation of ROS and detoxification could have resulted in this rise in protein carbonyl levels. Protein carbonylation (PC) is a type of protein oxidation that can be promoted by the production of ROS [81]. It usually results in the formation of reactive ketone groups or amino acid aldehydes that can lead to the degradation of protein functions [81]. This may increase PC expression in response to different stressors, such as *A. armata* exudate and PA-MPs, thus representing a form of oxidative damage. LPO occurs due to a chain of molecular reactions that can culminate in oxidative damage of lipids allowing toxic agents to penetrate cell membranes [76]. In this study, as LPO was not affected in any tissue, it is not expected that changes in the lipid bilayer's structure and function or in membrane permeability occurred. Furthermore, the absence of modifications in LPO suggests the efficiency in activation of ROS scavenging mechanisms to prevent oxidative damage at the lipid level [82].

Oxidative stress-related biomarkers were also assessed in the digestive gland, which is the main surface for PA-MP uptake after being filtered through the gills, as they are also recognised as an important detoxification organ [83]. CAT activity was inhibited in organisms exposed to the combined exposure to PA-MPs and *A. armata* exudate, and, as in the gills, it is hypothesised that the decrease in this enzymatic antioxidant is due to a strong response in the early stages of exposure leading to its inhibition. Depletion of CAT activity was also observed in the digestive tissue of *M. galloprovincialis* exposed to PE and PS for 7 days [14], *Mytilus* spp. exposed only to PS also for 7 days [73], and the clam *Scrobicularia plana* exposed to 1 mg PS/L for 14 days [72]. On the other hand, GST and tGSH were not altered along the different treatments, which may imply that the second phase of the biotransformation of ROS and detoxification was presumably not activated in the mussels' digestive glands, at least at the sampling point used. The absence of significant modifications in GST levels in the digestive tissues of mussels exposed to microplastics was previously demonstrated by Avio et al. [14], as well as the unaltered levels of LPO. Cole et al. [45] also did not find significant lipid peroxidation in the digestive gland of *Mytilus* spp. exposed to polyamide microfibers. In response to the PA-MP stress factor, which can trigger inflammation processes in the tissues of exposed organisms [84], there was oxidative damage in the form of protein carbonylation (PC) in the digestive glands of mussels exposed to the polyamide microplastic treatment.

LPO and PC levels remained unaltered in the muscle tissue in mussels exposed to all the treatments, suggesting that no oxidative damage occurred in this tissue. Although antioxidant defence-related biomarkers were not measured, the absence of effects at the protein and lipid levels allows us to infer that the antioxidant machinery was efficient in the muscle tissue.

AChE is generally used to evaluate the neurotoxic potential of various compounds in marine organisms [85] and has an important role in the regulation of cholinergic neurotransmissions [86]. Microplastics-induced neurotoxicity has been previously demonstrated in the mussel *Mytilus galloprovincialis* exposed to 1.5 g/L PE and PS (<100 μm) [14], the clam *Scrobicularia plana* exposed to 1 mg PS/L (20 μm) [72], and *Corbicula fluminea* after exposure to 0.2 mg/L red fluorescent microspheres (1–5 μm) [74]. Therefore, if this enzyme is adversely affected, the essential nervous system functions may be disrupted. However, in the present study, no alterations were detected in the AChE activity of either the gills or the digestive gland, which may indicate that the responses in these tissues were not related to neurotoxicity. On the other hand, the AChE activity exhibited an increase in the muscle tissues of mussels exposed to *A. armata* exudate. Silva et al. [87] discussed that exposure to this seaweed exudate followed by the induction of AChE activity may be related to an induced regulatory overcompensation by increasing AChE in the organisms' cholinergic system. Another possible explanation is when the AChE is released from the cellular membrane surface, which may trigger *de novo* synthesis to restore this enzyme [88]. Furthermore, this increase in AChE activity may signal an induction of inflammatory reactions, as AChE rise usually occurs in inflamed tissues or cells [89], and may be associated with cell-disrupting processes, especially apoptosis [85]. An AChE activation was previously observed in *G. umbilicalis* [87] and in the muscle tissue of *M. galloprovincialis* [45] exposed to lower concentrations of *A. armata* exudate. However, although previous studies have demonstrated neurotransmission impairment attributed to other MPs, in the present study no effect was observed under PA-MP exposure.

4.3. Energy Metabolism Biomarkers

LDH enzyme has an important role in the anaerobic pathway of energy production [90] and was not altered in exposed mussels. Thus, there are no indications of energy mobilisation through anaerobic metabolic vias to counteract stress caused by the metabolites released with the *A. armata* exudate and the presence of PA-MPs.

The energy reserves were measured as lipid, sugar and protein contents, which, in a normal situation, are used in trade-offs between the organisms' basal maintenance and physiological functions. Lipids and proteins were not altered in neither of the treatments. However, there was a significant increase in sugar levels in organisms exposed to the combined treatment of *A. armata* exudate and PA-MPs. The demand for additional cellular glucose may be related to the induction of gluconeogenesis and may imply a disruption in the energetic metabolism. Lacroix et al. [82] hypothesized that induction of gluconeogenesis could transduce a higher energy storage (in the form of glycogen) in the exposed mussels, but an increased need of glucose to fulfil alternative metabolic routes to combat oxidative stress could also explain this increase. Moreover, the increased gluconeogenesis can be correlated to an increase of reactive oxygen species, as ROS can be generated indirectly by increasing the aerobic metabolism so that organisms are apt to sustain energy costs of metabolic responses to stressful conditions, considering that the electron transport system is a primary site for ROS production [77]. Energy consumption was assessed by determining mitochondrial electron transport system (ETS) activity and may be used to measure the metabolic capacity in response to stress. Mussels exposed to PA-MPs demonstrated increased energy consumption, either with or without the exudate. The increased ETS activity, and consequent increment of aerobic energy production, can be associated with an increase in stress levels while the organisms try to maintain a state of physiological homeostasis [91] and may also support the gluconeogenesis hypothesis. Therefore, this metabolic activation demonstrates a transfer of resources to produce energy, allowing the mussels to cope with microplastics-induced stress. Moreover, a potential increment of non-enzymatic antioxidant capacity is suggested by the ETS increase [45] in the presence of PA-MPs. The increase in energy consumption was accompanied by a depletion of CEA activity in mussels exposed to PA-MPs during 4 days, which ultimately represents a significant decrease in the energy budget; this decline being most noticeable when both stressors are combined. CEA suppression implicates a lower amount of energy available for the mussels' growth, reproduction, defence, and byssus production, and thus is more susceptible to additional stress [92]. Shang et al. [93] also demonstrated a CEA decline in *Mytilus coruscus* exposed during 14 days to high concentrations (10^4 and 10^6 particles/L) of PS microspheres as well as an increased cellular energy demand (ETS activity). On the other hand, Van Cauwenberghe et al. [91] also detected increased ETS activity after exposing *M. edulis* for 14 days to 110 PS microspheres/mL (10, 30 and 90 μm), but this increased metabolism was not accompanied by any other alterations in the overall energy budget.

4.4. Byssal Thread Production

Byssus represent an extracellular and collagenous structure that allows mussels' attachment to the substratum, thus any interference in byssal threads production can diminish the capacity of mussels to firmly anchor to the surface [94], making them prone to dislodgement and more susceptible to natural stressors, such as tides, waves and predation [45]. Production of functional byssus declined in mussels exposed to PA-MPs, either in the presence or absence of the exudate, with a lower number of secreted byssal threads being found under stressor combination. Decreased byssal production was also observed in the mussels *Perna viridis* [71] and *Perna canaliculus* [79] exposed to polyvinyl chloride and polyethylene particles, respectively.

The exposure to PA-MPs and combined stressors led mussels to allocate more energy to cope with the oxidative stress, which, together with the high levels of protein oxidation, might have compromised the organisms' ability to invest in the growth and development of structures, such as the byssal threads. Thus, this study suggests that the presence of *A. armata* exudate combined with PA-MPs might increase the vulnerability of *M. galloprovincialis*, as byssal threads are crucial to anchor themselves to the rocky shores and to other mussels. This may consequently impair individuals' fitness, survival, the preservation of mussel beds, and their role in regulating macrofaunal and flora diversity [78].

5. Conclusions

In summary, the present findings suggest that 1 mg PA-MP/L in co-exposure with 2% *A. armata* exudate present a health hazard to *M. galloprovincialis*. In particular, the responses of oxidative stress biomarkers and the decrease in the final balance of the energy budget reflected the activation of antioxidant defences in exposed mussels, which prevented lipid peroxidation but not oxidative damage in proteins. Moreover, this was reflected in the impairment of byssus production under exposure to PA-MPs, which can compromise the attachment of mussels to the substratum and mussel bed stability. Thus, a potential amplification of the deleterious effects of the PA-MPs was observed in the presence of this invasive species exudate. This may anticipate that exposure to the secondary metabolites produced by *A. armata* may pose an additional impact to marine biota under the threat of MP pollution.

Supplementary Materials: The following are available online at https://www.mdpi.com/article/10.3390/toxics10020043/s1, Figure S1: Irregularly shaped polyamide microplastics (PA-MPs). Images taken at 10× magnification (Zeiss Primo Star light microscope, Jena, Germany), Table S1: Two-way ANOVA analysis results on oxidative stress-related biomarker responses in *Mytilus galloprovincialis* gills with *Asparagopsis armata* exudate and PA-MPs exposures as factors, Table S2: Results for two-way ANOVA analysis on biochemical biomarkers responses in the digestive gland of *M. galloprovincialis* with *A. armata* exudate exposure and water temperature as factors, Table S3: Results for two-way ANOVA analysis on biochemical biomarkers responses in the muscle of *M. galloprovincialis* with *A. armata* exudate exposure and water temperature as factors.

Author Contributions: Conceptualization, D.C., A.L.P.S., and M.D.B.; methodology, H.C.V., D.C., A.C.M.R., A.L.P.S., and M.D.B.; formal analysis, F.G.R., H.C.V., A.C.M.R., and M.D.B.; investigation, F.G.R., H.C.V., S.F.S.P., and M.D.B.; writing—original draft, F.G.R.; writing—review and editing, F.G.R., H.C.V., D.C., S.F.S.P., A.C.M.R., A.L.P.S., A.M.V.M.S., J.M.M.O., and M.D.B.; resources, A.L.P.S., A.M.V.M.S., and M.D.B.; supervision: D.C., A.L.P.S., J.M.M.O., and M.D.B.; project administration: A.L.P.S. and M.D.B.; funding acquisition: A.L.P.S. and M.D.B. All authors have read and agreed to the published version of the manuscript.

Funding: This research was funded through CESAM (UIDP/50017/2020+UIDB/50017/2020+LA/P/ 0094/2020), with financial support from FCT/MCTES through national funds. The authors are thankful for the financial support by the projects INSIDER (PTDC/CTA-AMB/30495/2017) and comPET (PTDC/CTA-AMB/30361/2017) funded by FEDER, through COMPETE2020-Programa Operacional Competitividade e Internacionalização (POCI), and by national funds (OE), through FCT/MCTES. The authors also thank FCT and POPH/FSE (Programa Operacional Potencial Humano/Fundo Social Europeu) for the doctoral grant of H.C.V. (PD/BD/127808/2016). A.L.P.S. is funded by a FCT research contract (CEECIND/01366/2018). M.D.B. is funded by national funds (OE), through FCT, I.P., in the scope of the framework contract foreseen in the numbers 4, 5 and 6 of the article 23, of the Decree-Law 57/2016, of 29 August, changed by Law 57/2017, of 19 July.

Institutional Review Board Statement: Not applicable.

Informed Consent Statement: Not applicable.

Data Availability Statement: The data presented in this study is available in the current manuscript, raw data is available on request from the corresponding author.

Conflicts of Interest: The authors declare no conflict of interest.

References

1. Pinteus, S.; Lemos, M.F.L.; Alves, C.; Neugebauer, A.; Silva, J.; Thomas, O.P.; Botana, L.M.; Gaspar, H.; Pedrosa, R. Marine invasive macroalgae: Turning a real threat into a major opportunity—The biotechnological potential of *Sargassum muticum* and *Asparagopsis armata*. *Algal Res.* **2018**, *34*, 217–234. [CrossRef]
2. Halpern, B.S.; Walbridge, S.; Selkoe, K.A.; Kappel, C.V.; Micheli, F.; D'Agrosa, C.; Bruno, J.F.; Casey, K.S.; Ebert, C.; Fox, H.E.; et al. A global map of human impact on marine ecosystems. *Science* **2008**, *319*, 948–952. [CrossRef]
3. PlasticsEurope. *Plastics–the Facts 2020: An analysis of European plastics production, Demand and Waste Data*; PlasticsEurope: Brussels, Belgium, 2020.
4. Li, J.; Lusher, A.L.; Rotchell, J.M.; Deudero, S.; Turra, A.; Bråte, I.L.N.; Sun, C.; Shahadat Hossain, M.; Li, Q.; Kolandhasamy, P.; et al. Using mussel as a global bioindicator of coastal microplastic pollution. *Environ. Pollut.* **2019**, *244*, 522–533. [CrossRef] [PubMed]
5. Cole, M.; Lindeque, P.; Halsband, C.; Galloway, T.S. Microplastics as contaminants in the marine environment: A review. *Mar. Pollut. Bull.* **2011**, *62*, 2588–2597. [CrossRef] [PubMed]
6. Alprol, A.E.; Gaballah, M.S.; Hassaan, M.A. Micro and Nanoplastics analysis: Focus on their classification, sources, and impacts in marine environment. *Reg. Stud. Mar. Sci.* **2021**, *42*, 101625. [CrossRef]
7. Beiras, R.; Schönemann, A.M. Currently monitored microplastics pose negligible ecological risk to the global ocean. *Sci. Rep.* **2020**, *10*, 22281. [CrossRef] [PubMed]
8. Lindeque, P.K.; Cole, M.; Coppock, R.L.; Lewis, C.N.; Miller, R.Z.; Watts, A.J.R.; Wilson-McNeal, A.; Wright, S.L.; Galloway, T.S. Are we underestimating microplastic abundance in the marine environment? A comparison of microplastic capture with nets of different mesh-size. *Environ. Pollut.* **2020**, *265*, 114721. [CrossRef]
9. Wright, S.L.; Thompson, R.C.; Galloway, T.S. The physical impacts of microplastics on marine organisms: A review. *Environ. Pollut.* **2013**, *178*, 483–492. [CrossRef]
10. Cole, M.; Liddle, C.; Consolandi, G.; Drago, C.; Hird, C.; Lindeque, P.K.; Galloway, T.S. Microplastics, microfibres and nanoplastics cause variable sub-lethal responses in mussels (*Mytilus* spp.). *Mar. Pollut. Bull.* **2020**, *160*, 111552. [CrossRef]
11. Li, J.; Qu, X.; Su, L.; Zhang, W.; Yang, D.; Kolandhasamy, P.; Li, D.; Shi, H. Microplastics in mussels along the coastal waters of China. *Environ. Pollut.* **2016**, *214*, 177–184. [CrossRef]
12. Kinjo, A.; Mizukawa, K.; Takada, H.; Inoue, K. Size-dependent elimination of ingested microplastics in the Mediterranean mussel *Mytilus galloprovincialis*. *Mar. Pollut. Bull.* **2019**, *149*, 110512. [CrossRef]
13. Von Moos, N.; Burkhardt-Holm, P.; Köhler, A. Uptake and effects of microplastics on cells and tissue of the blue mussel *Mytilus edulis* L. after an experimental exposure. *Environ. Sci. Technol.* **2012**, *46*, 11327–11335. [CrossRef] [PubMed]

14. Avio, C.G.; Gorbi, S.; Milan, M.; Benedetti, M.; Fattorini, D.; D'Errico, G.; Pauletto, M.; Bargelloni, L.; Regoli, F. Pollutants bioavailability and toxicological risk from microplastics to marine mussels. *Environ. Pollut.* **2015**, *198*, 211–222. [CrossRef] [PubMed]
15. Liu, L.; Xu, K.; Zhang, B.; Ye, Y.; Zhang, Q.; Jiang, W. Cellular internalization and release of polystyrene microplastics and nanoplastics. *Sci. Total Environ.* **2021**, *779*, 146523. [CrossRef] [PubMed]
16. Foley, C.J.; Feiner, Z.S.; Malinich, T.D.; Höök, T.O. A meta-analysis of the effects of exposure to microplastics on fish and aquatic invertebrates. *Sci. Total Environ.* **2018**, *631–632*, 550–559. [CrossRef]
17. Avio, C.G.; Gorbi, S.; Regoli, F. Plastics and microplastics in the oceans: From emerging pollutants to emerged threat. *Mar. Environ. Res.* **2017**, *128*, 2–11. [CrossRef] [PubMed]
18. Zhang, X.; Xia, M.; Zhao, J.; Cao, Z.; Zou, W.; Zhou, Q. Photoaging enhanced the adverse effects of polyamide microplastics on the growth, intestinal health, and lipid absorption in developing zebrafish. *Environ. Int.* **2022**, *158*, 106922. [CrossRef]
19. Lusher, A.; Hollman, P.; Mandoza-Hill, J. *Microplastics in fisheries and aquaculture: Status of Knowledge on Their Occurrence and Implications for Aquatic Organisms and Food Safety*; FAO: Rome, Italy, 2017; Volume 615.
20. Lusher, A.L.; McHugh, M.; Thompson, R.C. Occurrence of microplastics in the gastrointestinal tract of pelagic and demersal fish from the English Channel. *Mar. Pollut. Bull.* **2013**, *67*, 94–99. [CrossRef]
21. Thushari, G.G.N.; Senevirathna, J.D.M.; Yakupitiyage, A.; Chavanich, S. Effects of microplastics on sessile invertebrates in the eastern coast of Thailand: An approach to coastal zone conservation. *Mar. Pollut. Bull.* **2017**, *124*, 349–355. [CrossRef]
22. Pedrotti, M.L.; Petit, S.; Elineau, A.; Bruzaud, S.; Crebassa, J.C.; Dumontet, B.; Martí, E.; Gorsky, G.; Cózar, A. Changes in the floating plastic pollution of the mediterranean sea in relation to the distance to land. *PLoS ONE* **2016**, *11*, e0161581. [CrossRef]
23. Cincinelli, A.; Scopetani, C.; Chelazzi, D.; Martellini, T.; Pogojeva, M.; Slobodnik, J. Microplastics in the Black Sea sediments. *Sci. Total Environ.* **2021**, *760*, 143898. [CrossRef]
24. Erni-Cassola, G.; Zadjelovic, V.; Gibson, M.I.; Christie-Oleza, J.A. Distribution of plastic polymer types in the marine environment; A meta-analysis. *J. Hazard. Mater.* **2019**, *369*, 691–698. [CrossRef]
25. Davidson, A.D.; Campbell, M.L.; Hewitt, C.L.; Schaffelke, B. Assessing the impacts of nonindigenous marine macroalgae: An update of current knowledge. *Bot. Mar.* **2015**, *58*, 55–79. [CrossRef]
26. Thomsen, M.S.; Wernberg, T.; South, P.M.; Schiel, D.R. Non-native Seaweeds Drive Changes in Marine Coastal Communities Around the World. In *Seaweed Phylogeography*; Hu, Z.M., Fraser, C., Eds.; Springer: Dordrecht, The Netherlands, 2016; pp. 147–185. [CrossRef]
27. Katsanevakis, S.; Wallentinus, I.; Zenetos, A.; Leppäkoski, E.; Çinar, M.E.; Oztürk, B.; Grabowski, M.; Golani, D.; Cardoso, A.C. Impacts of invasive alien marine species on ecosystem services and biodiversity: A pan-European review. *Aquat. Invasions* **2014**, *9*, 391–423. [CrossRef]
28. Williams, S.L.; Smith, J.E. A global review of the distribution, taxonomy, and impacts of introduced seaweeds. *Annu. Rev. Ecol. Evol. Syst.* **2007**, *38*, 327–359. [CrossRef]
29. Molnar, J.L.; Gamboa, R.L.; Revenga, C.; Spalding, M.D. Assessing the global threat of invasive species to marine biodiversity. *Front. Ecol. Environ.* **2008**, *6*, 485–492. [CrossRef]
30. Simberloff, D.; Martin, J.L.; Genovesi, P.; Maris, V.; Wardle, D.A.; Aronson, J.; Courchamp, F.; Galil, B.; García-Berthou, E.; Pascal, M.; et al. Impacts of biological invasions: What's what and the way forward. *Trends Ecol. Evol.* **2013**, *28*, 58–66. [CrossRef] [PubMed]
31. Bonin, D.R.; Hawkes, M.W. Systematics and life histories of New Zealand Bonnemaisoniaceae (Bonnemaisoniales, Rhodophyta): I. The genus *Asparagopsis*. *N. Z. J. Bot.* **1987**, *25*, 577–590. [CrossRef]
32. Boudouresque, C.F.; Verlaque, M. Biological pollution in the Mediterranean Sea: Invasive versus introduced macrophytes. *Mar. Pollut. Bull.* **2002**, *44*, 32–38. [CrossRef]
33. Ní Chualáin, F.; Maggs, C.A.; Saunders, G.W.; Guiry, M.D. The invasive genus *Asparagopsis* (Bonnemaisoniaceae, Rhodophyta): Molecular systematics, morphology, and ecophysiology of *Falkenbergia* isolates. *J. Phycol.* **2004**, *40*, 1112–1126. [CrossRef]
34. Andreakis, N.; Procaccini, G.; Maggs, C.; Kooistra, W.H.C.F. Phylogeography of the invasive seaweed *Asparagopsis* (Bonnemaisoniales, Rhodophyta) reveals cryptic diversity. *Mol. Ecol.* **2007**, *16*, 2285–2299. [CrossRef]
35. Martins, G.M.; Cacabelos, E.; Faria, J.; Álvaro, N.; Prestes, A.C.L.; Neto, A.I. Patterns of distribution of the invasive alga *Asparagopsis armata* Harvey: A multi-scaled approach. *Aquat. Invasions* **2019**, *14*, 582–593. [CrossRef]
36. Rubal, M.; Costa-Garcia, R.; Besteiro, C.; Sousa-Pinto, I.; Veiga, P. Mollusc diversity associated with the non-indigenous macroalga *Asparagopsis armata* Harvey, 1855 along the Atlantic coast of the Iberian Peninsula. *Mar. Environ. Res.* **2018**, *136*, 1–7. [CrossRef] [PubMed]
37. Maggs, C.A.; Stegenga, H. Red algal exotics on North Sea coasts. *Helgol. Meeresunters.* **1999**, *52*, 243–258. [CrossRef]
38. Paul, N.A.; De Nys, R.; Steinberg, P.D. Seaweed-herbivore interactions at a small scale: Direct tests of feeding deterrence by filamentous algae. *Mar. Ecol. Prog. Ser.* **2006**, *323*, 1–9. [CrossRef]
39. Sala, E.; Boudouresque, C.F. The role of fishes in the organization of a Mediterranean sublittoral community. I: Algal communities. *J. Exp. Mar. Bio. Ecol.* **1997**, *212*, 25–44. [CrossRef]
40. Genovese, G.; Tedone, L.; Hamann, M.T.; Morabito, M. The Mediterranean red alga *Asparagopsis*: A source of compounds against *Leishmania*. *Mar. Drugs* **2009**, *7*, 361–366. [CrossRef]

41. Bansemir, A.; Blume, M.; Schröder, S.; Lindequist, U. Screening of cultivated seaweeds for antibacterial activity against fish pathogenic bacteria. *Aquaculture* **2006**, *252*, 79–84. [CrossRef]
42. Guerra-García, J.M.; Ros, M.; Izquierdo, D.; Soler-Hurtado, M.M. The invasive *Asparagopsis armata* versus the native *Corallina elongata*: Differences in associated peracarid assemblages. *J. Exp. Mar. Bio. Ecol.* **2012**, *416–417*, 121–128. [CrossRef]
43. Silva, C.O.; Lemos, M.F.L.; Gaspar, R.; Gonçalves, C.; Neto, J.M.; Silva, C.O. The effects of the invasive seaweed *Asparagopsis armata* on native rock pool communities: Evidences from experimental exclusion. *Ecol. Indic.* **2021**, *125*, 107463. [CrossRef]
44. Silva, C.O.; Novais, S.C.; Soares, A.M.V.M.; Barata, C.; Lemos, M.F.L. Impacts of the Invasive Seaweed *Asparagopsis armata* Exudate on Energetic Metabolism of Rock Pool Invertebrates. *Toxins* **2021**, *13*, 15. [CrossRef] [PubMed]
45. Coelho, S.D.; Vieira, H.C.; Oliveira, J.M.M.; Pires, S.F.S.; Rocha, R.J.M.; Rodrigues, A.C.M.; Soares, A.M.V.M.; Bordalo, M.D. How Does *Mytilus galloprovincialis* Respond When Exposed to the Gametophyte Phase of the Invasive Red Macroalga *Asparagopsis armata* Exudate? *Water* **2021**, *13*, 460. [CrossRef]
46. Vieira, H.C.; Rodrigues, A.C.M.; Pires, S.F.S.; Oliveira, J.M.M.; Rocha, R.J.M.; Soares, A.M.V.M.; Bordalo, M.D. Ocean warming may enhance biochemical alterations induced by an invasive seaweed exudate in the mussel *Mytilus galloprovincialis*. *Toxics* **2021**, *9*, 121. [CrossRef] [PubMed]
47. Beyer, J.; Green, N.W.; Brooks, S.; Allan, I.J.; Ruus, A.; Gomes, T.; Bråte, I.L.N.; Schøyen, M. Blue mussels (*Mytilus edulis* spp.) as sentinel organisms in coastal pollution monitoring: A review. *Mar. Environ. Res.* **2017**, *130*, 338–365. [CrossRef] [PubMed]
48. Abidli, S.; Pinheiro, M.; Lahbib, Y.; Neuparth, T.; Santos, M.M.; Trigui El Menif, N. Effects of environmentally relevant levels of polyethylene microplastic on *Mytilus galloprovincialis* (Mollusca: Bivalvia): Filtration rate and oxidative stress. *Environ. Sci. Pollut. Res.* **2021**, *28*, 26643–26652. [CrossRef]
49. Arribas, L.P.; Donnarumma, L.; Palomo, M.G.; Scrosati, R.A. Intertidal mussels as ecosystem engineers: Their associated invertebrate biodiversity under contrasting wave exposures. *Mar. Biodivers.* **2014**, *44*, 203–211. [CrossRef]
50. Oliveira, J.; Castilho, F.; Cunha, Â.; Pereira, M.J. Bivalve harvesting and production in Portugal: An overview. *J. Shellfish Res.* **2013**, *32*, 911–924. [CrossRef]
51. Pittura, L.; Avio, C.G.; Giuliani, M.E.; d'Errico, G.; Keiter, S.H.; Cormier, B.; Gorbi, S.; Regoli, F. Microplastics as vehicles of environmental PAHs to marine organisms: Combined chemical and physical hazards to the mediterranean mussels, Mytilus galloprovincialis. *Front. Mar. Sci.* **2018**, *5*, 103. [CrossRef]
52. *ASTM Standard E729-96*; Standard Guide for Conducting Acute Toxicity Tests on Test Materials with Fishes, Macroinvertebrates, and Amphibians. ASTM International: West Conshohocken, PA, USA, 2014; Voulume 11, 1–22. [CrossRef]
53. Prata, J.C.; Sequeira, I.F.; Monteiro, S.S.; Silva, A.L.P.; da Costa, J.P.; Dias-Pereira, P.; Fernandes, A.J.S.; da Costa, F.M.; Duarte, A.C.; Rocha-Santos, T. Preparation of biological samples for microplastic identification by Nile Red. *Sci. Total Environ.* **2021**, *783*, 147065. [CrossRef]
54. Campbell, S.H.; Williamson, P.R.; Hall, B.D. Microplastics in the gastrointestinal tracts of fish and the water from an urban prairie creek. *Facets* **2017**, *2*, 395–409. [CrossRef]
55. Bradford, M.M. A rapid and sensitive method for the quantitation of microgram quantities of protein utilizing the principle of protein-dye binding. *Anal. Biochem.* **1976**, *72*, 248–254. [CrossRef]
56. Ellman, G.L.; Courtney, K.D.; Andres, V.; Featherstone, R.M. A new and rapid colorimetric determination of acetylcholinesterase activity. *Biochem. Pharmacol* **1961**, *7*, 88–95. [CrossRef]
57. Domingues, I.; Gravato, C. Oxidative stress assessment in zebrafish larvae. *Methods Mol. Biol.* **2018**, *1797*, 477–486. [CrossRef] [PubMed]
58. Claiborne, A. Catalase activity. In *CRC Handbook of Methods in Oxygen Radical Research*; Greenwald, R.A., Ed.; CRC Press: Boca Raton, FL, USA, 1985; pp. 283–284.
59. Habig, W.H.; Pabst, M.J.; Jakoby, W.B. Glutathione S transferases. The first enzymatic step in mercapturic acid formation. *J. Biol. Chem.* **1974**, *249*, 7130–7139. [CrossRef]
60. Baker, M.A.; Cerniglia, G.J.; Zaman, A. Microtiter plate assay for the measurement of glutathione and glutathione disulfide in large numbers of biological samples. *Anal. Biochem.* **1990**, *190*, 360–365. [CrossRef]
61. Tietze, F. Enzymic method for quantitative determination of nanogram amounts of total and oxidized glutathione: Applications to mammalian blood and other tissues. *Anal. Biochem.* **1969**, *27*, 502–522. [CrossRef]
62. Rodrigues, A.C.M.; Gravato, C.; Quintaneiro, C.; Bordalo, M.D.; Barata, C.; Soares, A.M.V.M.; Pestana, J.L.T. Energetic costs and biochemical biomarkers associated with esfenvalerate exposure in *Sericostoma vittatum*. *Chemosphere* **2017**, *189*, 445–453. [CrossRef]
63. Bird, R.P.; Draper, H.H. Comparative Studies on Different Methods of Malonaldehyde Determination. *Methods Enzym.* **1984**, *105*, 299–305. [CrossRef]
64. Mesquita, C.S.; Oliveira, R.; Bento, F.; Geraldo, D.; Rodrigues, J.V.; Marcos, J.C. Simplified 2,4-dinitrophenylhydrazine spectrophotometric assay for quantification of carbonyls in oxidized proteins. *Anal. Biochem.* **2014**, *458*, 69–71. [CrossRef]
65. Vassault, A. Lactate dehydrogenase. In *Methods of Enzymatic Analysis—Enzymes: Oxireductases, Transferase*; Bergmyer, M.O., Ed.; Academic Press: New York, NY, USA, 1983; pp. 118–126.
66. Diamantino, T.C.; Almeida, E.; Soares, A.M.V.M.; Guilhermino, L. Lactate dehydrogenase activity as an effect criterion in toxicity tests with Daphnia magna straus. *Chemosphere* **2001**, *45*, 553–560. [CrossRef]

67. De Coen, W.; Janssen, C.R. The use of biomarkers in *Daphnia magna* toxicity testing. IV.Cellular Energy Allocation: A new methodology to assess the energy budget of toxicant-stressed Daphnia populations. *J. Aquat. Ecosyst. Stress Recover* **1997**, *6*, 43–45. [CrossRef]
68. Rodrigues, A.C.M.; Gravato, C.; Quintaneiro, C.; Golovko, O.; Žlábek, V.; Barata, C.; Soares, A.M.V.M.; Pestana, J.L.T. Life history and biochemical effects of chlorantraniliprole on *Chironomus riparius*. *Sci. Total Environ.* **2015**, *508*, 506–513. [CrossRef]
69. Gnaiger, E. Calculation of Energetic and Biochemical Equivalents of Respiratory Oxygen Consumption. In *Polarographic Oxygen Sensors*; Springer: Berlin/Heidelberg, Germany, 1983; pp. 337–375.
70. Browne, M.A.; Dissanayake, A.; Galloway, T.S.; Lowe, D.M.; Thompson, R.C. Ingested microscopic plastic translocates to the circulatory system of the mussel, *Mytilus edulis* (L.). *Environ. Sci. Technol.* **2008**, *42*, 5026–5031. [CrossRef]
71. Rist, S.E.; Assidqi, K.; Zamani, N.P.; Appel, D.; Perschke, M.; Huhn, M.; Lenz, M. Suspended micro-sized PVC particles impair the performance and decrease survival in the Asian green mussel *Perna viridis*. *Mar. Pollut. Bull.* **2016**, *111*, 213–220. [CrossRef]
72. Ribeiro, F.; Garcia, A.R.; Pereira, B.P.; Fonseca, M.; Mestre, N.C.; Fonseca, T.G.; Ilharco, L.M.; Bebianno, M.J. Microplastics effects in *Scrobicularia plana*. *Mar. Pollut. Bull.* **2017**, *122*, 379–391. [CrossRef] [PubMed]
73. Paul-Pont, I.; Lacroix, C.; González Fernández, C.; Hégaret, H.; Lambert, C.; Le Goïc, N.; Frère, L.; Cassone, A.L.; Sussarellu, R.; Fabioux, C.; et al. Exposure of marine mussels *Mytilus* spp. to polystyrene microplastics: Toxicity and influence on fluoranthene bioaccumulation. *Environ. Pollut.* **2016**, *216*, 724–737. [CrossRef] [PubMed]
74. Guilhermino, L.; Vieira, L.R.; Ribeiro, D.; Tavares, A.S.; Cardoso, V.; Alves, A.; Almeida, J.M. Uptake and effects of the antimicrobial florfenicol, microplastics and their mixtures on freshwater exotic invasive bivalve *Corbicula fluminea*. *Sci. Total Environ.* **2018**, *622–623*, 1131–1142. [CrossRef] [PubMed]
75. Thapa, H.R.; Lin, Z.; Yi, D.; Smith, J.E.; Schmidt, E.W.; Agarwal, V. Genetic and Biochemical Reconstitution of Bromoform Biosynthesis in *Asparagopsis* Lends Insights into Seaweed Reactive Oxygen Species Enzymology. *ACS Chem. Biol.* **2020**, *15*, 1662–1670. [CrossRef]
76. Prokić, M.D.; Radovanović, T.B.; Gavrić, J.P.; Faggio, C. Ecotoxicological effects of microplastics: Examination of biomarkers, current state and future perspectives. *TrAC-Trends Anal. Chem.* **2019**, *111*, 37–46. [CrossRef]
77. Regoli, F.; Giuliani, M.E. Oxidative pathways of chemical toxicity and oxidative stress biomarkers in marine organisms. *Mar. Environ. Res.* **2014**, *93*, 106–117. [CrossRef] [PubMed]
78. Gosling, E. *Marine Bivalve Molluscs*, 2nd ed.; John Wiley & Sons: Hoboken, NJ, USA, 2015.
79. Webb, S.; Gaw, S.; Marsden, I.D.; McRae, N.K. Biomarker responses in New Zealand green-lipped mussels *Perna canaliculus* exposed to microplastics and triclosan. *Ecotoxicol. Environ. Saf.* **2020**, *201*, 110871. [CrossRef]
80. Magara, G.; Elia, A.C.; Syberg, K.; Khan, F.R. Single contaminant and combined exposures of polyethylene microplastics and fluoranthene: Accumulation and oxidative stress response in the blue mussel, *Mytilus edulis*. *J. Toxicol. Environ. Health-Part A Curr. Issues* **2018**, *81*, 761–773. [CrossRef] [PubMed]
81. Suzuki, Y.J.; Carini, M.; Butterfield, D.A. Protein carbonylation. *Antioxid. Redox Signal.* **2010**, *12*, 323–325. [CrossRef]
82. Lacroix, C.; Richard, G.; Seguineau, C.; Guyomarch, J.; Moraga, D.; Auffret, M. Active and passive biomonitoring suggest metabolic adaptation in blue mussels (*Mytilus* spp.) chronically exposed to a moderate contamination in Brest harbor (France). *Aquat. Toxicol.* **2015**, *162*, 126–137. [CrossRef] [PubMed]
83. Dobal, V.; Suárez, P.; Ruiz, Y.; García-Martín, O.; Juan, F.S. Activity of antioxidant enzymes in *Mytilus galloprovincialis* exposed to tar: Integrated response of different organs as pollution biomarker in aquaculture areas. *Aquaculture* **2022**, *548*, 737638. [CrossRef]
84. Silva, C.J.M.; Patrício Silva, A.L.; Campos, D.; Machado, A.L.; Pestana, J.L.T.; Gravato, C. Oxidative damage and decreased aerobic energy production due to ingestion of polyethylene microplastics by *Chironomus riparius* (Diptera) larvae. *J. Hazard. Mater.* **2021**, *402*, 123775. [CrossRef]
85. Zhang, X.J.; Yang, L.; Zhao, Q.; Caen, J.P.; He, H.Y.; Jin, Q.H.; Guo, L.H.; Alemany, M.; Zhang, L.Y.; Shi, Y.F. Induction of acetylcholinesterase expression during apoptosis in various cell types. *Cell Death Differ.* **2002**, *9*, 790–800. [CrossRef]
86. Fonte, E.; Ferreira, P.; Guilhermino, L. Temperature rise and microplastics interact with the toxicity of the antibiotic cefalexin to juveniles of the common goby (*Pomatoschistus microps*): Post-exposure predatory behaviour, acetylcholinesterase activity and lipid peroxidation. *Aquat. Toxicol.* **2016**, *180*, 173–185. [CrossRef]
87. Silva, C.O.; Simões, T.; Félix, R.; Soares, A.M.V.M.; Barata, C.; Novais, S.C.; Lemos, M.F.L. *Asparagopsis armata* Exudate Cocktail: The Quest for the Mechanisms of Toxic Action of an Invasive Seaweed on Marine Invertebrates. *Biology* **2021**, *10*, 223. [CrossRef]
88. Espinoza, B.; Silman, I.; Arnon, R.; Tarrab-Hazdai, R. Phosphatidylinositol-specific phospholipase C induces biosynthesis of acetylcholinesterase via diacylglycerol in *Schistosoma mansoni*. *Eur. J. Biochem.* **1991**, *195*, 863–870. [CrossRef]
89. Gambardella, C.; Morgana, S.; Ferrando, S.; Bramini, M.; Piazza, V.; Costa, E.; Garaventa, F.; Faimali, M. Effects of polystyrene microbeads in marine planktonic crustaceans. *Ecotoxicol. Environ. Saf.* **2017**, *145*, 250–257. [CrossRef] [PubMed]
90. Shahriari, A.; Dawson, N.J.; Bell, R.A.V.; Storey, K.B. Stable suppression of lactate dehydrogenase activity during anoxia in the foot muscle of *Littorina littorea* and the potential role of acetylation as a novel posttranslational regulatory mechanism. *Enzyme Res.* **2013**, *2013*, 461374. [CrossRef]

91. Van Cauwenberghe, L.; Claessens, M.; Vandegehuchte, M.B.; Janssen, C.R. Microplastics are taken up by mussels (*Mytilus edulis*) and lugworms (*Arenicola marina*) living in natural habitats. *Environ. Pollut.* **2015**, *199*, 10–17. [CrossRef]
92. Erk, M.; Ivanković, D.; Strižak, Ž. Cellular energy allocation in mussels (*Mytilus galloprovincialis*) from the stratified estuary as a physiological biomarker. *Mar. Pollut. Bull.* **2011**, *62*, 1124–1129. [CrossRef] [PubMed]
93. Shang, Y.; Wang, X.; Chang, X.; Sokolova, I.M.; Wei, S.; Liu, W.; Fang, J.K.H.; Hu, M.; Huang, W.; Wang, Y. The Effect of Microplastics on the Bioenergetics of the Mussel *Mytilus coruscus* Assessed by Cellular Energy Allocation Approach. *Front. Mar. Sci.* **2021**, *8*, 1–8. [CrossRef]
94. Bell, E.C.; Gosline, J.M. Mechanical design of mussel byssus: Material yield enhances attachment strength. *J. Exp. Biol.* **1996**, *199*, 1005–1017. [CrossRef]

Article

Accumulation, Depuration, and Biological Effects of Polystyrene Microplastic Spheres and Adsorbed Cadmium and Benzo(a)pyrene on the Mussel *Mytilus galloprovincialis*

Rebecca von Hellfeld [1,2], María Zarzuelo [1], Beñat Zaldibar [1], Miren P. Cajaraville [1] and Amaia Orbea [1,*]

[1]   CBET Research Group, Department of Zoology and Animal Cell Biology, Biotechnology PiE and Science and Technology Faculty, Research Centre for Experimental Marine Biology, University of the Basque Country (UPV/EHU), Sarriena z/g, 48940 Leioa, Basque Country, Spain; rebecca.vonhellfeld@abdn.ac.uk (R.v.H.); mariazarzuelo1997@gmail.com (M.Z.); benat.zaldibar@ehu.eus (B.Z.); mirenp.cajaraville@ehu.eus (M.P.C.)

[2]   School of Biological Sciences, University of Aberdeen, 23 St. Machar Drive, Aberdeen AB24 3UU, UK

*   Correspondence: amaia.orbea@ehu.eus; Tel.: +34-946-012-735

Abstract: Filter feeders are target species for microplastic (MP) pollution, as particles can accumulate in the digestive system, disturbing feeding processes and becoming internalized in tissues. MPs may also carry pathogens or pollutants present in the environment. This work assessed the influence of polystyrene (PS) MP size and concentration on accumulation and depuration time and the role of MPs as vectors for metallic (Cd) and organic (benzo(a)pyrene, BaP) pollutants. One-day exposure to pristine MPs induced a concentration-dependent accumulation in the digestive gland (in the stomach and duct lumen), and after 3-day depuration, 45 μm MPs appeared between gill filaments, while 4.5 μm MPs also occurred within gill filaments. After 3-day exposure to contaminated 4.5 μm MPs, mussels showed increased BaP levels whilst Cd accumulation did not occur. Here, PS showed higher affinity to BaP than to Cd. Three-day exposure to pristine or contaminated MPs did not provoke significant alterations in antioxidant and peroxisomal enzyme activities in the gills and digestive gland nor in lysosomal membrane stability. Exposure to dissolved contaminants and to MP-BaP caused histological alterations in the digestive gland. In conclusion, these short-term studies suggest that MPs are ingested and internalized in a size-dependent manner and act as carriers of the persistent organic pollutant BaP.

Keywords: polystyrene microplastics; size-dependent uptake; vectors; cadmium; benzo(a)pyrene; mussels

Citation: von Hellfeld, R.; Zarzuelo, M.; Zaldibar, B.; Cajaraville, M.P.; Orbea, A. Accumulation, Depuration, and Biological Effects of Polystyrene Microplastic Spheres and Adsorbed Cadmium and Benzo(a)pyrene on the Mussel *Mytilus galloprovincialis*. *Toxics* **2022**, *10*, 18. https://doi.org/10.3390/toxics10010018

Academic Editors: Costanza Scopetani, Tania Martellini and Diana Campos

Received: 18 November 2021
Accepted: 29 December 2021
Published: 5 January 2022

Publisher's Note: MDPI stays neutral with regard to jurisdictional claims in published maps and institutional affiliations.

1. Introduction

In 2019, world plastic production reached 368 million tons [1], and the lack of efficient plastic management has led to severe consequences for ecosystems [2]. Moreover, the plethora of paths through which plastic enters the marine environment has allowed large quantities of plastic to accumulate [3]. The different types and sizes of plastic [2], such as water bottles, bags and industrially produced plastic pellets and microparticles [4], have been found to affect all trophic levels [5,6]. Microplastics (MPs) are defined as plastic particles with a diameter of less than 5 mm [7] and, according to the European Marine Strategy Framework Directive (MSFD) Technical Subgroup on Marine Litter, should be further classified as small MPs (<1 mm) and large MPs (1–5 mm) [8]. These minute particles easily disperse in the water column and are frequently found in sediment samples [9] and in biota [10].

Interactions with the environment alter the particles' structure, resulting in changing surface properties [11]. Over time, these processes increase their porosity, charge and roughness, leading to an increase in accumulation of other compounds present in the environment [12–14]. It has been found that the smaller a particle is, the larger its surface area-to-volume ratio will be, leading to greater contaminant adsorption [15]. It has further been reported that over 70% of chemicals listed as priority pollutants by the United States EPA bind with plastic debris [14]. Thus, the high capacity of plastic debris for adsorbing pollutants poses an additional threat to marine wildlife [16], as the adsorbed contaminants may desorb once the particle has been ingested [17]. The bioaccumulation potential of some of these potentially adsorbed contaminants can thus be seen throughout ecosystems [18], which may lead to the transfer of pollutants across generations [19,20] or alterations at subcellular level [21,22]. This underlines the importance of understanding the vector potential of MPs in order to accurately predict the risk of water borne contaminants in conjunction with the increasing pollution of marine waters with plastic particles [1].

Mussels are filter-feeding sessile organisms tolerant to salinity changes and other stressors. Moreover, due to their high water filtration rate and low metabolic activity, they accumulate dissolved and particulate pollutants at levels higher than those present in the water column [22]. This makes them an excellent species for MP research, allowing for comparability and transferability of results [23–26]. One widely distributed species is the Mediterranean mussel *Mytilus galloprovincialis*, found along almost every coastline worldwide [27], inhabiting the zone between the rocky shore and sandy bottom [28], and are thus widely used as sentinel organisms in pollution monitoring [24,29]. MPs were found to enter bivalves through the gill filaments, thus being the initial entry point for particulate pollutants and associated contaminants [4,23,30,31] and making it an organ of interest in biomarker studies. The ingested particles then move towards the mouth and enter the digestive gland [23,30,32] or even reach the gonad tissue [33]. Despite the relevant amount of data regarding MP particles entering bivalves, more information is needed on the retention time and depuration capacity. This is especially important, since the longer the retention time of these particles, the more likely it is that they will be transferred to the next trophic level upon consumption [34], as well as the more time chemicals and other compounds have to potentially desorb from the particles [23,32,35]. A multitude of studies have been conducted on the effects of pristine and contaminated plastic particles on the health of marine mussels [36,37], as well as the thus resulting health effect for humans [38] for different compounds.

Rios Mendoza et al. [39] assessed the concentration of pollutants sorbed to plastic debris in the North Pacific gyre and found that almost 80% of the debris they collected contained polycyclic aromatic hydrocarbons (PAHs), with concentrations ranging from a few to thousands of parts per billion (ppb). Benzo(a)pyrene (BaP), a PAH that originates from tar, burning wood, exhaust fumes and fumes from brunt organic material has been widely used as a model compound in aquatic toxicology [40] and, more recently, to assess the potential of MPs as carriers of hydrophobic pollutants. BaP reacts and binds to DNA, making it a highly efficient mutagen and carcinogen [41]. It is the only PAH classified as a recognized carcinogen by the International Agency for Research on Cancer (IARC) [42] and has routinely been employed as model contaminant for this group. It has also been classified as a candidate for being a substance "of very high concern" in the Registration, Evaluation, Authorization, and Restriction of Chemicals (REACH). Exposure has been found to induce CYP1A and morphological changes in gill tissue [42], as well as being found to accumulate through the food chain [43]. Direct exposure of mussels *M. galloprovincialis* to 20–25 µm low density PE (LDPE) MPs led to particle localization in the haemolymph and gills, as well as digestive tissue, whilst BaP contaminated MPs led to significant alterations of the immune system [44]. Dietary exposure to BaP contaminated polystyrene (PS) MPs also caused an exposure-time dependent increase in BaP concentration in mussels, particularly when sorbed to smaller MPs (0.5 µm versus 4.5 µm) [33]. Overall, BaP-contaminated MPs were more toxic than pristine MPs, according to haemocyte viability, catalase activity, and to the quantitative structure of digestive tubule epithelium.

Cadmium (Cd) is known to be persistent in the environment and to bioaccumulate up the food chain, similar to many lipophilic metals [45]. This makes it a suitable model contaminant to examine the vector potential of MPs for metals. Although metals readily adsorb to MPs, the co-exposure to copper or silver contaminated PE MPs was found to have no additional effect on marine microalgae [46] and zebrafish [47], whilst exposure to the metals alone had negative impacts on the individuals. However, synergistic sublethal toxicity of Cd and PS MPs at high levels (1, 5, 10 mg/L) was reported in zebrafish embryos [48]. In a study carried out in Vancouver (Canada), up to 7% of beached MPs were found to have adsorbed Cd [49].

Biomarkers, such as the activity of the antioxidant enzymes catalase (CAT) and superoxide dismutase (SOD), are often measured as indicators of the potential oxidative stress caused by pollutants, whilst the peroxisomal enzyme acyl-CoA oxidase (AOX) is assessed as a biomarker of exposure to organic contaminants [50]. The biomarker approach has also been applied to detecting deleterious effects caused by exposure to pristine and contaminated plastics on the health of aquatic organisms, such as mussels and copepods [33,44,51]. An established metal exposure biomarker is the quantification of lysosomal accumulation of metals in mussel tissues through autometallography [52]. Effects on cellular and tissue level can be determined through the assessment of the lysosomal membrane stability [32,44] and the histological structure of the digestive gland [33], respectively.

The present work aims (1) to examine the accumulation, depuration time and tissue distribution of 45 and 4.5 μm polystyrene MPs at different concentrations in the mussel *M. galloprovincialis* through histological analysis after short term exposure, and (2) to determine the fate and impact of adsorbed BaP and Cd on mussels through analytical chemistry and a battery of biomarkers.

2. Materials and Methods

2.1. Mussels

Mussels with a shell length between 3.5 and 4.5 cm were collected in the estuary of Plentzia, Basque Country (Bay of Biscay, 43°24′ N; 2°56′ W), considered as a reference site [50], during low tide in March (for experiment 1) and April 2016 (for experiment 2). Individuals were rinsed with water from the sampling location and transferred to the laboratory within the hour. Mussels for the experimental exposures to MPs were kept in an aerated tank with continuous filtered natural seawater supply for five days to acclimatize. Seawater from Plentzia was naturally filtered by sand in the uptake wells aided with a pump that sent the water to the Marine Station. Seawater gas balance was controlled in the station and then passed through a decantation/inertial tank and filtered (particle size ≤ 3 μm). Mussels were fed twice daily with Sera Marin "Coraliquid" (Sera, Heinsberg, Germany), and routine health checks were performed every morning, with no mortality observed during the acclimatization period.

2.2. Microplastics

PS spheres of 45 and 4.5 μm in diameter in a commercial solution (2.5% solids in deionised water with residual surfactant) were purchased from Polyscience Inc. (Badener, Germany). According to manufacturer's information, particles showed slight anionic charge and were monodispersed with a maximum coefficient of variation of 10% and 7% for the 45 and 4.5 μm particles, respectively.

2.3. Experiment 1: 1-Day Exposure and 3-Day Depuration of Pristine MPs

After the acclimatization period, mussels sampled in March were randomly distributed into 14 high density polyethylene containers (Deltalab, Barcelona, Spain) containing one litre of filtered natural seawater and exposed for 1 day to 1 (C1), 100 (C2) and 1000 (C3) particles/mL of PS microspheres of 45 and 4.5 μm in diameter, equivalent to 0.05, 5 and 50 mg/L for 45 μm MPs and 0.05, 5 and 50 μg/L for 4.5 μm MPs. In addition, a control group was maintained unexposed, and all treatments were run in duplicate. The selected MP particle concentrations of the present publication are within reported environmentally relevant concentrations (e.g., 1770 particles/L found in the southern North Sea [53]).

During the experiment, mussels were fed twice with "Coraliquid". After 1 day of exposure (E), 5 organisms per replicate of each treatment group were collected, cleaned, and processed for the histological localization of MPs. The remaining exposed mussels were then transferred back into uncontaminated water to allow depuration. After 1 (D1), 2 (D2) and 3 days (D3) of depuration 5 organisms per replicate of each exposure group were sampled. From each individual, a portion of the digestive gland and of the gill tissue was placed in histology cassettes and fixed in 10% buffered formalin for paraffin embedding. Tissue dehydration and infiltration steps were performed using n-butyl alcohol [33]. Paraffin embedded tissues were cut using a RM2125RT microtome (Leica Microsystems GmbH, Wetzlar, Germany) into 5 μm thickness sections. Three histological sections with a distance between them of at least 15 μm were collected onto microscopy slides from each individual and tissue. Sections were dewaxed utilizing n-butyl alcohol and stained with hematoxylin and eosin (H&E). Slides were mounted with Kaiser's glycerin gelatin (Merck KGaA, Darmstadt, Germany). Sections were examined for MP localization and photographed using an Olympus BX50 microscope (Olympus, Tokyo, Japan).

2.4. Experiment 2: 3-Day Exposure to Pristine and Contaminant Adsorbed MPs

Particles of 4.5 μm were contaminated with BaP or Cd after Batel et al. [43]. The procedure was repeated daily prior to dosing. The MPs were incubated in the dark in 10 mL of a 1 μM BaP or Cd water solution (252.3 and 112.4 μg/L, respectively). BaP was initially dissolved in DMSO and then diluted in MilliQ water to reach a final DMSO concentration in the incubation medium of 0.01%, a concentration that was found to induce no alterations in biomarker responses in mussels [54]. Cd 1 μM was prepared from CdCl$_2$. After 1 day in the orbital shaker (Rotabit, Selecta, Barcelona, Spain), the MP suspension was filtered through a 0.45 μm sterile filter (Merck Millipore, Darmstadt, Germany). MPs retained in the filter were washed twice with dH$_2$O and recovered with 10 mL dH$_2$O. The MPs were then resuspended in 40 mL dH$_2$O and added to the aquaria.

After the initial acclimatization period, mussels sampled in April were distributed in glass aquaria of 10 L. For 3 days, organisms were either exposed to 1000 particles/mL pristine 4.5 μm MPs, 1000 particles/mL plastic particles previously exposed to 1 μM Cd (MP-Cd) or BaP (MP-BaP), 1 μM dissolved Cd or BaP without plastic particles, or filtered natural seawater as control group. Particle concentration was based on the outcome of experiment 1. For the dissolved pollutant exposure, the glass tanks were pre-exposed for 24 h to allow for saturation. The water in the aquaria was fully renewed fully every 24 h prior to redosing. Mussels were fed and monitored as described above.

Every day, 30 min and 24 h after dosing, water samples were collected from the aquaria to monitor Cd and BaP concentrations. After 3 days, mussel samples were cleaned and collected for (1) chemical analyses of Cd and BaP concentrations for which whole mussels were frozen and stored at −20 °C until analysis; (2) activity of antioxidant and peroxisomal enzymes catalase (CAT), superoxide dismutase (SOD) and acyl-CoA oxidase (AOX), for which the digestive gland and gills of mussels were dissected, frozen in liquid nitrogen and stored at −80 °C until analysis; (3) the evaluation of the lysosomal membrane stability, for which half of the digestive gland was frozen in liquid nitrogen and stored at −80 °C until cryo-sectioning, and (4) MP localization, quantitative assessment of the structure of the digestive gland, and metal localization and distribution after autometallographical staining, for which the other half of the digestive gland was placed in histology cassettes and fixed in 10% buffered formalin for paraffin embedding, as described above.

2.4.1. Chemical Analysis of the Mussel and Water Samples

Chemical analyses of water and mussel samples were carried out in the General Research Services (SGIker) at the University of the Basque Country. Sixty mussels sampled in March were used for chemical analysis of PAHs and metals to ascertain the background concentration of contaminants.

For the analysis of metal body burdens, mussel samples were dried in pools (5 replicates) at 120 °C for 48 h, weighted and digested in HNO_3 Tracepur® 69% (Panreac, Barcelona, Spain). Once the concentrated acid was evaporated, pellets were resuspended in 0.01 M HNO_3 Tracepur® and quantified. The metal analysis was carried out by inductively coupled plasma atomic emission spectrometry (ICP-AES, Horiba Yobin Yvon Activa, Horiba Japan Domestic Group, Kyoto, Japan) for Fe and Zn and by ICP-mass spectrometry (ICP-MS; Agilent 7700, Agilent Technologies, Santa Clara, CA, USA) for Cr, Ni, Cu, Cd and Pb. The certified reference material NIST 2976 was used for quality control. A detection limit of 13 ng/g for Fe, Cr, Ni, Cu and Cd; 0.1 µg/g for Pb, and 2.0 µg/g for Zn was determined.

The analysis of the 16 EPA PAHs was performed by gas chromatography and mass spectrometry in 5 replicates. Approximately 1 g freeze-dried samples were extracted with acetone in a microwave oven (MARX, CEM, Matthews, NC, USA) and cleaned up by solid phase extraction (SPE) using Millipore cartridges (Merck Millipore). Six deuterated PAHs were added to the samples to monitor the recovery efficiency and two blank samples were run in parallel. The extracts were analysed in a 6890 Agilent gas chromatograph coupled to a 5975C Agilent mass spectrometer (Agilent Technologies, Avondale, PA, USA). A detection limit of 1 ng/g was determined for all PAHs, except for acenaphthalene (0.1 ng/g) and naphthalene (10 ng/g).

During the second experiment, water samples were collected for chemical analyses. For the measurement of Cd concentration, 50 mL of water from the MP, MP-Cd and Cd groups (3 replicates of each) were collected. All water samples were filtered through a PES membrane (0.2 µm), acidified with ultra-pure hydrochloric acid (1% v/v) and stored at 4 °C for no longer than two days before analysis. Analyses were carried out in an ICP-MS Agilent 7700 spectrophotometer as mentioned above. For the analysis of BaP, 500 mL of water were collected (3 replicates of each) in glass bottles from the MP, MP-BaP and BaP groups and stored at 4 °C in the dark until being analysed. These water samples were mixed with propanol and, after adding deuterated BaP as internal standard, samples were extracted by SPE and analysed using the same equipment described above. At the end of the exposure, 10 mussels were also collected and frozen whole, to assess Cd and BaP concentrations in the same groups mentioned for the water samples. Five pooled samples were used for the assessment of Cd concentration and two pooled samples for BaP concentration. Analyses of mussel samples were performed as described above.

2.4.2. Biochemical Analysis of the Antioxidant and Peroxisomal Enzyme Activity

Digestive glands or gills of six individuals per experimental group were homogenized in 3 mL of TVBE buffer (1 mM sodium bicarbonate, 1 mM EDTA, 0.1% ethanol and 0.01% Triton X-100, pH 7.6) per gram of tissue using a glass-Teflon® homogenizer (Potter S, B. Braun Melsungen AG, Melsungen, Germany) in an ice water-cooled bath. Homogenized samples were centrifuged at $500 \times g$ for 15 min in a Beckman Coulter Allegra 25R Centrifuge (Beckman Coulter Life Sciences, Indianapolis, IN, USA). The pellet was discarded, and 50 µL aliquots of the supernatant were frozen and stored for the measurement of AOX activity and protein concentration. The remaining supernatant was centrifuged at $12,000 \times g$ for 45 min. The pellet (mitochondrial fraction) was resuspended in 1 mL homogenization buffer per gram of initial tissue and frozen for later determination of CAT activity and protein concentration. The supernatant (S12 fraction) was divided in three aliquots and frozen for the measurement of CAT and SOD activity, and protein concentration.

Peroxisomal AOX activity was measured as described by Small et al. [55]. The assay is based on the H_2O_2-dependent oxidation of dichlorofluorescein catalysed by an exogenous peroxidase using 30 mM palmitoyl-CoA as substrate. CAT activity was calculated as the sum of the activities assessed in the mitochondrial and S12 fractions by measuring the disappearance of H_2O_2 at 240 nm (extinction coefficient 40 M^{-1} cm^{-1}) in a Shimadzu UV-1800 spectrophotometer (Shimadzu, Columbia, SC, USA) using 50 mM H_2O_2 as substrate in 80 mM potassium phosphate buffer (pH 7) [56]. SOD activity was determined in the S12 fraction at 550 nm by measuring the inhibition of cytochrome c reduction by superoxide generated by the xanthine oxidase/hypoxanthine system in an assay mixture that contained 50 mM potassium phosphate buffer plus 0.1 mM EDTA (pH 7.8), 50 mM hypoxanthine, 1.87 mU mL^{-1} xanthine oxidase and 10 mM cytochrome c [57]. One SOD unit was defined as the amount of enzyme that inhibits the rate of cytochrome c reduction by 50%. Protein concentration was measured in all fractions using the Quick Start™ Bradford Protein Assay Kit 3 (Bio Rad Life Sciences, Hercules, CA, USA).

2.4.3. Lysosomal Membrane Stability

The lysosomal membrane stability (LMS) test was performed according to a standardized protocol [58]. Serial tissue sections (10 μm thick) of 10 individuals per experimental group were cut in a Leica CM 3050S cryostat (Leica) and stored at −40 °C until required for staining. Briefly, the lysosomal membrane was destabilized at 37 °C for different periods of time (0, 3, 5, 10, 15, 20, 30 and 40 min) using 0.1 M sodium citrate buffer (pH 4.5) plus 2.5% NaCl. Then, sections were incubated for 20 min at 37 °C in 0.1 M citrate buffer (pH 4.5) containing 2.5% NaCl, 0.04% naphthol AS-BI N-acetyl-β-D-glucosaminide dissolved in 2-methoxiethanol (Merck KGaA, Darmstadt, Germany) and 7% Polypep® (Merck KGaA) as a section stabilizer. After incubation, sections were rinsed in a saline solution (3% NaCl) at 37 °C for 2 min and introduced into 0.1 M phosphate buffer (pH 7.4) containing 0.1% diazonium dye Fast Violet B salt, at room temperature for 10 min. Slides were rinsed in running tap water for 5 min, fixed for 10 min in 10% formaldehyde containing 2% calcium acetate at 4 °C and rinsed in distilled water. Finally, slides were mounted in Kaiser's glycerol gelatine. The determination of lysosomal membrane stability was based on the time of acid labialization required to produce maximum lysosomal staining. The labialization period (LP) was assessed under an Olympus BX-50 light microscope using an objective lens of 40× magnification. Each digestive gland was divided into four sections for the analysis to obtain the mean value of LP.

2.4.4. Tissue Metal Accumulation after Autometallography

A set of paraffin sections (10 individuals per experimental group) was stained with the BBI Solutions Silver enhancer kit (TAAB Laboratories Equipment, Aldermaston, UK) to assess the presence of metals in histological sections of the gills and digestive gland shown as black silver deposits (BSDs). Five fields of each section were photographed using the 40× magnification objective and the percentage of the digestive tissue area occupied by BSDs was measured by image analysis with the aid of ImageJ software (version1.50i, National Institutes of Health, USA).

2.4.5. Quantitative Histological Analysis

Changes in digestive gland structure of 10 individuals per experimental group were assessed by means of quantitative histology in paraffin sections stained with H&E. Volume density of basophilic cells (*VvBAS*), mean epithelium thickness (*MET*), mean luminal radius (*MLR*) and mean diverticular radius (*MDR*) of digestive gland tubules were determined applying a stereological procedure [59,60]. A M-168 Weibel multipurpose test system was superimposed to microscopic images (20× objective) with the aid of a drawing tube attached to an Olympus BX51 microscope and hits on basophilic cells (b) digestive cells (d), diverticular lumen (l) and interstitial connective tissue (c) were recorded. The following equations were applied:

$$VvBAS = \frac{b}{(d+b)} \tag{1}$$

$$MET = \frac{2d\sqrt{\pi}}{\left(\sqrt{((b+d)+\sqrt{1})}\right)} \tag{2}$$

$$MLR = \sqrt{\frac{1}{\pi}} \tag{3}$$

$$MDR = \sqrt{\frac{(b + d + 1)}{\pi}} \tag{4}$$

MLR/MET and MET/MDR ratios were calculated as well, along with connective to diverticula (CTD) ratio, which was calculated as CTD $= c/(b + d + 1)$ [59].

2.5. Statistics

The normal distribution and homogeneity of variances of each dataset was assessed with the Shapiro test and the Levene's test, respectively. For data following a normal distribution and with homogeneous variances, one-way ANOVA was applied followed by the Tukey's HSD post hoc test. The non-normal/non-homogenous data were assessed using one-way Kruskal–Wallis test followed by Dunn's test. Analyses were performed using SPSS Standard (version 21.0.0 for Mac OS X) and statistical significance was established at $p < 0.05$.

3. Results

3.1. Accumulation and Depuration of MPs in Mussel Tissue

None of the unexposed control organisms showed any plastic particle in the histological assessment. In experiment 1, after 1 day of exposure, the abundance of the 45 μm sized particles in the digestive gland increased in a concentration dependent manner, with particles present in the digestive gland of the 50% of the individuals exposed to the lowest concentration (1 particle/mL) and in the 100% of organisms exposed to 100 and 1000 particles/mL (Table 1). Moreover, increasing concentrations also led to longer retention times within the mussels. Amounts of 20% and 40% of the mussels exposed to the two highest concentrations retained MPs in the digestive gland by the third depuration day (Table 1). At the lowest concentration, no particles were found in any structure of the digestive gland after ≤ 3 days of depuration.

In mussels exposed to 100 particles/mL, most particles appeared in the lumen of the stomach lumen (Figure 1A), duct, and tubule, and the connective tissue. Exposure to 1000 particles/mL led to a higher abundance in all sample regions (Figure 1B,C, Table 1). By the third day of depuration, MPs remained mostly in the stomach lumen, with few observed in digestive duct lumen (Table 1). MPs of 45 μm were observed in the gills less frequently than in the digestive gland, with 40% being the highest observed prevalence prior to depuration of the highest exposure concentration group (1000 particles/mL), and particles were rarely observed within the gills after 1 day of depuration. However, they were found both within and outside the gill filaments (Figure 1D). Exposure to 1 particle/mL led to some particles observed after 1 day of exposure outside of the filaments, whilst no particles were observed within the structure.

Particles of 4.5 μm were found in both the digestive gland and the gills in almost all treated groups, even after the full depuration period (Table 1). The highest prevalence of 4.5 μm MPs in mussels was observed after 1-day exposure. After 2 days of depuration, 40 to 60% of organisms exposed to 100 and 1000 particles/mL still showed particles in the gill and digestive gland samples. Overall, a concentration-dependent increase in abundance and dispersal was found in the digestive gland The digestive gland samples exhibited a steeper decrease in affected individuals with depuration time than the gill samples, with a reduction of 50 to 60% when exposed to 1–1000 particles/mL. In the digestive gland, the 4.5 μm particles were found exclusively in the stomach and duct lumen at the lowest exposure concentration and the organisms had depurated completely by the final day. When exposed to 1000 particles/mL, MPs were found in stomach, duct, and tubule lumen by the end of the experiment.

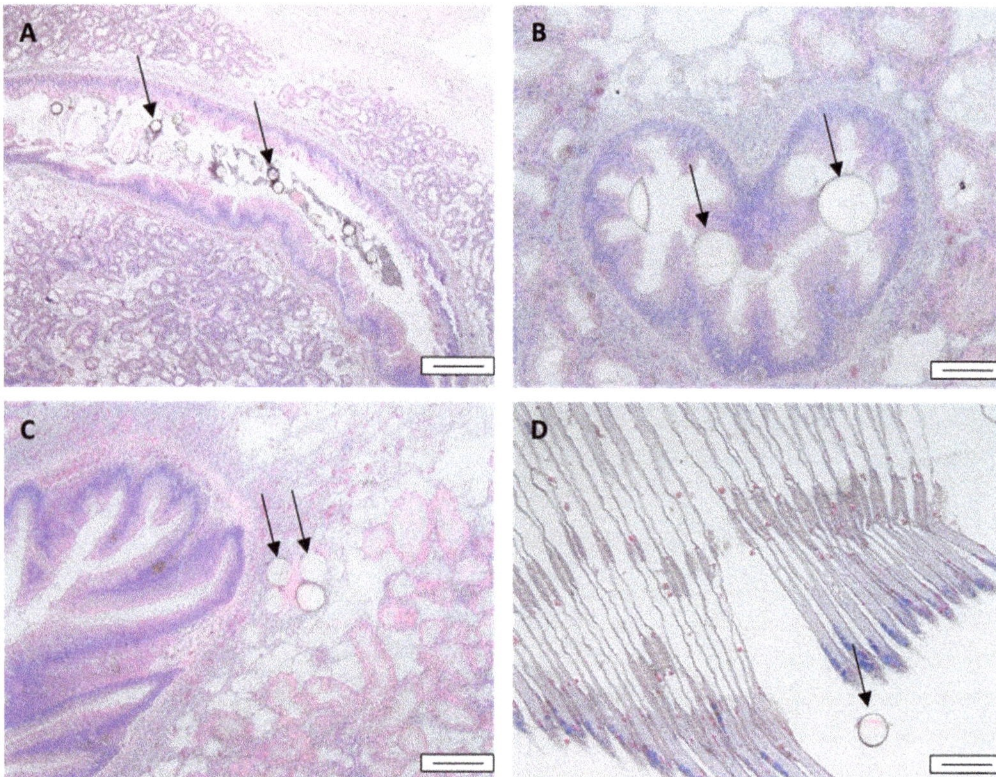

Figure 1. Micrographs of H&E-stained sections of digestive gland and gills of mussels after 1 day of exposure to pristine 45 μm MPs. (**A**) MPs in the lumen of the stomach after exposure to 100 particles/mL; (**B**) MPs in the lumen of a duct after exposure to 1000 particles/mL; (**C**) MPs in the connective tissue after exposure to 1000 particles/mL; (**D**) MPs outside a gill filament after exposure to 100 particles/mL. Black arrows point to MP particles. Scale bars: (**A**) 200 μm, (**B**) 50 μm, (**C,D**) 100 μm.

Gill depuration for organisms exposed to 4.5 μm particles was between 30 and 50% with increasing concentration. Here, the particles were mainly located between the filaments or in the frontal area of the gill filaments. Throughout all exposure concentrations, some particles were observed within the gill filaments, however, with slight decreases noted over the depuration time.

In experiment 2, mussels exposed for 3 days to pristine 4.5 μm MPs and to MPs contaminated with Cd and BaP showed the same tissue distribution of MPs described above (Figure 2) but, in this case, some particles were also observed within the stomach epithelium (Figure 2A).

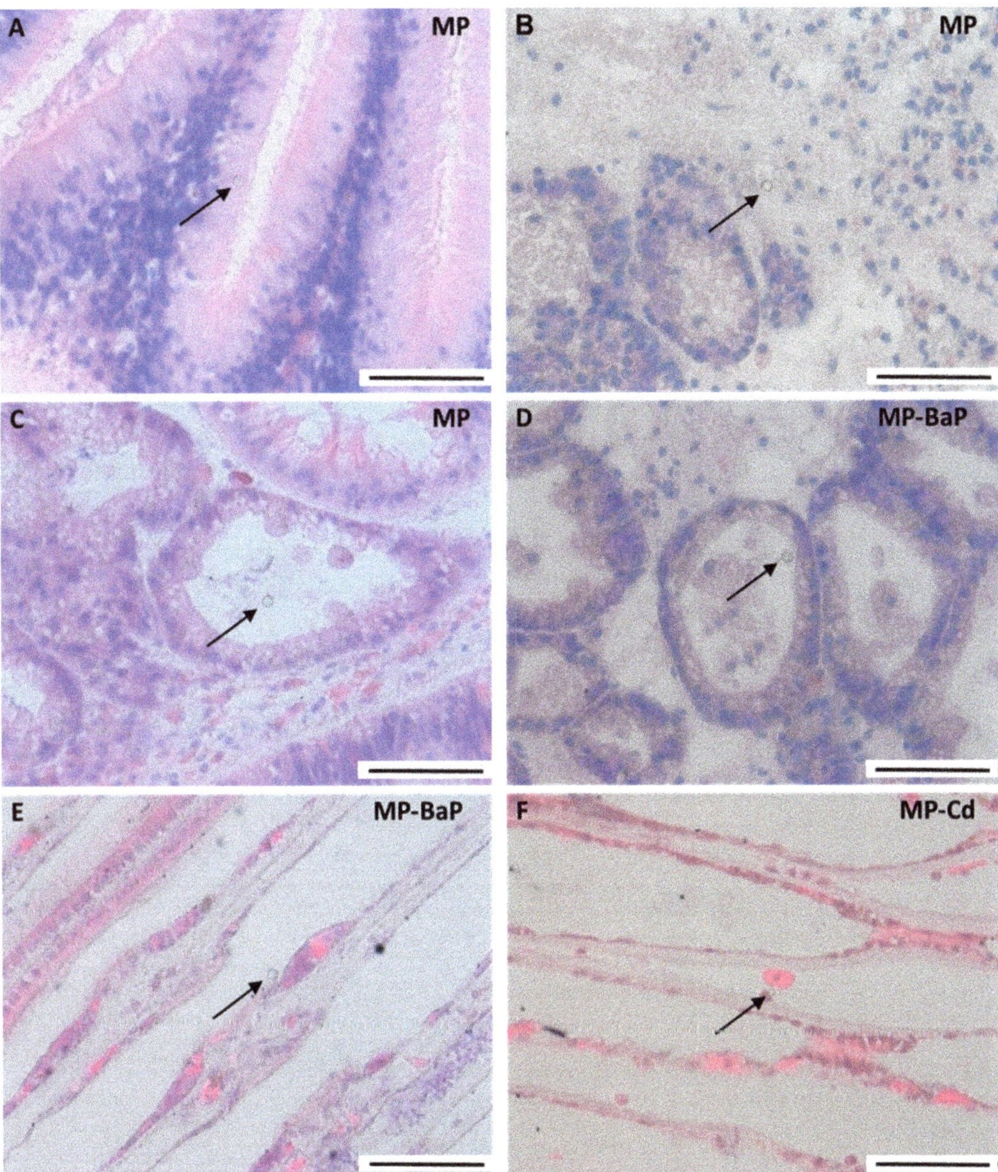

Figure 2. Micrographs of H&E-stained sections of digestive gland and gills of mussels after 3-day exposure to 1000 particles/mL pristine and contaminated 4.5 μm MPs. (**A**) MP in the stomach epithelium after exposure to pristine particles; (**B**) MP in the connective tissue surrounding the digestive tubules after exposure to pristine particles; (**C**) MP in the lumen of a digestive tubule after exposure to pristine particles; (**D**) MP in the lumen of a digestive tubule after exposure to MP-BaP; (**E**) MP over a gill filament after exposure to MP-BaP; (**F**) MP inside a gill filament after exposure to MP-Cd. Black arrows point to MP particles. Scale bars: 50 μm.

Table 1. Prevalence of mussels presenting MPs and abundance of particles found in the different structures of the digestive gland and in the gills. Data are expressed as mean ± standard deviation.

					Digestive Gland						Gills	
	Group		*n*	F	Stomach Lumen	Duct Lumen	Tubule Lumen	Connective Tissue	*n*	F	within Filaments	outside Filaments
45 µm particles	C1	E	10	50	0.3 ± 0.67	n.o.	n.o.	0.1 ± 0.3	10	20	n.o.	0.3 ± 0.67
		D1	10	10	0.3 ± 0.95	0.1 ± 0.32	n.o.	n.o.	9	n.o.	n.o.	n.o.
		D2	10	n.o.	n.o.	n.o.	n.o.	n.o.	10	n.o.	n.o.	n.o.
		D3	10	n.o.	n.o.	n.o.	n.o.	n.o.	10	n.o.	n.o.	n.o.
	C2	E	8	100	8.1 ± 7.41	3.5 ± 8.75	0.25 ± 0.71	1.37 ± 2.39	10	30	n.o.	0.8 ± 1.62
		D1	10	n.o.	n.o.	n.o.	n.o.	n.o.	10	20	n.o.	0.2 ± 0.42
		D2	10	20	1.1 ± 2.33	0.1 ± 0.32	n.o.	n.o.	10	n.o.	n.o.	n.o.
		D3	10	20	0.2 ± 0.63	n.o.	n.o.	n.o.	10	10	n.o.	0.1 ± 0.32
	C3	E	10	100	83.6 ± 112.32	13 ± 27.02	1.1 ± 1.91	11.7 ± 31.60	10	40	n.o.	1.1 ± 2.18
		D1	10	90	16.7 ± 21.71	2.4 ± 3.47	n.o.	0.3 ± 0.67	10	20	0.1 ± 0.32	0.4 ± 0.97
		D2	10	60	9.1 ± 21.75	3.2 ± 9.77	0.7 ± 2.21	2.3 ± 5.66	10	20	n.o.	0.3 ± 0.67
		D3	10	40	5.1 ± 11.73	0.1 ± 0.32	n.o.	n.o.	10	10	n.o.	0.1 ± 0.32
4.5 µm particles	C1	E	10	50	0.4 ± 0.63	0.3 ± 0.67	n.o.	n.o.	10	50	0.4 ± 0.7	1 ± 1.05
		D1	10	30	0.3 ± 0.48	0.1 ± 0.32	n.o.	n.o.	10	40	0.1 ± 0.32	0.9 ± 1.29
		D2	10	20	0.1 ± 0.32	0.1 ± 0.32	n.o.	n.o.	10	20	0.1 ± 0.32	0.5 ± 0.97
		D3	9	n.o.	n.o.	n.o.	n.o.	n.o.	10	20	0.1 ± 0.32	0.1 ± 0.32
	C2	E	9	70	1.7 ± 1.41	0.9 ± 0.93	0.7 ± 0.71	n.o.	10	80	0.5 ± 0.71	2.7 ± 1.95
		D1	10	50	0.6 ± 0.84	0.3 ± 0.48	0.2 ± 0.42	0.1 ± 0.32	10	70	0.4 ± 0.70	1.9 ± 1.59
		D2	10	50	0.2 ± 0.42	0.2 ± 0.42	0.2 ± 0.42	n.o.	10	40	0.3 ± 0.67	0.6 ± 0.97
		D3	10	20	0.1 ± 0.32	n.o.	n.o.	n.o.	10	40	0.2 ± 0.42	0.4 ±0.70
	C3	E	10	90	1 ± 0.82	0.9 ± 0.74	0.9 ± 0.74	0.3 ± 0.48	10	90	0.8 ± 0.79	2.8 ± 1.75
		D1	10	80	0.2 ± 0.42	0.7 ± 0.82	0.5 ± 0.53	0.1 ± 0.32	10	70	0.8 ± 0.79	1.4 ± 1.43
		D2	10	50	0.4 ± 0.52	0.3 ± 0.48	0.2 ± 0.42	n.o.	10	60	0.3 ± 0.48	1.1 ± 1.20
		D3	10	30	0.1 ± 0.32	0.1 ± 0.32	0.1 ± 0.32	n.o.	9	40	0.6 ± 1.01	0.6 ± 0.88

E: exposure group; D: depuration groups; F: prevalence (%); n: number of examined individuals; n.o: no particles observed.

3.2. Metal and PAH Accumulation in Mussels and Concentration in Exposure Media

The background contamination by metals and PAHs of the mussels sampled in Plentzia can be found in Table 2. PAHs such as acenaphthalene, indenopyrene and dibenzo(a,h)anthracene were found at levels below the detection limit (bdl) in some of the analysed samples. The Cd concentration detected in field mussels was similar to the Cd concentration measured in mussels exposed to pristine MPs and in mussels exposed to MP-Cd (Tables 2 and 3). However, organisms exposed to dissolved Cd for 3 days showed a Cd concentration 50 times greater than that of mussels exposed to pristine plastics. Field mussels showed slightly higher concentration of BaP than mussels exposed to pristine MPs, possibly due to the acclimatization period the exposed organisms were given after sampling, which the organisms from the field did not have. Whilst plastic-bound Cd did not increase the tissue Cd concentration, plastic-bound BaP notably increased the BaP concentration in the tissue samples, indicating that the plastic particles acted as vehicles for BaP to mussels. The highest concentration of BaP was observed in mussels exposed to BaP dissolved in water (Table 3).

Table 2. Results of the chemical analyses of mussels sampled in Plentzia (Bay of Biscay) in March 2016. Data are expressed as mean ± standard deviation.

Metal	µg/g dw
Fe	141 ± 22.06
Zn	446.6 ± 93.75
Cr	1.72 ± 0.16
Ni	1.52 ± 0.34
Cu	4.62 ± 0.128
Cd	0.65 ± 0.07
Pb	1.95 ± 0.21
PAH	**ng/g dw**
Naphthalene	1555.8 ± 253.56
Acenaphthylene	5.6 ± 2.07
Acenaphthalene	bdl
Fluorene	5 ± 2
Phenanthrene	25.6 ± 1.52
Anthracene	111.6 ± 168.45
Fluoranthene	54 ± 5.48
Pyrene	63.6 ± 9.96
Benzo(a)anthracene	31 ± 3.16
Chrysene	57.2 ± 4.44
Benzo(b)fluoranthene	33 ± 4.64
Benzo(k)fluoranthene	26.2 ± 4.33
Benzo(a)pyrene	81.8 ± 35.05
Indeno pyrene	<6 ± 2 *
Dibenzo(a,h)anthracene	<3.67 ± 1.5 *
Benzo(ghi)perylene	14.4 ± 4.16
TOTAL PAHs	**<2071.5**

bdl: below detection limit; dw: dry weight; * values for some of the replicates were bdl and those samples were not used to calculate mean values.

Table 3. Results of the chemical analyses of Cd and BaP concentrations in mussels exposed to pristine or contaminated MPs, or to dissolved contaminants. Data are expressed as mean ± standard deviation.

Group	Cd (µg/g dw)	BaP (ng/g dw)
MP	0.59 ± 0.09	18.5 ± 16.25
MP-Cd	0.60 ± 0.16	nm
Cd	33 ± 7.85	nm
MP-BaP	nm	3050 ± 777.82
BaP	nm	192,450 ± 11,101.58

dw: dry weight; nm: not measured for this sample set.

The analysis of the water samples collected from the exposure tanks (Table 4) showed that the Cd concentration was below the detection limit in the aquaria containing pristine MPs and MP-Cd. Water samples from the tanks of the Cd-exposed organisms indicated that the actual Cd concentration 30 min after adding the contaminant reflected the nominal exposure concentration (1 µM = 112 µg/L), and the value decreased after 1 day. BaP concentration measured in the aquaria containing pristine MPs and MP-BaP was low. In the tank containing dissolved BaP, although markedly below the nominal exposure concentration (1 µM = 252 µg/L), BaP concentration was high 30 min after adding the contaminant and dropped notably after 1 day (Table 4).

Table 4. Results of the chemical analyses of Cd and BaP concentrations in water samples of the second experiment. Samples were collected 30 min and 1 day after each dosing. Data are expressed as mean ± standard deviation.

Group	Cd (µg/L)		BaP (µg/L)	
	30 min	1 day	30 min	1 day
MP	bdl	bdl	0.03 ± 0.03	0.12 ± 0.10
MP-Cd	bdl	bdl	nm	nm
Cd	112.33 ± 2.08	74.37 ± 7.99	nm	nm
MP-BaP	nm	nm	0.30 ± 0.29	0.29 ± 0.13
BaP	nm	nm	40.60 ± 31.71	1.72 ± 1.45

bdl: below detection limit; nm: not measured for this sample set.

3.3. Activity of Antioxidant and Peroxisomal Enzymes

The highest CAT activity in the digestive gland was measured in organisms exposed to dissolved BaP and to MP-BaP (Figure 3A), with values of 6.431 ± 2.020 and 5.402 ± 1.497 mmol/min mg^{-1} protein, respectively. The lowest CAT activity was found in organisms exposed to dissolved Cd, as well as in the control organisms, with the mean values being 4.072 ± 0.873 and 4.523 ± 0.649 mmol/min mg^{-1} protein, respectively. Gill samples (Figure 3B) of organisms exposed to MP-BaP showed the highest CAT activity (4.053 ± 1.797 mmol/min mg^{-1} protein), whilst groups treated with dissolved BaP and MP-Cd expressed the lowest activity (1.868 ± 0.903 and 2.236 ± 0.720 mmol/min mg^{-1} protein, respectively). No statistically significant differences were found among the CAT activities measured in the digestive gland or gills of control and treated mussels.

The lowest SOD activity in the digestive gland (Figure 3C) was measured in organisms exposed to MP-Cd and in control organisms, with mean values of 0.659 ± 0.127 and 0.751 ± 0.409 units/min mg^{-1} protein, respectively. The highest activity was measured after BaP exposure, with 1.013 ± 0.437 units/min mg^{-1} protein, followed by those exposed to dissolved Cd, with 0.948 ± 0.171 units/min mg^{-1} protein. Assessing the gill samples (Figure 3D), the lowest SOD activity was also measured in organisms exposed to MP-Cd (0.973 ± 0.13 units mg^{-1} protein), followed by those exposed to pristine MPs, with 2.582 ± 0.967 units mg^{-1} protein. Here, the highest mean activity was also observed in mussels exposed to MP-BaP, with 4.276 ± 3.557 units mg^{-1} protein. The SOD activity measured in the gills was found to be significantly influenced by the treatment ($\chi^2(5) = 15.656$, $p = 0.008$). Post hoc testing determined that mussels exposed to MP-Cd presented significantly lower activity than mussels exposed to pristine MPs ($p = 0.019$) and mussels exposed to MP-BaP ($p = 0.003$).

The lowest mean AOX activity in the digestive gland samples (Figure 3E) was measured in mussels exposed to pristine MPs (0.149 ± 0.056 mU mg^{-1} protein), while the highest activity was observed in organisms exposed to the contaminated MPs. Significant differences were obtained (F (5,28) = 3.048 and $p = 0.025$), caused by the difference between the treatment groups exposed to pristine MPs and MP-BaP (Tukey HSD post hoc: $p = 0.043$).

3.4. Lysosomal Membrane Stability

Overall, all experimental groups, including control mussels, showed low labilization period (LP) values (Figure 4). The mean LP value measured in BaP-exposed mussels (9.16 ± 3.06 min) was the lowest of all groups. The longest LP was found in organisms exposed to pristine MPs (11.25 ± 2.43 min). No statistically significant differences were found among experimental groups.

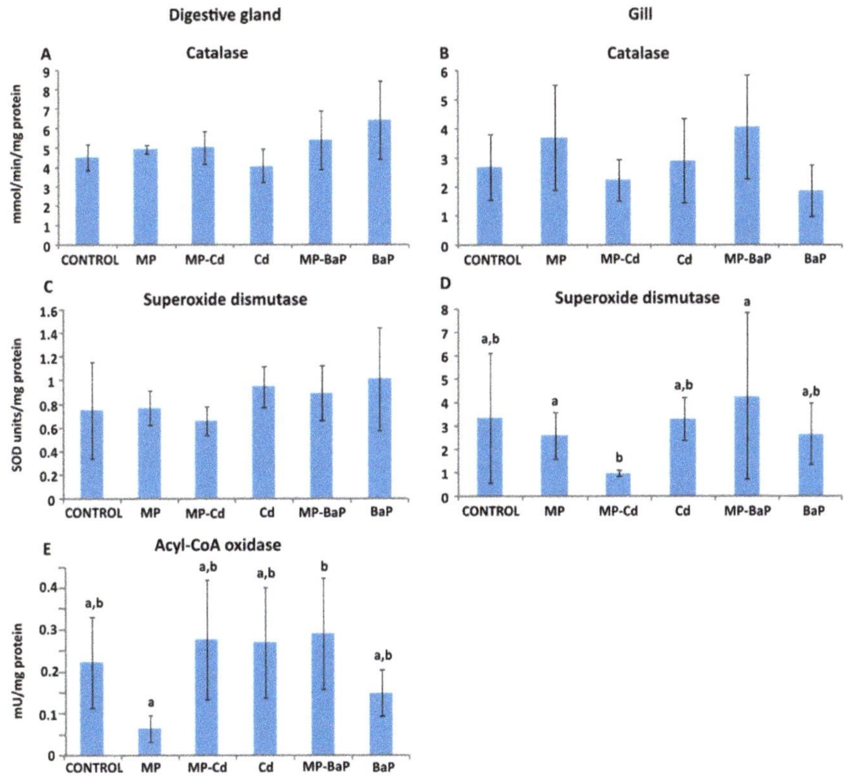

Figure 3. Activity of the antioxidant enzymes catalase in the digestive gland (**A**) and gills (**B**), superoxide dismutase in the digestive gland (**C**) and gills (**D**), and activity of acyl-CoA oxidase in the digestive gland (**E**) of mussels, presented as mean ± standard deviation ($n = 6$). Different letters indicate statistically significant differences ($p < 0.05$) according to the Tukey's post hoc test after one-way ANOVA.

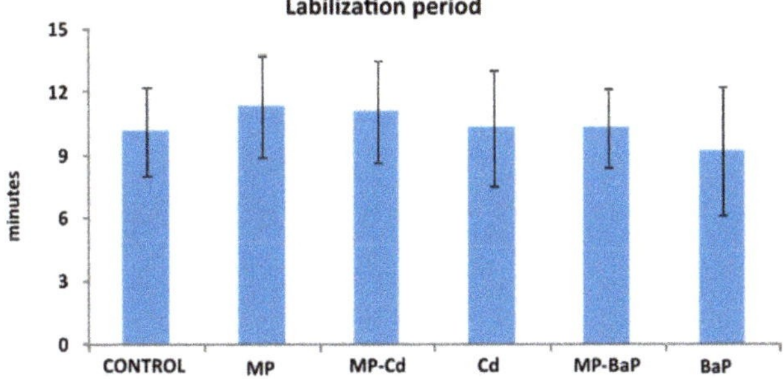

Figure 4. Labilization period (LP) of the digestive cell lysosomes. Mean ± standard deviation ($n = 10$). Statistically significant differences were not found according to the Kruskal–Wallis test ($p < 0.05$).

3.5. Tissue Metal Distribution and Accumulation after Autometallography

Metals revealed as BSDs after autometallographical staining (Figure 5) were mainly detected in the frontal zone of the gill filaments (Figure 5A) as well as in the digestive gland epithelium (Figure 5B–F). Occasionally, metals were also detected in the digestive gland haemocytes (Figure 5D). Since the main area for metal accumulation was the epithelium of the digestive tubules, the measurement of the percentage of tissue area that showed BSDs was focused in that area. As expected, results indicated that highest values were observed in mussels exposed to dissolved Cd (Figures 5D and 6), while the lowest were observed in mussels exposed to dissolved BaP (Figure 6). The Kruskal–Wallis test showed a $\chi^2(4) = 27.449$ with $p = 0.000$, and the post hoc Dunn's test showed significant differences between mussels exposed to dissolved Cd and those exposed to dissolved BaP and to MP-BaP ($p = 0.000$ and 0.016, respectively).

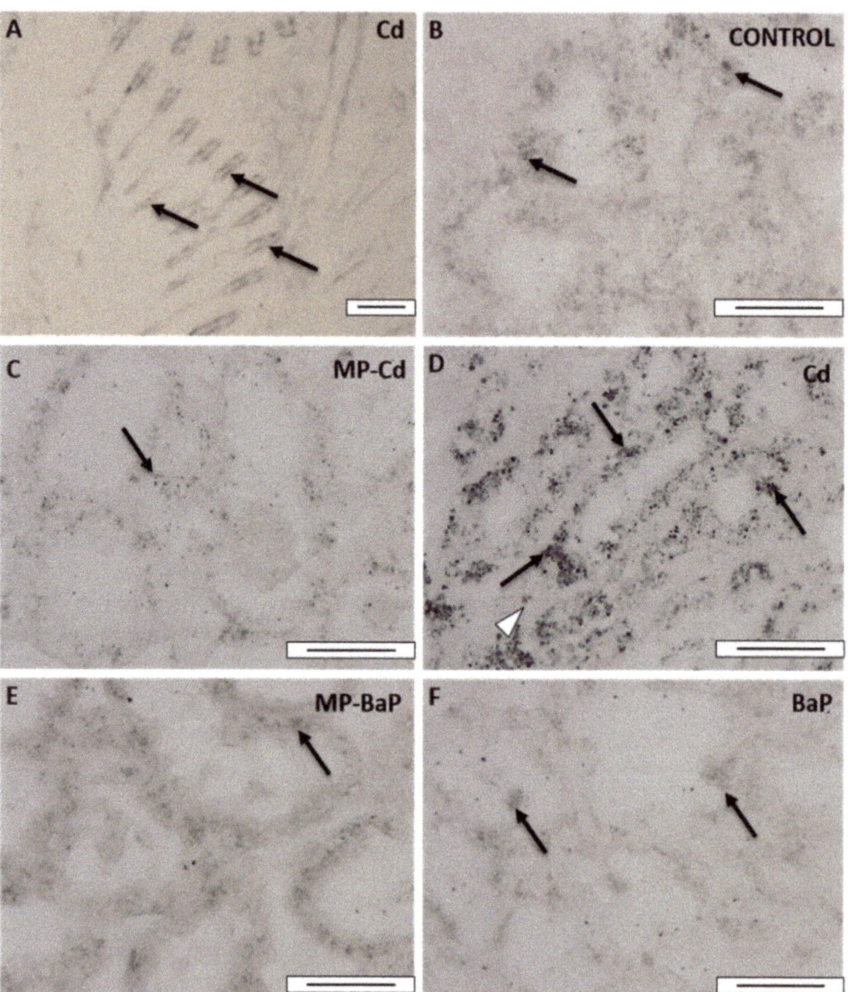

Figure 5. Micrographs of the gills (**A**) and digestive gland (**B–F**) of mussels after autometallographical staining. (**A**) Mussel exposed to 1 μM Cd for 3 days; (**B**) control mussel; (**C**) mussel exposed to 1000 particles/mL 4.5 μm MP-Cd for 3 days; (**D**) mussel exposed to 1 μM Cd for 3 days; (**E**) mussel exposed to 1000 particles/mL 4.5 μm MP-BaP for 3 days; (**F**) mussel exposed to 1 μM BaP for 3 days. Black silver deposits indicate the presence of metals in the gill cells (black arrows in **A**), in the digestive tissue (black arrows in **B–F**) and haemocytes (white triangle in **D**). Scale bars: 50 μm.

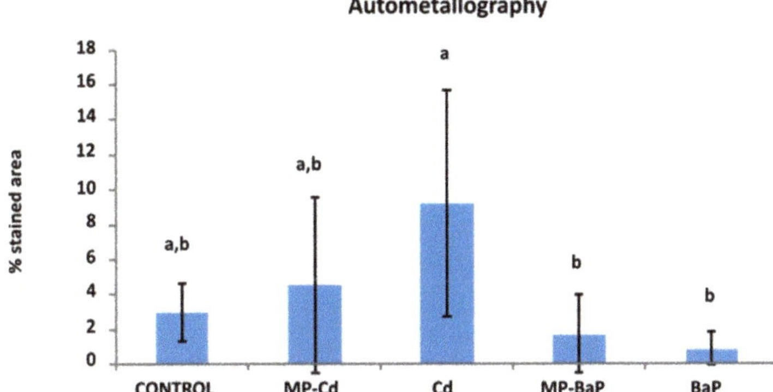

Figure 6. Results of the quantitative analysis of the autometallographical staining of the digestive gland. Mean $\pm$ standard deviation (n = 7–10). Different letters indicate statistically significant differences ($p < 0.05$), according to the Dunn's post hoc test after performing a one-way Kruskal–Wallis test.

3.6. Quantitative Histological Analysis

The volume density of basophilic cells (Figure 7A) had the highest values in mussels exposed to Cd (0.199 ± 0.024 µm^3/µm^3) and lowest in individuals exposed to MP-Cd (0.13 ± 0.026 µm^3/µm^3). One-way ANOVA indicated significant differences among experimental groups ($\chi^2(5) = 13.422$ with $p = 0.000$). The post hoc test showed that control mussels and mussels exposed to pristine MPs and to MP-Cd presented significantly lower values of VvBAS than the rest of exposed mussels. Mussels exposed to Cd showed significantly higher values than the other treatments. Similarly, tissue integrity (CTD) presented a similar trend to that shown by VvBAS (Figure 7B) with the lowest values observed in control mussels and those exposed to pristine MPs and MP-Cd, while mussels exposed to MP-BaP displayed intermediate values and mussels exposed to Cd and to BaP showed significantly higher values ($\chi^2(5) = 5.955$ with $p = 0.000$). The MLR/MET and MET/MDR ratios (Figure 7C,D) also presented a similar trend to that observed in VvBAS and CTD. Overall, control mussels and those exposed to pristine MPs and MP-Cd presented the lowest MLR/MET and highest MET/MDR values. The highest values for MLR/MET (2.04 ± 0.26 µm/µm) and lowest values in MET/MDR (0.33 ± 0.027 µm/µm) were measured in Cd exposed mussels. In both cases, significant differences were observed among experimental groups. In the case of MLR/MET ($\chi^2(5) = 10.311$ with $p = 0.000$), two statistical groups were distinguished with control mussels, mussels exposed to pristine MPs and to MP-Cd in one and mussels exposed to Cd and to BaP in other group, while mussels exposed to MP-BaP presented intermediate values. In the case of MET/MDR ($\chi^2(5) = 9.563$ with $p = 0.000$), control mussels, mussels exposed to pristine MP and to MP-Cd presented significantly higher values than mussels exposed to Cd and to BaP, and mussels treated with MP-BaP presented intermediate values (Figure 7D).

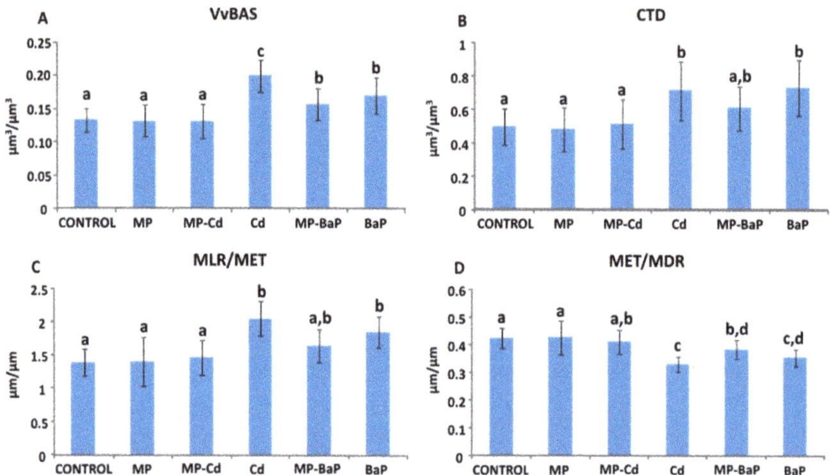

Figure 7. Results of the quantitative histological analysis of the structure of the digestive gland. (**A**) Volume density of basophilic cells; (**B**) connective-to-diverticula ratio; (**C**) mean luminal radius to mean epithelium thickness; (**D**) mean epithelium thickness to mean diverticular radius. Mean ± standard deviation (n = 10). Different letters indicate statistically significant differences ($p < 0.05$) according to the Tukey's post hoc test after one-way ANOVA.

4. Discussion

A recent review regarding the applicability of mussels as global indicators for the coastal contamination by MPs concluded that the mussel provides great potential for global biomonitoring of both spatial and temporal international trends [26]. At least two ways of MP uptake, dependent on the particle size, have been described in mussels: via the gills involving microvilli and endocytosis and via the cilia that transferred MPs to the stomach and digestive gland [30]. A recent review further assessed the viability of the mussel digestive gland in terms of assessing anthropogenic pollutants, concluding that it is a reliable organ for cellular, molecular and biochemical assessment [61].

4.1. Quantitative Histological Analysis

Throughout the experiments, the selected MP particle concentrations were within reported environmentally relevant concentrations (e.g., 1770 particles/L found in the southern North Sea [53]), with the second experiment being designed based on the accumulation observed in the first experiment. This allows for the hypothesis that the results obtained in the present study may resemble natural occurrences. Most previous works are based on higher test concentrations to allow establishing effect concentrations [62–64], which however makes drawing conclusions for the aquatic ecosystem health more difficult. Studies have determined the uptake of MPs by marine mussels, as well as the organism's ability to retain these particles for a length of time [31,33]. It was further shown that MPs are capable of being transferred through the food web [43,65,66], leading to increased concentrations in organisms higher in the trophic system, such as baleen whales [67] and the thorough assessment of the possible cellular and molecular effects of MP ingestion are thus of paramount importance.

The results of the first experiment determined that the digestive gland retained both 4.5 and 45 µm particles, even after a 3-day depuration period, whilst the 4.5 µm particles were observed in the gills more prominently than the 45 µm ones. In terms of digestive gland retention, these findings are in accordance with those of Gonçalves et al. [25], showing that 10 µm particles could be observed within the gut lumen but not the gills after 15 min of exposure. Long-term MP exposure (21 days) further showed that particles accumulated in the diverticula of the stomach and digestive gland, whilst no particles were found in other organs, even after a 7-day depuration period. The study further found that ingested MPs passed through the entire digestive tract and were expelled with the organism's faeces. The present findings of 45 μm particles being more evident in the stomach lumen and connective tissue were also supported by previous studies [32].

These findings overall indicate that MP ingestion occurs in a concentration and depuration-time dependent manner, concurrent with previous work [31]. Here it was also determined that the gills have a higher affinity to small particles, also concurrent with previous findings [23,33]. The longer prevalence of the 4.5 μm particles in the gills indicates a longer overall exposure time, as the trapped particles may be ingested even after the direct exposure has ended. It should thus be considered that, at environmentally relevant concentrations, mussels are able to rapidly (1 day) ingest the particles and translocate them into various tissues before depurating them over time. Moreover, 4.5 μm MPs crossing the digestive gland epithelia have been observed in this study after 3 days of exposure as well as in previous studies [33].

4.2. Accumulation and Effects of Contaminated MPs

Research carried out in the North Pacific Gyre found that different samples of seawater contained between 0.4 and 9 ng/L of PAHs, whilst sampled MP fragments contained PAH concentrations of 6 to 249 ng/g of plastic [68], indicating an accumulation of PAHs on plastic particles. The contaminant measurements within the mussel tissue samples carried out in the present work further suggest that BaP was more easily adsorbed to the PS plastic particles than cadmium. Similarly, contaminated MPs have been reported as negligible vector for mercury bioaccumulation in clams [69]. The mussel tissue contained Cd and BaP, according to the respective exposure groups and the concentration of contaminants within the aquaria water, decreased with exposure time, allowing the assumption that the removed quantity was, at least partially, taken up by the mussels. This statement was further supported by previous research, indicating that MPs exposed to pyrene over 6 days showed a concentration and time dependent adsorption, as well as then significantly increasing the pyrene body burden in exposed mussels by more than 13-fold [32]. Similarly, González-Soto et al. [33] reported an exposure time- and MP size-dependent accumulation of BaP in mussels after exposure for 7 and 26 days to BaP contaminated PS MPs of 0.5 and 4.5 μm. Pittura et al. [44] further showed that BaP readily adsorbed to LDPE MPs and increased the measured BaP concentration in the digestive gland after a 7-day exposure.

Having shown that mussels successfully accumulated MPs, this study further investigated the variable effect that both pristine and contaminated MPs can have on the organism. First, several enzyme activities were assessed as a response to environmental stressors. Lowered antioxidant activities have been considered an indicator for overwhelmed antioxidant defences or an inability to remove reactive oxygen species [70]. A recently published review stated that MP ingestion frequently challenged the oxidative state of invertebrates and seemingly required an upregulation of the antioxidant system in response [71], further supporting the application of these markers in studies on the impact of MPs. However, the exposure to pristine MP particles has frequently failed to induce significant responses in antioxidant levels [32,44,72,73].

In the present study, catalase activity in the digestive gland and gills was not significantly impacted. A previous study found, however, that a 7-day exposure of mussels to 500 μg/L BaP decreased catalase activity, followed by an increase in activity after 21 days [74]. Furthermore, a study assessing the effects of MP pollution and ocean acidification on mussels determined that of the assessed antioxidant biomarkers only catalase activity was significantly increased with increased MP concentration [73]. Revel et al. [75] also showed that exposure to 10 μg/L MPs for 26 days significantly increases both catalase and SOD activity in the digestive gland of mussels. In the present work, SOD activity only varied significantly for the gill samples of organisms exposed to MP-Cd (lowered activity) in comparison to those exposed to both pristine and MP-BaP. This result suggested that a longer exposure time could be needed to affect the overall SOD activity in mussels.

In the present study, peroxisomal AOX activity decreased significantly in the digestive gland of organisms exposed to pristine MPs compared to those exposed to MP-BaP. Even though lipid metabolism, where peroxisomes play a key role, has been highlighted as a relevant target for MPs pollution [76], recent work has shown that no significant difference in AOX activity in the digestive gland was measured when exposed to either pristine or MP-BaP as well as dissolved BaP over 7 days [44]. These findings were further supported with a recent study conducted with oysters (*Crassostrea gigas*) [72]. Work performed by Orbea and Cajaraville [50] found that mussels inhabiting or transplanted to sites polluted by PAHs showed increased AOX activity. However, lab studies where mussels were exposed to BaP yielded controversial results. Orbea et al. [74] found a significant decrease of AOX activity in the digestive gland of mussels waterborne exposed for 1 day, while no changes were seen after 7 and 21 days of treatment. Cancio et al. [77] did not observe alteration of AOX activity after 1 day of BaP injection, but increased activity was registered after 7 days.

Lysosomal membrane stability (LMS) can be used not only as a diagnostic biomarker for lysosomal stress, but also for the prognosis of the animals' health status [78]. The measured labilization periods (LPs) were overall lower than expected indicating, a possible disturbed health status of all mussels. The presence of relevant concentrations of some pollutants, such as Zn or PAHs (mainly naphthalene), at levels that have been described in moderately polluted areas [79] may be responsible of the relatively low stability of the lysosome membrane. Moreover, the data may also indicate that other stressing factors, such reproductive status, could be triggering the response, or that Plentzia is not as clean site as previously thought. Other pollutants in addition to PAHs and metals should be considered in future works. Similar levels of LP have been previously described in mussels from the same area exposed in the lab during similar exposure time (96 h [80]) The influence of feeding during the experimental period in lysosomal compartment that could lead into changes in both lysosomal size and lysosomal membrane stability [81] should also be considered. No differences in LP were found among exposure groups whereas longer exposure conditions to BaP contaminated MPs along with dissolved BaP led to significantly decreased membrane stability in a time dependent manner [44]. These differences could be due to differences in exposure periods or to different lysosome population measurement.

Regarding tissue metal distribution and accumulation after autometallography, it became evident that Cd accumulated mainly in the gills and the digestive gland. The percentage of BSDs in the cells followed the expected trend, where samples of organisms exposed to dissolved Cd showed a larger positively stained area followed by those exposed to MP-Cd, in agreement with results obtained by analytical chemistry. These results indicate that autometallography can be a suitable technique to detect the exposure to metal-contaminated MPs.

The present study demonstrated that the structure of the digestive gland was only impacted by the 3-day exposure to both dissolved BaP and MP-BaP, as well as Cd. The VvBAS significantly increased in mussels exposed to BaP and MP-BaP, whilst the two dissolved contaminants caused a significant increase of the CTD and MLR/MET ratios. Moreover, Cd exposure significantly reduced MET/MDR ratio, which also occurred, to a lesser degree, after exposure to dissolved BaP and to MP-BaP. The increase in VvBAS has previously been determined as an indicator of environmental stressors [82]. The effects exerted on cell type composition of the digestive gland, however, can be reversed, as shown by a study conducted after the Prestige oil spill in 2002. It was found that mussels affected by the contamination showed signs of recovery after two years [83]. Current results indicated that, although control mussels presented moderate levels of stress [60], in agreement with LP data, the presence of Cd, BaP and MP-BaP induced higher levels of stress. Moreover, the CTD values indicative of the structural integrity of the digestive gland tissue, with a high ratio indicating reduced digestive tissue [59], suggest that both Cd and BaP and the MP-BaP induced a reduction of digestion capability, which may disrupt the normal functioning of the organism in the long-run. This was further supported by the MLR/MET and MET/MDR values, where an increase of the first parameter and decrease on the second is indicative of epithelial thinning due to stressors [59,60], a response that has previously been observed in mussels exposed to the water accommodated fraction of oils [84], metals [85] or MPs [33]. The fact that control mussels present some altered biomarker responses (relatively low LP and high VvBAS) could be indicative that selected season (spring; developing gametes) and site (Plentzia) could be optimized for future research, as commented before. Longer exposure-times could confirm whether present alterations are transitory of are confirmed and increased after exposure. Conversely, although generally for dissolved contaminants biochemical changes precede histological ones [29], higher alterations were observed at tissue level compared with biochemical measurements. Similarly, the Manila clam (*Ruditapes philippinarum*) was found to ingest MPs and whilst none of the assessed biochemical biomarkers showed significant responses after 7-day exposure, the histological assessment of individuals exposed to MPs alone or co-exposed with Hg indicated deterioration of the gill epithelial tissue along with haemocyte infiltration [69].

5. Conclusions

Results of the present work demonstrated that marine mussels ingest MPs of various sizes (4.5 and 45 μm) and that these particles can further be accumulated in the digestive gland in a concentration and depuration time dependent manner. Furthermore, it was found that BaP body burdens increased notably in mussels exposed to MP-BaP, making it evident that plastic debris with adsorbed contaminants are posing a threat to the marine wildlife. This research was carried out over a 3-day period, which would only indicate initial impacts of exposure. Many factors may influence the impact that plastics and contaminants may have on organisms, of which not all are known or fully understood yet. Activity of the antioxidant and peroxisomal enzymes did not show a clear response to MP exposure, but autometallography appeared as a suitable technique to detect the exposure to metal-contaminated MPs. Quantitative histological analysis allowed for the determination of stress caused by the exposure to BaP and Cd and to MP-BaP by determination of changes in the basophilic cells volume density and the connective-to-diverticula ratio, as well as two ratios indicating digestive tubule structure. The results of this research make it evident that more work is needed in this field, as there are still knowledge gaps in the understanding of contaminants and their association with plastic debris, as well as their impact on marine organisms at long-term.

Author Contributions: Conceptualization, methodology, and writing—review and editing, B.Z., M.P.C. and A.O.; formal analysis, R.v.H., M.Z. and B.Z.; investigation: R.v.H., M.Z., B.Z. and A.O.; writing—original draft preparation, R.v.H., B.Z. and A.O.; visualization, R.v.H., M.Z., B.Z., M.P.C. and A.O.; supervision, B.Z. and A.O.; project administration and funding acquisition, M.P.C. and A.O. All authors have read and agreed to the published version of the manuscript.

Funding: This work was performed in the framework of the European project "PLASTOX: Direct and indirect ecotoxicological impacts of microplastics on marine organisms" (JPI Oceans) and Centre for Advanced Studies (CAS) project "H2020 CAS6 Nanoplastics" funded by the European Commission- Joint Research Centre (JRC/A/05). It was funded by Spanish MINECO (NACE project CTM2016-81130-R), UPV/EHU (VRI grant PLASTOX) and the Basque Government through a grant to consolidated research groups (IT810-13 and IT302-19).

Institutional Review Board Statement: Not applicable.

Informed Consent Statement: Not applicable.

Data Availability Statement: Data are contained within the article.

Acknowledgments: Thank you for the technical and human support provided by SGIker (UPV/EHU).

Conflicts of Interest: The authors declare no conflict of interest.

References

1. Plastics Europe Plastics—The Facts 2020. Available online: https://plasticseurope.org/wp-content/uploads/2021/09/Plastics_the_facts-WEB-2020_versionJun21_final.pdf (accessed on 12 March 2020).
2. Sharma, S.; Chatterjee, S. Microplastic pollution, a threat to marine ecosystem and human health: A short review. *Environ. Sci. Pollut. Res.* **2017**, *24*, 21530–21547. [CrossRef]
3. Andrady, A.L. Microplastics in the marine environment. *Mar. Pollut. Bull.* **2011**, *62*, 1596–1605. [CrossRef]
4. Browne, M.A. Sources and pathways of microplastics to Habitats. In *Marine Anthropogenic Litter*; Springer International Publishing: Cham, Germany, 2015; pp. 229–244. ISBN 978-3-319-16510-3.
5. Besseling, E.; Foekema, E.M.; Van Franeker, J.A.; Leopold, M.F.; Kühn, S.; Bravo Rebolledo, E.L.; Heße, E.; Mielke, L.; IJzer, J.; Kamminga, P.; et al. Microplastic in a macro filter feeder: Humpback whale Megaptera novaeangliae. *Mar. Pollut. Bull.* **2015**, *95*, 248–252. [CrossRef]
6. Zhang, F.; Wang, X.; Xu, J.; Zhu, L.; Peng, G.; Xu, P.; Li, D. Food-web transfer of microplastics between wild caught fish and crustaceans in East China Sea. *Mar. Pollut. Bull.* **2019**, *146*, 173–182. [CrossRef]
7. Barnes, D.K.A.; Milner, P. Drifting plastic and its consequences for sessile organism dispersal in the Atlantic Ocean. *Mar. Biol.* **2005**, *146*, 815–825. [CrossRef]
8. Galgani, F.; Hanke, G.; Werner, S.; De Vrees, L. Marine litter within the European Marine Strategy Framework Directive. *ICES J. Mar. Sci.* **2013**, *70*, 1055–1064. [CrossRef]
9. Vermeiren, P.; Muñoz, C.; Ikejima, K. Microplastic identification and quantification from organic rich sediments: A validated laboratory protocol. *Environ. Pollut.* **2020**, *262*, 114298. [CrossRef]
10. Ugwu, K.; Herrera, A.; Gómez, M. Microplastics in marine biota: A review. *Mar. Pollut. Bull.* **2021**, *169*, 112540. [CrossRef]
11. Ziccardi, L.M.; Edgington, A.; Hentz, K.; Kulacki, K.J.; Kane Driscoll, S. Microplastics as vectors for bioaccumulation of hydrophobic organic chemicals in the marine environment: A state-of-the-science review. *Environ. Toxicol. Chem.* **2016**, *35*, 1667–1676. [CrossRef]
12. Fotopoulou, K.N.; Karapanagioti, H.K. Surface properties of beached plastic pellets. *Mar. Environ. Res.* **2012**, *81*, 70–77. [CrossRef]
13. Mato, Y.; Isobe, T.; Takada, H.; Kanehiro, H.; Ohtake, C.; Kaminuma, T. Plastic resin pellets as a transport medium for toxic chemicals in the marine environment. *Environ. Sci. Technol.* **2001**, *35*, 318–324. [CrossRef] [PubMed]
14. Rochman, C.M.; Hoh, E.; Hentschel, B.T.; Kaye, S. Long-term field measurement of sorption of organic contaminants to five types of plastic pellets: Implications for plastic marine debris. *Environ. Sci. Technol.* **2013**, *47*, 1646–1654. [CrossRef] [PubMed]
15. Velzeboer, I.; Kwadijk, C.J.A.F.; Koelmans, A.A. Strong sorption of PCBs to nanoplastics, microplastics, carbon nanotubes, and fullerenes. *Environ. Sci. Technol.* **2014**, *48*, 4869–4876. [CrossRef]
16. Lee, H.; Shim, W.J.; Kwon, J.-H. Sorption capacity of plastic debris for hydrophobic organic chemicals. *Sci. Total Environ.* **2014**, *470–471*, 1545–1552. [CrossRef]
17. Cole, M.; Lindeque, P.; Halsband, C.; Galloway, T.S. Microplastics as contaminants in the marine environment: A review. *Mar. Pollut. Bull.* **2011**, *62*, 2588–2597. [CrossRef]
18. Atwell, L.; Hobson, K.A.; Welch, H.E. Biomagnification and bioaccumulation of mercury in an arctic marine food web: Insights from stable nitrogen isotope analysis. *Can. J. Fish. Aquat. Sci.* **1998**, *55*, 1114–1121. [CrossRef]
19. Robinson, K.J.; Hall, A.J.; Debier, C.; Eppe, G.; Thomé, J.-P.P.; Bennett, K.A. Persistent organic pollutant burden, experimental POP exposure, and tissue properties affect metabolic profiles of blubber from Gray Seal pups. *Environ. Sci. Technol.* **2018**, *52*, 13523–13534. [CrossRef]
20. Shaw, S.D.; Brenner, D.; Bourakovsky, A.; Mahaffey, C.A.; Perkins, C.R. Polychlorinated biphenyls and chlorinated pesticides in harbor seals (Phoca vitulina concolor) from the northwestern Atlantic coast. *Mar. Pollut. Bull.* **2005**, *50*, 1069–1084. [CrossRef]
21. Buckman, A.H.; Veldhoen, N.; Ellis, G.; Ford, J.K.B.; Helbing, C.C.; Ross, P.S. PCB-associated changes in mRNA expression in Killer Whales (Orcinus orca) from the NE Pacific Ocean. *Environ. Sci. Technol.* **2011**, *45*, 10194–10202. [CrossRef]
22. Vandermeersch, G.; Van Cauwenberghe, L.; Janssen, C.R.; Marques, A.; Granby, K.; Fait, G.; Kotterman, M.J.J.; Diogène, J.; Bekaert, K.; Robbens, J.; et al. A critical view on microplastic quantification in aquatic organisms. *Environ. Res.* **2015**, *143*, 46–55. [CrossRef] [PubMed]
23. Browne, M.A.; Dissanayake, A.; Galloway, T.S.; Lowe, D.M.; Thompson, R.C. Ingested microscopic plastic translocates to the circulatory system of the mussel, *Mytilus edulis* (L.). *Environ. Sci. Technol.* **2008**, *42*, 5026–5031. [CrossRef] [PubMed]
24. Beyer, J.; Green, N.W.; Brooks, S.; Allan, I.J.; Ruus, A.; Gomes, T.; Bråte, I.L.N.; Schøyen, M. Blue mussels (*Mytilus edulis* spp.) as sentinel organisms in coastal pollution monitoring: A review. *Mar. Environ. Res.* **2017**, *130*, 338–365. [CrossRef] [PubMed]
25. Gonçalves, C.; Martins, M.; Sobral, P.; Costa, P.M.; Costa, M.H. An assessment of the ability to ingest and excrete microplastics by filter-feeders: A case study with the Mediterranean mussel. *Environ. Pollut.* **2019**, *245*, 600–606. [CrossRef]
26. Li, J.; Lusher, A.L.; Rotchell, J.M.; Deudero, S.; Turra, A.; Bråte, I.L.N.; Sun, C.; Shahadat Hossain, M.; Li, Q.; Kolandhasamy, P.; et al. Using mussel as a global bioindicator of coastal microplastic pollution. *Environ. Pollut.* **2019**, *244*, 522–533. [CrossRef]
27. Branch, G.M.; Nina Steffani, C. Can we predict the effects of alien species? A case-history of the invasion of South Africa by *Mytilus galloprovincialis* (Lamarck). *J. Exp. Mar. Biol. Ecol.* **2004**, *300*, 189–215. [CrossRef]
28. Ceccherelli, V.U.; Rossi, R. Settlement, growth and production of the mussel Mytilus galloprovincialis. *Mar. Ecol. Prog. Ser.* **1984**, *16*, 173–184. [CrossRef]

29. Cajaraville, M.P.; Bebianno, M.J.; Blasco, J.; Porte, C.; Sarasquete, C.; Viarengo, A. The use of biomarkers to assess the impact of pollution in coastal environments of the Iberian Peninsula: A practical approach. *Sci. Total Environ.* **2000**, *247*, 295–311. [CrossRef]

30. Von Moos, N.; Burkhardt-Holm, P.; Köhler, A. Uptake and effects of microplastics on cells and tissue of the Blue Mussel *Mytilus edulis L.* after an experimental exposure. *Environ. Sci. Technol.* **2012**, *46*, 11327–11335. [CrossRef]

31. Fernández, B.; Albentosa, M. Insights into the uptake, elimination and accumulation of microplastics in mussel. *Environ. Pollut.* **2019**, *249*, 321–329. [CrossRef]

32. Avio, C.G.; Gorbi, S.; Milan, M.; Benedetti, M.; Fattorini, D.; D'Errico, G.; Pauletto, M.; Bargelloni, L.; Regoli, F. Pollutants bioavailability and toxicological risk from microplastics to marine mussels. *Environ. Pollut.* **2015**, *198*, 211–222. [CrossRef] [PubMed]

33. González-Soto, N.; Hatfield, J.; Katsumiti, A.; Duroudier, N.; Lacave, J.M.; Bilbao, E.; Orbea, A.; Navarro, E.; Cajaraville, M.P. Impacts of dietary exposure to different sized polystyrene microplastics alone and with sorbed benzo[a]pyrene on biomarkers and whole organism responses in mussels *Mytilus galloprovincialis*. *Sci. Total Environ.* **2019**, *684*, 548–566. [CrossRef] [PubMed]

34. Au, S.Y.; Lee, C.M.; Weinstein, J.E.; van den Hurk, P.; Klaine, S.J. Trophic transfer of microplastics in aquatic ecosystems: Identifying critical research needs. *Integr. Environ. Assess. Manag.* **2017**, *13*, 505–509. [CrossRef]

35. Koelmans, A.A.; Bakir, A.; Burton, G.A.; Janssen, C.R. Microplastic as a vector for chemicals in the aquatic environment: Critical review and model-supported reinterpretation of empirical studies. *Environ. Sci. Technol.* **2016**, *50*, 3315–3326. [CrossRef] [PubMed]

36. Tang, Y.; Zhou, W.; Sun, S.; Du, X.; Han, Y.; Shi, W.; Liu, G. Immunotoxicity and neurotoxicity of bisphenol A and microplastics alone or in combination to a bivalve species, *Tegillarca granosa*. *Environ. Pollut.* **2020**, *265*, 115115. [CrossRef]

37. Shi, W.; Han, Y.; Sun, S.; Tang, Y.; Zhou, W.; Du, X.; Liu, G. Immunotoxicities of microplastics and sertraline, alone and in combination, to a bivalve species: Size-dependent interaction and potential toxication mechanism. *J. Hazard. Mater.* **2020**, *396*, 122603. [CrossRef] [PubMed]

38. Zhou, W.; Han, Y.; Tang, Y.; Shi, W.; Du, X.; Sun, S.; Liu, G. Microplastics aggravate the Bioaccumulation of two waterborne veterinary antibiotics in an edible bivalve species: Potential mechanisms and implications for human health. *Environ. Sci. Technol.* **2020**, *54*, 8115–8122. [CrossRef]

39. Rios Mendoza, L.M.; Jones, P.R.; Moore, C.; Narayan, U.V. Quantification of persistent organic pollutants adsorbed on plastic debris from the North Pacific Gyre's "eastern garbage patch". *J. Environ. Monit.* **2010**, *12*, 2226–2236. [CrossRef]

40. Banni, M.; Negri, A.; Dagnino, A.; Jebali, J.; Ameur, S.; Boussetta, H. Acute effects of benzo[a]pyrene on digestive gland enzymatic biomarkers and DNA damage on mussel *Mytilus galloprovincialis*. *Ecotoxicol. Environ. Saf.* **2010**, *73*, 842–848. [CrossRef]

41. Baan, R.; Grosse, Y.; Straif, K.; Secretan, B.; El Ghissassi, F.; Bouvard, V.; Benbrahim-Tallaa, L.; Guha, N.; Freeman, C.; Galichet, L.; et al. A review of human carcinogens—Part F: Chemical agents and related occupations. *Lancet Oncol.* **2009**, *10*, 1143–1144. [CrossRef]

42. Speciale, A.; Zena, R.; Calabrò, C.; Bertuccio, C.; Aragona, M.; Saija, A.; Trombetta, D.; Cimino, F.; Lo Cascio, P. Experimental exposure of blue mussels (*Mytilus galloprovincialis*) to high levels of benzo[a]pyrene and possible implications for human health. *Ecotoxicol. Environ. Saf.* **2018**, *150*, 96–103. [CrossRef]

43. Batel, A.; Linti, F.; Scherer, M.; Erdinger, L.; Braunbeck, T. Transfer of benzo[a]pyrene from microplastics to *Artemia nauplii* and further to zebrafish via a trophic food web experiment: CYP1A induction and visual tracking of persistent organic pollutants. *Environ. Toxicol. Chem.* **2016**, *35*, 1656–1666. [CrossRef]

44. Pittura, L.; Avio, C.G.; Giuliani, M.E.; D'Errico, G.; Keiter, S.H.; Cormier, B.; Gorbi, S.; Regoli, F. Microplastics as vehicles of environmental PAHs to marine organisms: Combined chemical and physical hazards to the Mediterranean mussels, *Mytilus galloprovincialis*. *Front. Mar. Sci.* **2018**, *5*, 103. [CrossRef]

45. Li, J.; Chapman, E.C.; Shi, H.; Rotchell, J.M. PVC does not influence cadmium uptake or effects in the mussel (*Mytilus edulis*). *Bull. Environ. Contam. Toxicol.* **2020**, *104*, 315–320. [CrossRef] [PubMed]

46. Davarpanah, E.; Guilhermino, L. Single and combined effects of microplastics and copper on the population growth of the marine microalgae *Tetraselmis chuii*. *Estuar. Coast. Shelf Sci.* **2015**, *167*, 269–275. [CrossRef]

47. Khan, F.R.; Syberg, K.; Shashoua, Y.; Bury, N.R. Influence of polyethylene microplastic beads on the uptake and localization of silver in zebrafish (*Danio rerio*). *Environ. Pollut.* **2015**, *206*, 73–79. [CrossRef] [PubMed]

48. Zhang, R.; Wang, M.; Chen, X.; Yang, C.; Wu, L. Combined toxicity of microplastics and cadmium on the zebrafish embryos (*Danio rerio*). *Sci. Total Environ.* **2020**, *743*, 140638. [CrossRef]

49. Munier, B.; Bendell, L.I. Macro and micro plastics sorb and desorb metals and act as a point source of trace metals to coastal ecosystems. *PLoS ONE* **2018**, *13*, e0191759. [CrossRef]

50. Orbea, A.; Cajaraville, M.P. Peroxisome proliferation and antioxidant enzymes in transplanted mussels of four Basque estuaries with different levels of polycyclic aromatic hydrocarbon and polychlorinated biphenyl pollution. *Environ. Toxicol. Chem.* **2006**, *25*, 1616–1626. [CrossRef]

51. Jeong, C.-B.; Kang, H.-M.; Lee, M.-C.; Kim, D.-H.; Han, J.; Hwang, D.-S.; Souissi, S.; Lee, S.-J.; Shin, K.-H.; Park, H.G.; et al. Adverse effects of microplastics and oxidative stress-induced MAPK/Nrf2 pathway-mediated defense mechanisms in the marine copepod *Paracyclopina nana*. *Sci. Rep.* **2017**, *7*, 41323. [CrossRef]

52. Soto, M.; Marigómez, I. BSD Extent, an index for metal pollution screening based on the metal content within digestive cell lysosomes of mussels as determined by autometallography. *Ecotoxicol. Environ. Saf.* **1997**, *37*, 141–151. [CrossRef]

53. Dubaish, F.; Liebezeit, G. Suspended microplastics and black carbon particles in the jade system, southern north sea. *Water Air Soil Pollut.* **2013**, *224*, 1352. [CrossRef]

54. Marigómez, I.; Izagirre, U.; Lekube, X. Lysosomal enlargement in digestive cells of mussels exposed to cadmium, benzo[a]pyrene and their combination. *Comp. Biochem. Physiol. Part C Toxicol. Pharmacol.* **2005**, *141*, 188–193. [CrossRef]

55. Small, G.M.; Burdett, K.; Connock, M.J. A sensitive spectrophotometric assay for peroxisomal acyl-CoA oxidase. *Biochem. J.* **1985**, *227*, 205–210. [CrossRef]

56. Aebi, H. *Catalase Methods of Enzymatic Analysis*, 2nd ed.; Academic Press: New York, NY, USA, 1974.

57. Porte, C.; Sole, M.; Albaigés, J.; Livingstone, D.R. Responses of mixed-function oxygenase and antioxidase enzyme system of *Mytilus sp.* to organic pollution. *Comp. Biochem. Physiol. Part C Comp. Pharmacol.* **1991**, *100*, 183–186. [CrossRef]

58. UNEP. *UNEP/RAMOGE: Manual on the Biomarkers Recommended for the MED POL Biomonitoring Programme*; UNEP: Athens, Greece, 1999.

59. Garmendia, L.; Soto, M.; Vicario, U.; Kim, Y.; Cajaraville, M.P.; Marigómez, I. Application of a battery of biomarkers in mussel digestive gland to assess long-term effects of the *Prestige oil* spill in Galicia and Bay of Biscay: Tissue-level biomarkers and histopathology. *J. Environ. Monit.* **2011**, *13*, 915. [CrossRef] [PubMed]

60. Bignell, J.; Cajaraville, M.P.; Marigómez, I. Background Document: Histopathology of Mussels (*Mytilus* spp.) for Health Assessment in Biological Effects Monitoring. In *Integrated Marine Environmental Monitoring of Chemicals and Their Effects*; Davies, I.M., Vethaak, A.D., Eds.; ICES Cooperative Research Report No. 315; ICEM: Copenhagen, Denmark, 2012; p. 277. Available online: https://www.ices.dk/sites/pub/Publication%20Reports/Cooperative%20Research%20Report%20(CRR)/CRR315.pdf (accessed on 12 March 2020).

61. Faggio, C.; Tsarpali, V.; Dailianis, S. Mussel digestive gland as a model tissue for assessing xenobiotics: An overview. *Sci. Total Environ.* **2018**, *636*, 220–229. [CrossRef]

62. Rochman, C.M.; Browne, M.A.; Underwood, A.J.; van Franeker, J.A.; Thompson, R.C.; Amaral-Zettler, L.A. The ecological impacts of marine debris: Unraveling the demonstrated evidence from what is perceived. *Ecology* **2016**, *97*, 302–312. [CrossRef] [PubMed]

63. Koelmans, A.A.; Besseling, E.; Foekema, E.; Kooi, M.; Mintenig, S.; Ossendorp, B.C.; Redondo-Hasselerharm, P.E.; Verschoor, A.; van Wezel, A.P.; Scheffer, M. Risks of plastic debris: Unravelling fact, opinion, perception, and belief. *Environ. Sci. Technol.* **2017**, *51*, 11513–11519. [CrossRef] [PubMed]

64. Baroja, E.; Christoforou, E.; Lindström, J.; Spatharis, S. Effects of microplastics on bivalves: Are experimental settings reflecting conditions in the field? *Mar. Pollut. Bull.* **2021**, *171*, 112696. [CrossRef]

65. Setälä, O.; Fleming-Lehtinen, V.; Lehtiniemi, M. Ingestion and transfer of microplastics in the planktonic food web. *Environ. Pollut.* **2014**, *185*, 77–83. [CrossRef]

66. Cedervall, T.; Hansson, L.-A.; Lard, M.; Frohm, B.; Linse, S. Food chain transport of nanoparticles affects behaviour and fat metabolism in fish. *PLoS ONE* **2012**, *7*, e32254. [CrossRef]

67. Fossi, M.C.; Coppola, D.; Baini, M.; Giannetti, M.; Guerranti, C.; Marsili, L.; Panti, C.; de Sabata, E.; Clò, S. Large filter feeding marine organisms as indicators of microplastic in the pelagic environment: The case studies of the Mediterranean basking shark (*Cetorhinus maximus*) and fin whale (*Balaenoptera physalus*). *Mar. Environ. Res.* **2014**, *100*, 17–24. [CrossRef]

68. Rios Mendoza, L.M.; Jones, P.R. Characterisation of microplastics and toxic chemicals extracted from microplastic samples from the North Pacific Gyre. *Environ. Chem.* **2015**, *12*, 611–617. [CrossRef]

69. Sıkdokur, E.; Belivermiş, M.; Sezer, N.; Pekmez, M.; Bulan, Ö.K.; Kılıç, Ö. Effects of microplastics and mercury on manila clam *Ruditapes philippinarum*: Feeding rate, immunomodulation, histopathology and oxidative stress. *Environ. Pollut.* **2020**, *262*, 114247. [CrossRef] [PubMed]

70. Fernandes, A.M.; Fero, K.; Arrenberg, A.B.; Bergeron, S.A.; Driever, W.; Burgess, H.A. Deep brain photoreceptors control light-seeking behavior in zebrafish larvae. *Curr. Biol.* **2012**, *22*, 2042–2047. [CrossRef]

71. Trestrail, C.; Nugegoda, D.; Shimeta, J. Invertebrate responses to microplastic ingestion: Reviewing the role of the antioxidant system. *Sci. Total Environ.* **2020**, *734*, 138559. [CrossRef]

72. Revel, M.; Châtel, A.; Perrein-Ettajani, H.; Bruneau, M.; Akcha, F.; Sussarellu, R.; Rouxel, J.; Costil, K.; Decottignies, P.; Cognie, B.; et al. Realistic environmental exposure to microplastics does not induce biological effects in the Pacific oyster *Crassostrea gigas*. *Mar. Pollut. Bull.* **2020**, *150*, 110627. [CrossRef] [PubMed]

73. Wang, X.; Huang, W.; Wei, S.; Shang, Y.; Gu, H.; Wu, F.; Lan, Z.; Hu, M.; Shi, H.; Wang, Y. Microplastics impair digestive performance but show little effects on antioxidant activity in mussels under low pH conditions. *Environ. Pollut.* **2020**, *258*, 113691. [CrossRef]

74. Orbea, A.; Ortiz-Zarragoitia, M.; Cajaraville, M.P. Interactive effects of benzo(a)pyrene and cadmium and effects of di(2-ethylhexyl) phthalate on antioxidant and peroxisomal enzymes and peroxisomal volume density in the digestive gland of mussel Mytilus galloprovincialis Lmk. *Biomarkers* **2002**, *7*, 33–48. [CrossRef]

75. Revel, M.; Lagarde, F.; Perrein-Ettajani, H.; Bruneau, M.; Akcha, F.; Sussarellu, R.; Rouxel, J.; Costil, K.; Decottignies, P.; Cognie, B.; et al. Tissue-specific biomarker responses in the Blue Mussel *Mytilus* spp. exposed to a mixture of microplastics at environmentally relevant concentrations. *Front. Environ. Sci.* **2019**, *7*, 33. [CrossRef]

76. Kögel, T.; Bjorøy, Ø.; Toto, B.; Bienfait, A.M.; Sanden, M. Micro- and nanoplastic toxicity on aquatic life: Determining factors. *Sci. Total Environ.* **2020**, *709*, 136050. [CrossRef]

77. Cancio, I.; Orbea, A.; Völkl, A.; Fahimi, H.D.; Cajaraville, M.P. Induction of peroxisomal oxidases in mussels: Comparison of effects of lubricant oil and benzo(a)pyrene with two typical peroxisome proliferators on peroxisome structure and function in *Mytilus galloprovincialis*. *Toxicol. Appl. Pharmacol.* **1998**, *149*, 64–72. [CrossRef] [PubMed]

78. Moschino, V.; Da Ros, L. Biochemical and lysosomal biomarkers in the mussel *Mytilus galloprovincialis* from the Mar Piccolo of Taranto (Ionian Sea, Southern Italy). *Environ. Sci. Pollut. Res.* **2016**, *23*, 12770–12776. [CrossRef] [PubMed]

79. Kimbrough, K.L.; Johnson, W.E.; Lauenstein, G.G.; Christensen, J.D.; Apeti, D.A. *An Assessment of Two Decades of Contaminant Monitoring in the Nation's Coastal Zone*; NOAA Technical Memorandum NOS NCCOS 74; NCCOS: Silver Spring, MD, USA, 2008; 105p. Available online: https://coastalscience.noaa.gov/data_reports/an-assessment-of-two-decades-of-contaminant-monitoring-in-the-nations-coastal-zone/ (accessed on 12 March 2020).

80. Blanco-Rayón, E.; Ivanina, A.V.; Sokolova, I.M.; Marigómez, I.; Izagirre, U. Food-type may jeopardize biomarker interpretation in mussels used in aquatic toxicological experimentation. *PLoS ONE* **2019**, *14*, e0220661. [CrossRef] [PubMed]

81. Izagirre, U.; Ramos, R.; Marigómez, I. Natural variability in size and membrane stability of lysosomes in mussel digestive cells: Seasonal and tidal zonation. *Mar. Ecol. Prog. Ser.* **2008**, *372*, 105–117. [CrossRef]

82. De los Ríos, A.; Pérez, L.; Ortiz-Zarragoitia, M.; Serrano, T.; Barbero, M.C.; Echavarri-Erasun, B.; Juanes, J.A.; Orbea, A.; Cajaraville, M.P. Assessing the effects of treated and untreated urban discharges to estuarine and coastal waters applying selected biomarkers on caged mussels. *Mar. Pollut. Bull.* **2013**, *77*, 251–265. [CrossRef]

83. Cajaraville, M.P.; Garmendia, L.; Orbea, A.; Werding, R.; Gómez-Mendikute, A.; Izagirre, U.; Soto, M.; Marigómez, I. Signs of recovery of mussels health two years after the Prestige oil spill. *Mar. Environ. Res.* **2006**, *62*, S337–S341. [CrossRef]

84. Cajaraville, M.P.; Marigómez, J.A.; Díez, G.; Angulo, E. Comparative effects of the water accommodated fraction of three oils on mussels—2. Quantitative alterations in the structure of the digestive tubules. *Comp. Biochem. Physiol. Part C Comp. Pharmacol.* **1992**, *102*, 113–123. [CrossRef]

85. Jimeno-Romero, A.; Bilbao, E.; Valsami-Jones, E.; Cajaraville, M.P.; Soto, M.; Marigómez, I. Bioaccumulation, tissue and cell distribution, biomarkers and toxicopathic effects of CdS quantum dots in mussels, Mytilus galloprovincialis. *Ecotoxicol. Environ. Saf.* **2019**, *167*, 288–300. [CrossRef]

Article

Variable Fitness Response of Two Rotifer Species Exposed to Microplastics Particles: The Role of Food Quantity and Quality

Claudia Drago [1,*] and **Guntram Weithoff** [1,2]

1   Department for Ecology and Ecosystem Modelling, University of Potsdam, 14469 Potsdam, Germany; weithoff@uni-potsdam.de
2   Berlin-Brandenburg Institute of Advanced Biodiversity Research (BBIB), 14195 Berlin, Germany
*   Correspondence: drago@uni-potsdam.de

Abstract: Plastic pollution is an increasing environmental problem, but a comprehensive understanding of its effect in the environment is still missing. The wide variety of size, shape, and polymer composition of plastics impedes an adequate risk assessment. We investigated the effect of differently sized polystyrene beads (1-, 3-, 6-μm; PS) and polyamide fragments (5–25 μm, PA) and non-plastics items such as silica beads (3-μm, SiO_2) on the population growth, reproduction (egg ratio), and survival of two common aquatic micro invertebrates: the rotifer species *Brachionus calyciflorus* and *Brachionus fernandoi*. The MPs were combined with food quantity, limiting and saturating food concentration, and with food of different quality. We found variable fitness responses with a significant effect of 3-μm PS on the population growth rate in both rotifer species with respect to food quantity. An interaction between the food quality and the MPs treatments was found in the reproduction of *B. calyciflorus*. PA and SiO_2 beads had no effect on fitness response. This study provides further evidence of the indirect effect of MPs in planktonic rotifers and the importance of testing different environmental conditions that could influence the effect of MPs.

Keywords: microplastics; population growth rate; polystyrene; polyamide; silica beads; fitness response; rotifers; *Brachionus fernandoi*; *Brachionus calyciflorus*; egg ratio

Citation: Drago, C.; Weithoff, G. Variable Fitness Response of Two Rotifer Species Exposed to Microplastics Particles: The Role of Food Quantity and Quality. *Toxics* **2021**, *9*, 305. https://doi.org/10.3390/toxics9110305

Academic Editors: Costanza Scopetani, Tania Martellini and Diana Campos

Received: 21 October 2021
Accepted: 9 November 2021
Published: 13 November 2021

Publisher's Note: MDPI stays neutral with regard to jurisdictional claims in published maps and institutional affiliations.

1. Introduction

Plastic pollution is continuously increasing and without effective control, it will become more and more serious in the future. Currently, about 60 to 80% of the litter material in the environment is plastic [1].Plastic litter has a broad size, ranging from large plastic fishing nets and fragments of containers to very small particles in the millimeter or micrometer range and down to nanoparticles below 1 μm. Microplastics (MPs) have been found virtually everywhere in both terrestrial and aquatic ecosystems such as rivers, lakes, and oceans [2,3]. Plastics can enter aquatic systems from waste water treatment plants [4], through surface runoff [5–7], or from being deposited through the air [8]. Many studies have reported that microplastics harm a wide variety of aquatic organisms: the ingestion of large amounts of microplastics by aquatic organisms can reduce energy reserves and can affect growth and reproduction, which consequently increases the mortality of, for example, crustaceans [9], fish, mollusca, anellida[10]. The uptake of MPs from even smaller zooplankton can make them more available to larger taxa [11]. However, evidence supporting a quantitative risk assessment for microplastics is still missing due to a lack of method standardization and result ambiguity [12].A study from Sun et al. [13] showed that small-sized microplastics (0.07 μm; 0.05 μm) decreased rotifer survival and reproduction, whereas large-sized microplastics (0.7 and 7 μm) had no effect on rotifer life history traits. In contrast, Xue et al., [14] showed that larger microplastics (10–22 μm), in association with the algal food of similar size, suppressed the reproduction of rotifer, and this negative effect could be alleviated by increasing the food supply. Similar discrepancies have been found in studies conducted with the microcrustacean *Daphnia* [15,16]. Such discrepancies can result from different experimental set-ups, different shapes and types of plastics, and their relationship with food availability or food-size selection. Because of the shapes, size, and polymer composition of microplastics, there is still a necessity to better understand the effect of microplastics on aquatic organisms. Representative forms of microplastics in the environment are fragments and fibers, while microspheres are found less often [17,18]. Fragments and fibers accounted for 60% of all types of MPs, even in remote areas such as Lake Hovsgol in Mongolia [19]. One relevant component of shape is "spikiness". It was shown that spiky particles (e.g., filaments) and irregularly shaped particles (e.g., fragments) had showed a greater potential to harm animals than smooth particles such as spheres did, because spiky particles are more difficult to egest than smooth particles [20].

Rotifers are a widely distributed group of zooplankton that is present in all types of freshwater and brackish water bodies. They play an important role in aquatic food webs at the interface between primary producers and secondary consumers. As filter feeding organisms, rotifers have a very limited capability for food particle selection. Thus, rotifers cannot avoid the ingestion of plastic particles while they are feeding on natural food, such as algae. Therefore, rotifers are good model organisms for the study of and to understand how microplastic pollution influences aquatic ecosystems. Since field populations of rotifers are often resource limited [21–24], resource availability and natural fluctuation of algal growth should also be taken into account when estimating the risk of plastic pollution. We tested two closely related rotifers species, which were previously considered as one species, *Brachionus calyciflorus* and *Brachionus fernandoi*. These two species, even though they have a very similar morphology, exhibit different ecology and life history traits [25–27].

We used 1-, 3-, 6-, µm polystyrene beads (PS) because they are commonly used in toxicological studies of other organisms [28,29]. In addition, we used polyamide nylon fragments (PA) that were 5–25 µm in length because they are relevant in the field. As a non-plastic control, we used silica beads (SiO_2) (3 µm), and as the positive control, we used a treatment without artificial particles (only food algae). The different artificial beads were offered together with food algae at limiting and saturating food concentrations [30]. Moreover, the effects of the different microplastics were tested in *B. calyciflorus* in association with a different algal diet of *Monoraphidium minutum* and *Cryptomonas* sp., which is considered to be a high-quality food that can be ingested by rotifers [31,32].

The aim of this study was to quantify and compare the effect of differently sized and shaped particles made of different materials. We hypothesized that (1) the ingested beads could induce a decrease in the growth rate and reproduction of brachionids, acting as non-nutritional particles and that (2) the effect of microplastics is influenced by the food quantity and food quality.

2. Materials and Methods

2.1. Cultivation of Organisms

We used two species of pelagic rotifers, *Brachionus calyciflorus* s.s. (strain USA) and *B. fernandoi* (strain A10; [26]). Rotifers were raised in six well microtiter plates with sterile and vitamin-supplemented Woods Hole Culture Medium (WC) with saturating densities of *Monoraphidium minutum* (SAG 243-1, Culture Collection of Algae, University of Göttingen, Germany; ESD = 3.5 µm) as food. The phytoplankton species *Cryptomonas* sp. (Culture collection Göttingen, strain SAG-26-80; ESD = 5.9 µm [33]) was used as additional food in the food quality experiments [26]. Cultures were kept at 20 °C in a light–dark cycle of 14:10 h and at a light intensity of 35 µM photon s^{-1} m^{-2} photosynthetic active radiation (300–700 nm). Prior to the experiment, the rotifers were sieved through a mesh (30 µm) and were rinsed with sterile culture medium in order to separate them from their food. The carbon content was determined by an elemental analyzer (Euro EA 3000, HEKAtech Gmbh, Wegberg, Germany).

2.2. Microplastics

We used polystyrene microspheres (PS) of three different diameters as the microplastic beads in this study: 1.03, 3.06, and 5.73 µm (Polysciences, Inc. Fluoresbrite® YG Polystyrene Microspheres, Warrington, USA); for convenience, we refer to them as 1-, 3- and 6-PS. A stock solution was prepared with deionized MilliQ water under sterile conditions to minimize bacterial growth. To keep the beads as singular particles, each stock solution was sonicated for 30 min and was mixed using a vortexer. Stock suspensions of silica (SiO_2) beads in the size of 3.0 (cat. #SiO_2-F-3.0) were purchased from microParticles GmbH (Berlin, Germany). The stock solution was prepared using the same methods as the one prepared for the PS beads. Nylon fragments (5–25 µm) were prepared by size fractionating polyamide nylon-6 powder (nylon, PA) (Goodfellow; AM306010) with 25 µm cellulose filter (Whatman® qualitative filter paper, Grade 4) and 5 µm nylon mesh under a laminar flow hood. Prior to use, the microplastics were exposed to UV-light for 20 min to avoid bacterial contamination. For quantification, the fragments were suspended in ultrapure water and were analyzed with an electronic particle counter (CASY Schärfe System GmbH, Reutlingen, Germany) to assess the concentration and the total volume; moreover, a subsample was inspected using microscope, and the stock concentration and size range was assessed (Figure S2). The PS microbeads, the silica beads, and the PA fragments used in the present study have been previously used in numerous studies determining the effect and the ingestion of microplastics in pelagic and benthic organisms [28,29,34,35].

2.3. Experimental Procedure

For the population growth experiments, the two rotifer species fed on two carbon concentrations (0.5 mg C L^{-1}, "Limiting food concentration" LF and 2 mg C L^{-1} "Saturating food concentration" HF, Table S1) of *M. minutum* in combination with 1, 3, 6 PS beads, three types of SiO_2 beads, and 2 mg/L PA fragments with four replicates (Table S2). In this study, we used the same total amount of plastic (or silica) material, i.e., smaller particles were provided in higher numbers than larger particles.

In the second experiment, only the rotifer species *B.calyciflorus* was fed with a mix of algae species: *M. minutum* and *Cryptomonas* sp. Two carbon concentrations (0.5 "LF" and 2 mg C L^{-1} "HF") were used. Both algal species were supplied in 0.25 mg C L^{-1} for LF and 1 mg C L^{-1} for HF, respectively. *B. fernandoi* was not exposed to the mixture of algal food because it became mictic, i.e., it switched to sexual reproduction when fed with the mixed diet.

The experiment was conducted in 6-well microtiter plates at 20 °C in the dark to avoid additional algal growth. In the beginning, 10 individuals were randomly chosen from the stock culture and were pipetted into each well filled with 10 mL of the respective food suspension. At intervals of 24 h, the animals (live and dead) and their eggs were counted in each well. When the populations increased, 10 live individuals were randomly picked and transferred into new wells daily, receiving fresh food suspensions. In a case where less than 10 individuals survived, all of the remaining animals were transferred. The experiment lasted for 10 days (there was the exception of one replicate from *B. fernandoi* at low food concentration that got lost). Microtiter plates were placed on a rocker (Bio-Rad, Double Rocker, Labnet International Inc., Woodbridge, NJ, USA) to reduce the particle sedimentation. For each replicate the intrinsic growth rate (*r*), the egg ratio (*m*; eggs/female), and the survival (*l*) per day (*t*) were calculated on a daily basis using the following equations [36–38]:

$$r = ln(N_t) - ln(N_{t-1}) \tag{1}$$

$$m = \frac{H_t}{N_t} \tag{2}$$

$$l = 1 - \frac{D_t}{N_{t-1}} \tag{3}$$

where $N_{(t-1)}$ is the initial number of individuals and where N_t, H_t, and D_t are the final numbers of individuals, total eggs, and dead, respectively, on consecutive experimental days. The population growth rate (d^{-1}) of each replicate as well as reproduction (eggs $ind^{-1} d^{-1}$) and the probability of survival (d^{-1}) were calculated by averaging r, m, or l of consecutive experimental days.

2.4. Statistical Analysis

To compare the results from different experiments, we used the intensity of growth rate reduction (Δr) relative to the control group. The intensity of the growth rate reduction (Δr) was expressed as the difference in the per capita population growth rates with and without microbeads; a measure often used in food limitation experiments follows [21,23,24,39,40]:

$$\Delta r = r_c - r_s \tag{4}$$

where r_c is the per capita population growth rate in the experiment without microbeads (control), and r_s is the growth rate with the microbeads. A statistically significant growth reduction was present if the 95% confidence limits did not include zero and if the confidence intervals did not overlap. The effect of plastics and the interaction of food quantity, food quality, and plastics on the egg ratio and percentage of survival was analyzed using three-way ANOVAs and a pairwise comparison (Emmeans test) grouped by food against the reference group "control" with Bonferroni adjustment. The egg ratio was square-root transformed, and the percentage of survival was Yeo–Johnson transformed (lambda = 4.99) with the R-package "bestNormalize". Normality was assessed graphically using QQ-plot, and the homogeneity of variances was assessed using Levene's test. All of the statistical analyses were performed, and graphs were generated using R software (version 1.1.383).

3. Results

3.1. Effect of the MP Beads on Population Growth Rate

Brachionus calyciflorus and *B. fernandoi* experienced significant population growth rate reductions when exposed to the PS beads (Figure S5). Otherwise, there were no significant growth rate reductions in the treatments using PA fragments and silica beads (Figures S1, S3, and S4 showing ingested polymers).

In detail, we found a significant growth rate reduction when *B. calyciflorus* was only fed on the *M. minutum* algae with the 1-µm PS beads ($\Delta r = 0.14$; CI = 0.061) and 3- ($\Delta r = 0.16$; CI = 0.079) at the saturating food concentration. For the limiting food concentration, we found significant growth reductions with the 3- ($\Delta r = 0.31$; CI = 0.072) and 6-µm beads ($\Delta r = 0.19$; CI = 0.067). Contrarily, when a mixed algal diet was provided to *B. calyciflorus*, no growth rate reduction was found at the saturating food concentration, and the rotifers showed a significant decrease in growth rate for the limiting food concentration for particles that were 3 µm in size (PS: $\Delta r = 0.25$; CI = 0.171; silicate $\Delta r = 0.14$; CI = 0.103). In a similar manner, *B. fernandoi* exhibited no growth rate reductions at the saturating food concentrations, and only exhibited reductions when exposed to the limiting food concentration and to the 3-µm PS beads ($\Delta r = 0.20$; CI = 0.071), where we found a significant decrease in growth rate (Figure 1).

3.2. Effect of the MP Beads on Reproduction

Brachionus calyciflorus and *B. fernandoi* responded similarly regarding the production of eggs per individual ($F_{1137} = 1.3$, $p = 0.26$; Table 1 and Figure 2).

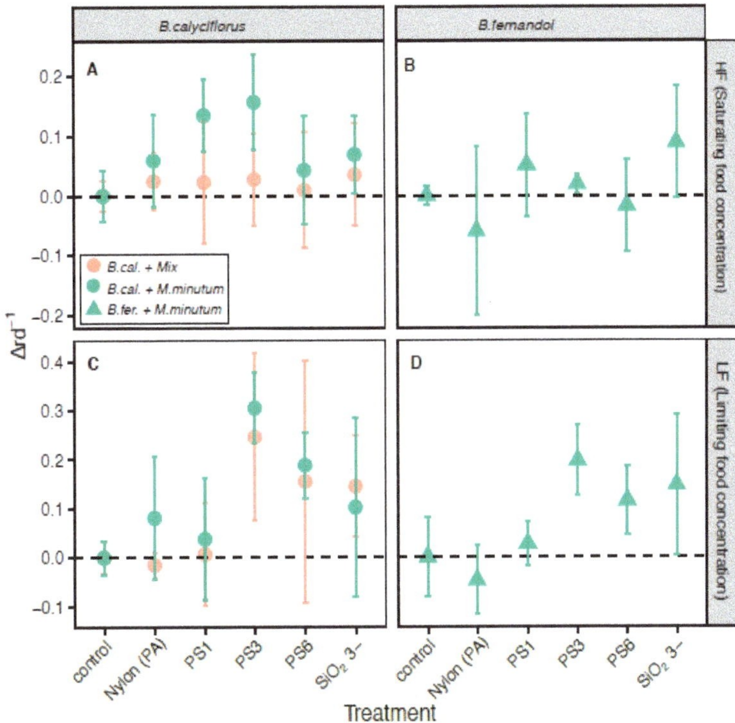

Figure 1. Intensity of food reduction ($\Delta r \pm 95\%$ confidence interval (CI)) of the rotifer *B. calyciflorus* and *B. fernandoi* at high and low food concentrations; (**A–C**) the red circles refer to the experiment with *B. calyciflorus* and the mixed algal diet (*M. minutum* and *Cryptomonas* sp.), and the green circles refers to the experiment with *B. calyciflorus* and one algal species (*M. minutum*); (**B–D**) the green triangle refers to *B. fernandoi*.

Table 1. Results of three-way ANOVAs using square-root transformed data on the egg ratio and Yeo–Johnson transformed data on survival (lambda = 4.99) for the two rotifer species (*Brachionus calyciflorus* and *Brachionus fernandoi*) and the two algal diets (*Monoraphidium minutum*; *Monoraphidium minutum* + *Cryptomonas* sp.). The two species were provided with two quantities (0.5 and 2.0 mg C L^{-1}) of *Monoraphidium minutum*. *B. calyciflorus* was provided with the same food quantities of a mixture of *Monoraphidium minutum* and *Cryptomonas* sp. as food.

	Egg-Ratio			Probability of Survival		
Independent variables	Df	F-Value	*p*-Value	Df	F-Value	*p*-Value
Alg	1137	125.5	<0.0001	1137	0.4	0.534
food	1137	997.0	<0.0001	1137	28.6	<0.0001
food × Alg	1137	33.5	<0.0001	1137	2.8	0.099
food × Treatment	5137	1.0	0.422	5137	3.9	<0.01
Specie	1137	1.3	0.258	1137	20.2	<0.0001
Specie × food	1137	16.6	<0.0001	1137	2.4	0.126
Specie × food × Treatment	5137	1.5	0.190	5137	0.6	0.699
Specie × Treatment	5137	0.3	0.907	5137	3.3	<0.01
Treatment	5137	20.3	<0.0001	5137	5.6	<0.001
Treatment × Alg	5137	4.2	<0.01	5137	3.2	<0.01

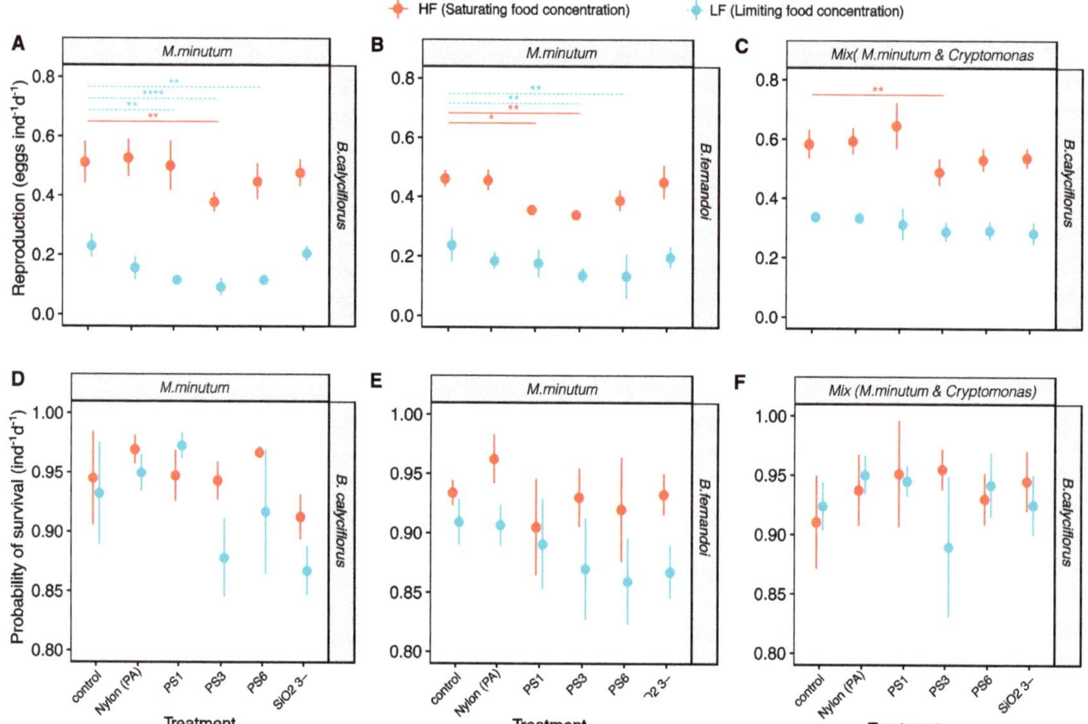

Figure 2. A−B−C egg ratio of *B. calyciflorus* and *B. fernandoi* exposed to the microbeads (mean ± SD); (**A**) egg ratio from *B. calyciflorus* fed on one algal species (*M. minutum*), with a statistically significant difference between the control group and the microbead treatment group; (**B**) egg ratio from *B. fernandoi* fed on one algal species (*M. minutum*), with a statistically significant difference between the control group and the microbead treatment group; (**C**) egg ratio from *B. calyciflorus* fed on mix algal diet (*M. minutum* and *Cryptomonas* sp.), with a statistically significant difference between the control group and the microbead treatment group; D−E−F percentage of survival of *B. calyciflorus* and *B. fernandoi* exposed to the microbeads (mean ± SD); (**D**) survival of *B. calyciflorus* fed on one algal species (*M. minutum*), with a statistically significant difference between the control group and the microbead treatment group; (**E**) survival from *B. fernandoi* feeding on one algal specie (*M. minutum*); (**F**) survival from *B. calyciflorus* fed on mix algal diet (*M. minutum* and *Cryptomonas* sp.), with a statistically significant difference between the control group and the microbead treatment group.

The egg productions were affected by the food concentration (F_{1137} = 997.0, $p < 0.0001$; Table 1), the different algal diets (F_{1137} = 125.5, $p < 0.0001$; Table 1), and the plastic treatments (F_{5137} = 20.3, $p < 0.0001$; Table 1). Moreover, the effect of the food concentrations on the egg ratio differed between the two rotifer species (F_{1137} = 16.6, $p < 0.0001$; Table 1) and between the two algal diets within the same species (F_{1137} = 33.5, $p < 0.0001$; Table 1). Regarding the effect of the plastic treatments, in general, we did not find significant changes after limiting the saturating food concentration (F_{5137} = 1.0, $p = 0.42$; Table 1); on the contrary, the effect varied between the two algal diets (F_{5137} = 4.23, $p < 0.01$; Table 1). The rotifers responded differently depending on the plastic treatments, but no significantly different effect was found between the control group and the rotifers exposed to PA fragments and silica beads. A reduction in egg production was mostly found with the 3-μm PS beads, with the exception of the experiment with *B. calyciflorus* when limiting then food concentration in the mixed algal diet. *B. calyciflorus* was more vulnerable to a decrease in the egg ratio when fed on a monoculture diet and with PS beads when the food concentration was limited (LF: PS1, $p < 0.01$; PS3, $p < 0.0001$; PS6, $p < 0.01$; Table S3), and a minor vulnerability was also shown with the saturating food concentration (HF: PS3, $p < 0.01$; Table S3). When the mixed algal diet was provided, *B. calyciflorus* exhibited a less pronounced decrease in the egg ratio, with the only significant reduction only being seen with the 3-µm PS beads (HF: PS3, $p < 0.01$; Table S3). Similarly, *B. fernandoi* showed an eggs ratio reduction with PS beads at the saturating (HF: PS1, $p < 0.05$; PS3, $p < 0.01$; Table S3) and limiting food concentrations (LF: PS3, $p < 0.01$; PS6, $p < 0.01$; Table S3).

3.3. Effect of the MP Beads on Survival

The probability of survival was affected by the food quantity ($F_{1137} = 28.6$, $p < 0.0001$; Table 1) and plastic treatments ($F_{5137} = 5.6$, $p < 0.001$; Table 1) and differed between the two species ($F_{1137} = 20.2$, $p < 0.0001$; Table 1). The effect of the beads changed depending on the food concentration ($F_{5137} = 3.9$, $p < 0.01$; Table 1), on the algal diet ($F_{5137} = 3.2$, $p < 0.01$; Table 1), and on the species ($F_{5137} = 3.3$, $p < 0.01$; Table 1). Nevertheless, for the two species and the different algal diets, no significant differences were found between the control group and the beads.

4. Discussion

The aim of this research was to investigate and compare the effects of different sizes and types of microbeads and the role of food quantity and quality in a freshwater rotifer population. In this study, we highlighted the decrease of the population growth rate and reproduction (egg ratio) of two freshwater rotifer species, *Brachionus calyciflorus* and *Brachionus fernandoi*, in response to exposure to PS beads at the limiting food concentration. Moreover, *B. calyciflorus* exhibited reduced fitness when exposed to MPs with a single algal food species at the saturating food concentration. In contrast, the (PA) nylon fragments and the silicate beads had no effect on the population growth rate, egg ratio, and survival.

4.1. The Role of Food Quantity and Food Quality on Microplastics Effect

Our experiments showed that the population growth rates of the two rotifers species and with both algal diets were more affected at the limiting food concentration with the presence of the 3-µm PS beads. Only *B. calyciflorus* showed a reduction in the population growth rate at a high food concentration with the monoculture algal diet. In fact, the population growth rate of *B. calyciflorus* did not decline when a mixed algal diet was provided at the saturating food concentration; similarly, *B. fernandoi* only exhibited a reduced population growth rate at the limiting food concentration. In addition, the growth rate reduction was less pronounced in *B. calyciflorus* with the mixed algal diet than it was with the monoculture algal diet (Figure S1). The egg production was also mostly affected mostly by the PS beads; the effect of the microplastics, if present, was not influenced by the different food concentration but instead depended more on the algal diet provided to the rotifers. For instance, *B. calyciflorus* and *B. fernandoi* showed a reduced egg ratio at the limiting and saturating food concentrations, with different intensities, but when a mix algal diet was provided, *B. calyciflorus* only exhibited a reduced egg ratio with the 3-µm PS beads at the saturating food concentration and had no effect at the limiting food concentration. For *B. calyciflorus* at the limiting food concentration, we found an inverse relation between the population growth rate and the number of eggs produced, where the number of individuals decreased but not the number of eggs; in contrast, at the saturating food concentration, the number of eggs per individual declined, but not the number of individuals. Although the population growth rate and egg ratio are expected to be linked to each other, they do not match perfectly. On the one hand, at low food levels, animals can increase their life span at the expense of reproduction. In our experimental set up, this led to a lower growth rate reduction but to a strong decline in the egg ratio. On the other hand, at the maximal growth rates, a high number of not yet reproducing juveniles are part of the population, leading to sub-maximal egg ratios. Our findings are in accordance with Korez et al., [41] where a marine isopod was not affected by microplastics when they received a sufficient amount of food with a high nutritional quality. A surplus in the microplastics at a low food concentration caused a significant reduction in food uptake and digestive enzyme activities. One likely explanation for the decrease in egg ratio in rotifers that is connected to microbeads exposure, is the food dilution effects, which have been found in nematodes and crustacea [12,29]. Microbeads, which are mostly of the same size of the supplied food, interfere with normal food ingestion, and in addition, the particles act as a non-food item, providing no energy resource. Thus, the microbeads occupy space in the digestive tract, decreasing the available space for algal food. A similar study on cladocerans determined that chronic exposure to PS beads led to a reduction in the number of offspring, which could be explained by the downregulation of several digestive enzymes that can interfere with the animal´s nutrient supply and that can affect their fitness [42].

Food quality may be more important in the explanation of the variation in zooplankton fitness than food quantity [43]. The food quality acts on consumer physiology through morphological traits such as the shape as well as the nutritional value. This is evident for organisms such as rotifers, who strongly depend on dietary nutrient supply. A decrease in food supply may lead to a shift in energy allocation and less available energy, resulting in a decrease fitness response [44–46]. Our findings indicate no differences between the two species in terms of the egg ratio, but as in previous studies, the food quantity influenced the reproduction differently [38]. Previous studies demonstrated the importance of food quality effects on the population growth rate, fecundity, and survival [47] as well as the differences in the life history traits between *B. calyciflorus* and *B. fernandoi* feeding on different algal foods [38]. Divergence in other life history traits were found [27] between *B. fernandoi* and *B. calyciflorus* by Zhang et al. since *B. fernandoi* invests less in sexual reproduction and has a higher population growth rate than the others brachionids. In addition, *B. calyciflorus* has a higher heat tolerance than *B. fernandoi* [26].These findings support the finding that *B. fernandoi* and *B. calyciflorus* differ in their ecology and react to stressors in a different way.

4.2. Size Particles Effect

The population growth rate and reproduction of the two rotifer species was significantly reduced when exposed to 3-µm PS beads. The size of the 3-µm PS beads is close to the size of the food alga and is at the lower end of the efficiently used food-size spectrum in *Brachionus* species [48–51]. This can explain why an effect was only found for the 3- and 6-µm beads. Our results are in accordance with Xue et al., [14], who showed that the reproduction of rotifers was suppressed when they were exposed to polyethylene microbeads (10–20 µm) along with algal food of a similar size. In our experiment, the survival percentage was not affected by the presence of microbeads, even when exposed to 3-µm PS, which had the strongest negative fitness response.

Different results were found by testing very small, nano-sized PS particles (37 nm, 0.07 µm) in marine brachionids, where the population growth rate decreased by more than 50%. On the contrary, large-sized PS beads had no effect on the population growth rate and reproduction [13]. The different results could be related to the different feeding efficiencies of the rotifer species. Furthermore, the nano-sized plastic beads mostly interfered at the cellular level. Micro- to medium-sized particles, similar to those in the present study, and particles that are up to 20 µm in size might interfere with the feeding and may dilute the food; in addition, large particles seem to have no effect on micro-zooplankton because they are non-edible food for them [48–51].

4.3. Silica and (PA) Nylon Microbeads

No effect on the fitness response was found when the rotifers were exposed to silica beads and polyamide fragments. The concentration and the specific density of the material play an important role in the uptake of particles in rotifers and could be a likely explanation for our findings. In fact, silica beads and the polyamide (PA) have a higher specific weight and a higher sinking velocity than PS. To prevent sedimentation, we applied agitation, but the ingestion process itself might have been affected by the weight. One may speculate that heavy particles are difficult to ingest. In the natural environments, animals are exposed to particles along with other suspended solids. A number of studies found no negative effects on the fitness of rotifers when they were exposed to suspended clay, whereas cladocerans were affected by clay particles [52,53]. Although rotifers and cladocerans are typical filter feeders, rotifers can feed more selectively, and they were able to avoid ingesting clay particles [52,53]. These results suggest that rotifers might be less affected by plastic pollution than cladocerans. Studying the effect of irregularly shaped MPs, *D. magna* was more affected by MPs than by mineral particles of a similar size, potentially leading to extinction within one and four generations [44,54,55]. A mechanism counteracting the ingestion of fragments is aggregation, which leads to particle sizes that are unable to be digested [20,49]. Until now, no general conclusion can be drawn as to which factors drive the ingestion and impact the size, shape, weight, and type of plastics on animals: Klein et al. [56] have recently found that the ingestion of beads and fragments in freshwater shrimp was more influenced by the size of the particles than by their shape, whereas the ingestion was not influenced by the presence of the food. Copepods, instead, ingest more fragments than beads or fibers [57]. Marine off-shore zooplankton ingested more fragments than the ones close to the urban coast [58]. These findings suggest a strong particle type and a species-specific role.

4.4. Ecological Relevance

A crucial issue in the research on plastic pollution is that the detection of particles becomes more and more difficult with decreasing size. At the moment, there is no method available that can reliably quantify microplastics in the size range used in this study in natural water samples with algae, bacteria, and detritus. The concentration of the smallest MPs size (<10 μm) cannot be estimated at present, but from modelling studies, it is likely that the number of MPs in the environment increases when the size decreases [59]. For instance, the number of particles in marine environment and freshwater sediment has been underestimated due to technical limitation [60,61]. At the time of the study, the concentrations of microbeads were, most likely, higher than the ones in the field; however, with increasing production and fragmentation, the amount of small microplastics will increase continuously.

Typically, laboratory conditions are chosen to match the needs of the test species as well as possible. In contrast, in the field, environmental conditions are highly variable over time and are often suboptimal in terms of temperature or food supply. In particular, food supply can vary strongly from low to high and vice versa over the course of mere days [62]. Under such suboptimal conditions, when animals are already stressed, the effects of pollutants can be stronger than they would be under ideal conditions, as demonstrated in the present study. Furthermore, the PS beads used for the experiment do not contain plasticizer or additives since they are used for standard tests. In fact, the polymer type and the chemicals that they contain can contribute to the toxicity of microplastics, creating an additional stress [63]. Indeed, one single plastic product can contain hundreds of chemicals [64]. These include additives such as antioxidants, flame retardants, plasticizers, and colorants as well as residual monomers and oligomers and side products of polymerization and compounds and impurities [65]. Once taken up, these plastic chemicals can have negative impacts. For instance, aqueous leachates from epoxy resin or PVC plastic products can induce acute toxicity [66] and alter life history traits [67] in *Daphnia magna*. Still, studies on the contribution of plastic chemicals to microplastic toxicity are scarce. Studies testing for the combined effects of more than two factors are generally rare [68]. In a study with *Daphnia*, Hiltunen et al. [69] tested for temperature, food quality, and microplastics. Using lower plastic concentrations, as was also the case in our study, they found that decreased food quality had the biggest effect on life history, and the low plastic concentrations had no effect. In another study, increasing the food quantity disproportionately reduced the uptake of MP, and no effect on *Daphnia* life history was found [70]. However, some results only become apparent after long-term exposure [71]. Combining these results, food quantity and quality have a strong impact on consumer life history that can be enhanced by high microplastic pollution.

5. Conclusions

Our study reveals that the negative effect of microplastics on a common freshwater invertebrate depends on the environmental conditions, which in this study, were food quality and quantity. This is one reason for the differing results in microplastic research and requires more attention in terms of plastic risk assessment. In addition, although standardized toxicological tests provide useful information on the toxic potential of pollutants, more realistic studies with various environmental conditions are needed to obtain deeper and more comprehensive insights on the problem of plastic pollution.

Supplementary Materials: The following are available online at https://www.mdpi.com/article/10.3390/toxics9110305/s1, Figure S1: Population growth rate, Figure S2: Size range distribution of PA nylon beads, Figures S3 and S4: PA beads ingested by *B. calyciflorus*, Figure S5A,B: PS beads ingested by *B. calyciflorus*. Table S1: Concentration of food algae, Table S2: Concentration of microbeads, Table S3: Results from the Emmeans' test.

Author Contributions: Conceptualization, methodology, investigation, C.D. and G.W.; writing—original draft preparation, C.D.; writing—review and editing, C.D. and G.W.; supervision, G.W.; funding acquisition, G.W. All authors have read and agreed to the published version of the manuscript.

Funding: This research was supported by the BMBF project MikroPlaTaS (02WPL1448C).

Institutional Review Board Statement: Not applicable.

Informed Consent Statement: Not applicable.

Data Availability Statement: The data presented in this study are available upon request from the corresponding author.

Acknowledgments: We thank Claudia Wahl for the technical support, S. Schälicke for her valuable advice on the experimental design, and Rico Leiser for the support during the data analysis. We also thank S. Bolius, Julia Pawlak, and Markus Stark for their writing assistance. We acknowledge the support of the Deutsche Forschungsgemeinschaft and Open Access Publishing Fund of University of Potsdam. We also thank the reviewers for their comments that improved the manuscript.

Conflicts of Interest: The authors declare no conflict of interest.

References

1. Moore, C.J. Synthetic polymers in the marine environment: A rapidly increasing, long-term threat. *Environ. Res.* **2008**, *108*, 131–139. [CrossRef]
2. Eerkes-Medrano, D.; Thompson, R.C.; Aldridge, D.C. Microplastics in freshwater systems: A review of the emerging threats, identification of knowledge gaps and prioritisation of research needs. *Water Res.* **2015**, *75*, 63–82. [CrossRef]
3. Li, J. Microplastics in freshwater systems: A review on occurrence, environmental effects, and methods for microplastics detection. *Water Res.* **2018**, *137*, 362–374. [CrossRef]
4. Murphy, F.; Ewins, C.; Carbonnier, F.; Quinn, B. Wastewater treatment works (WwTW) as a source of microplastics in the aquatic environment. *Environ. Sci. Technol.* **2016**, *50*, 5800–5808. [CrossRef]
5. Kole, P.J.; Löhr, A.J.; Van Belleghem, F.G.; Ragas, A.M. Wear and Tear of Tyres: A Stealthy Source of Microplastics in the Environment. *Int. J. Environ. Res. Public Health* **2017**, *14*, 1265. [CrossRef]
6. Corradini, F.; Meza, P.; Eguiluz, R.; Casado, F.; Huerta-Lwanga, E.; Geissen, V. Evidence of Microplastic Accumulation in Agricultural Soils from Sewage Sludge Disposal. *Sci. Total Environ.* **2019**, *671*, 411–420. [CrossRef]
7. Liu, K.; Wang, X.; Wei, N.; Song, Z.; Li, D. Accurate Quantification and Transport Estimation of Suspended Atmospheric Microplastics in Megacities: Implications for Human Health. *Environ. Int.* **2019**, *132*, 105127. [CrossRef]
8. Loppi, S.; Roblin, B.; Paoli, L.; Aherne, J. Accumulation of Airborne Microplastics in Lichens from a Landfill Dumping Site (Italy). *Sci. Rep.* **2021**, *11*, 4564. [CrossRef] [PubMed]
9. Cole, M.; Webb, H.; Lindeque, P.K.; Fileman, E.S.; Halsband, C.; Galloway, T.S. Isolation of microplastics in biota-rich seawater samples and marine organisms. *Sci. Rep.* **2015**, *4*, 4528. [CrossRef] [PubMed]
10. De Sá, L.C.; Oliveira, M.; Ribeiro, F.; Rocha, T.L.; Futter, M.N. Studies of the Effects of Microplastics on Aquatic Organisms: What Do We Know and Where Should We Focus Our Efforts in the Future? *Sci. Total Environ.* **2018**, *645*, 1029–1039. [CrossRef] [PubMed]

11. Bermúdez, J.R.; Metian, M.; Oberhänsli, F.; Taylor, A.; Swarzenski, P.W. Preferential Grazing and Repackaging of Small Polyethylene Microplastic Particles (≤5 Mm) by the Ciliate *Sterkiella* sp. *Mar. Environ. Res.* **2021**, *166*, 105260. [CrossRef]

12. De Ruijter, V.N.; Redondo-Hasselerharm, P.E.; Gouin, T.; Koelmans, A.A. Quality Criteria for Microplastic Effect Studies in the Context of Risk Assessment: A Critical Review. *Environ. Sci. Technol.* **2020**, *54*, 11692–11705. [CrossRef] [PubMed]

13. Sun, Y.; Xu, W.; Gu, Q.; Chen, Y.; Zhou, Q.; Zhang, L.; Yang, Z. Small-Sized Microplastics Negatively Affect Rotifers: Changes in the Key Life-History Traits and Rotifer–*Phaeocystis* Population Dynamics. *Environ. Sci. Technol.* **2019**, *53*, 9241–9251. [CrossRef] [PubMed]

14. Xue, Y.-H.; Sun, Z.-X.; Feng, L.-S.; Jin, T.; Xing, J.-C.; Wen, X.-L. Algal Density Affects the Influences of Polyethylene Microplastics on the Freshwater Rotifer *Brachionus Calyciflorus*. *Chemosphere* **2021**, *270*, 128613. [CrossRef] [PubMed]

15. Canniff, P.M.; Hoang, T.C. Microplastic Ingestion by *Daphnia Magna* and Its Enhancement on Algal Growth. *Sci. Total Environ.* **2018**, *633*, 500–507. [CrossRef]

16. Beiras, R.; Bellas, J.; Cachot, J.; Cormier, B.; Cousin, X.; Engwall, M.; Gambardella, C.; Garaventa, F.; Keiter, S.; Le Bihanic, F.; et al. Ingestion and Contact with Polyethylene Microplastics Does Not Cause Acute Toxicity on Marine Zooplankton. *J. Hazard. Mater.* **2018**, *360*, 452–460. [CrossRef]

17. Christensen, N.D.; Wisinger, C.E.; Maynard, L.A.; Chauhan, N.; Schubert, J.T.; Czuba, J.A.; Barone, J.R. Transport and Characterization of Microplastics in Inland Waterways. *J. Water Process Eng.* **2020**, *38*, 101640. [CrossRef]

18. Kruse, J.; Laermanns, H.; Stock, F.; Foeldi, C.; Schaefer, D.; Scherer, C.; Bogner, C. Proceedings of the Microplastic in fluvial environments - an example of the Elbe river near Dessau-Roßlau, Germany, EGU General Assembly 2021, online, 19–30 April 2021. EGU21-2686. [CrossRef]

19. Free, C.M.; Jensen, O.P.; Mason, S.A.; Eriksen, M.; Williamson, N.J.; Boldgiv, B. High-Levels of Microplastic Pollution in a Large, Remote, Mountain Lake. *Mar. Pollut. Bull.* **2014**, *85*, 156–163. [CrossRef]

20. Frydkjær, C.K.; Iversen, N.; Roslev, P. Ingestion and Egestion of Microplastics by the Cladoceran *Daphnia Magna*: Effects of Regular and Irregular Shaped Plastic and Sorbed Phenanthrene. *Bull. Environ. Contam. Toxicol.* **2017**, *99*, 655–661. [CrossRef]

21. Merriman, J.L.; Kirk, K.L. Temporal patterns of resource limitation in natural populations of rotifers. *Ecology* **2000**, *81*, 141–149. [CrossRef]

22. Cordova, S.E.; Giffin, J.; Kirk, K.L. Food Limitation of Planktonic Rotifers: Field Experiments in Two Mountain Ponds: Food Limitation. *Freshw. Biol.* **2001**, *46*, 1519–1527. [CrossRef]

23. Weithoff, G. Vertical Niche Separation of Two Consumers (Rotatoria) in an Extreme Habitat. *Oecologia* **2004**, *139*, 594–603. [CrossRef] [PubMed]

24. Ortega-Mayagoitia, E.; Ciros-Perez, J.; Sanchez-Martinez, M. A Story of Famine in the Pelagic Realm: Temporal and Spatial Patterns of Food Limitation in Rotifers from an Oligotrophic Tropical Lake. *J. Plankton Res.* **2011**, *33*, 1574–1585. [CrossRef]

25. Michaloudi, E.; Papakostas, S.; Stamou, G.; Neděla, V.; Tihlaříková, E.; Zhang, W.; Declerck, S.A.J. Reverse Taxonomy Applied to the *Brachionus Calyciflorus* Cryptic Species Complex: Morphometric Analysis Confirms Species Delimitations Revealed by Molecular Phylogenetic Analysis and Allows the (Re)Description of Four Species. *PLoS ONE* **2018**, *13*, e0203168. [CrossRef] [PubMed]

26. Paraskevopoulou, S.; Tiedemann, R.; Weithoff, G. Differential Response to Heat Stress among Evolutionary Lineages of an Aquatic Invertebrate Species Complex. *Biol. Lett.* **2018**, *14*, 20180498. [CrossRef] [PubMed]

27. Zhang, W.; Lemmen, K.D.; Zhou, L.; Papakostas, S.; Declerck, S.A. Patterns of Differentiation in the Life History and Demography of Four Recently Described Species of the *Brachionus Calyciflorus* Cryptic Species Complex. *Freshw. Biol.* **2019**, *64*, 1994–2005. [CrossRef]

28. Jeong, C.-B.; Won, E.-J.; Kang, H.-M.; Lee, M.-C.; Hwang, D.-S.; Hwang, U.-K.; Zhou, B.; Souissi, S.; Lee, S.-J.; Lee, J.-S. Microplastic Size-Dependent Toxicity, Oxidative Stress Induction, and p-JNK and p-P38 Activation in the Monogonont Rotifer (*Brachionus Koreanus*). *Environ. Sci. Technol.* **2016**, *50*, 8849–8857. [CrossRef]

29. Rauchschwalbe, M.-T.; Fueser, H.; Traunspurger, W.; Höss, S. Bacterial Consumption by Nematodes Is Disturbed by the Presence of Polystyrene Beads: The Roles of Food Dilution and Pharyngeal Pumping. *Environ. Pollut.* **2021**, *273*, 116471. [CrossRef]

30. Ramos-Rodríguez, E.; Conde-Porcuna, J.M. Nutrient Limitation on a Planktonic Rotifer: Life History Consequences and Starvation Resistance. *Limnol. Oceanogr.* **2003**, *48*, 933–938. [CrossRef]

31. Stemberger, R.S. A General Approach to the Culture of Planktonic Rotifers. *Can. J. Fish. Aquat. Sci.* **1981**, *38*, 721–724. [CrossRef]

32. Bogdan, K.G.; Gilbert, J.J. Body Size and Food Size in Freshwater Zooplankton. *Proc. Natl. Acad. Sci. USA* **1984**, *81*, 6427–6431. [CrossRef]

33. Pagano, M. Feeding of Tropical Cladocerans (*Moina Micrura, Diaphanosoma Excisum*) and Rotifer (*Brachionus Calyciflorus*) on Natural Phytoplankton: Effect of Phytoplankton Size-Structure. *J. Plankton Res.* **2008**, *30*, 401–414. [CrossRef]

34. Coppock, R.L.; Galloway, T.S.; Cole, M.; Fileman, E.S.; Queirós, A.M.; Lindeque, P.K. Microplastics alter feeding selectivity and faecal density in the copepod, *Calanus Helgolandicus*. *Sci. Total Environ.* **2019**, *687*, 780–789. [CrossRef]

35. Mueller, M.-T.; Fueser, H.; Trac, L.N.; Mayer, P.; Traunspurger, W.; Ho, S. Surface-Related Toxicity of Polystyrene Beads to Nematodes and the Role of Food Availability. *Environ. Sci. Technol.* **2020**, *54*, 1790–1798. [CrossRef] [PubMed]

36. Rothhaupt, K.O. Algal Nutrient Limitation Affects Rotifer Growth Rate but Not Ingestion Rate. *Limnol. Oceanogr.* **1995**, *40*, 1201–1208. [CrossRef]

37. Sarma, S.S.S.; Gulati, R.D.; Nandini, S. Factors Affecting Egg-Ratio in Planktonic Rotifers. *Hydrobiologia* **2005**, *546*, 361–373. [CrossRef]

38. Schälicke, S.; Teubner, J.; Martin-Creuzburg, D.; Wacker, A. Fitness Response Variation within and among Consumer Species Can Be Co-Mediated by Food Quantity and Biochemical Quality. *Sci. Rep.* **2019**, *9*, 16126. [CrossRef] [PubMed]

39. Osenberg, C.W.; Mittelbach, G.G. *The Relative Importance of Resource Limitation and Predator Limitation in Food Chains*; Polis, G.A., Winemiller, K.O., Eds.; Food Webs; Springer: Boston, MA, USA, 1996. [CrossRef]

40. Devetter, M.; Sed'a, J. Decline of Clear-Water Rotifer Populations in a Reservoir: The Role of Resource Limitation. *Hydrobiologia* **2005**, *546*, 509–518. [CrossRef]

41. Korez, Š.; Gutow, L.; Saborowski, R. Feeding and Digestion of the Marine Isopod Idotea Emarginata Challenged by Poor Food Quality and Microplastics. *Comp. Biochem. Physiol. Part C Toxicol. Pharmacol.* **2019**, *226*, 108586. [CrossRef]

42. Trotter, B.; Wilde, M.V.; Brehm, J.; Dafni, E.; Aliu, A.; Arnold, G.J.; Fröhlich, T.; Laforsch, C. Long-Term Exposure of *Daphnia Magna* to Polystyrene Microplastic (PS-MP) Leads to Alterations of the Proteome, Morphology and Life-History. *Sci. Total Environ.* **2021**, *795*, 148822. [CrossRef]

43. Müller-Navarra, D.C.; Brett, M.T.; Liston, A.M.; Goldman, C.R. A Highly Unsaturated Fatty Acid Predicts Carbon Transfer between Primary Producers and Consumers. *Nature* **2000**, *403*, 74–77. [CrossRef]

44. Ogonowski, M.; Schür, C.; Jarsén, Å.; Gorokhova, E. The Effects of Natural and Anthropogenic Microparticles on Individual Fitness in *Daphnia Magna*. *PLoS ONE* **2016**, *11*, e0155063. [CrossRef] [PubMed]

45. Imhof, H.K.; Rusek, J.; Thiel, M.; Wolinska, J.; Laforsch, C. Do Microplastic Particles Affect *Daphnia Magna* at the Morphological, Life History and Molecular Level? *PLoS ONE* **2017**, *12*, e0187590. [CrossRef] [PubMed]

46. Guilhermino, L.; Martins, A.; Cunha, S.; Fernandes, J.O. Long-Term Adverse Effects of Microplastics on *Daphnia Magna* Reproduction and Population Growth Rate at Increased Water Temperature and Light Intensity: Combined Effects of Stressors and Interactions. *Sci. Total Environ.* **2021**, *784*, 147082. [CrossRef] [PubMed]

47. Schälicke, S.; Sobisch, L.; Martin-Creuzburg, D.; Wacker, A. Food Quantity–Quality Co-limitation: Interactive Effects of Dietary Carbon and Essential Lipid Supply on Population Growth of a Freshwater Rotifer. *Freshw Biol* **2019**, *64*, 903–912. [CrossRef]

48. Rothhaupt, K. Differences in Particle Size-Dependent Feeding Efficiencies of Closely Related Rotifer Species. *Limnol. Oceanogr.* **1990**, *35*, 16–23. [CrossRef]

49. Drago, C.; Pawlak, J.; Weithoff, G. Biogenic Aggregation of Small Microplastics Alters Their Ingestion by a Common Freshwater Micro-Invertebrate. *Front. Environ. Sci.* **2020**, *8*, 264. [CrossRef]

50. Starkweather, P.L. Aspects of the Feeding Behavior and Trophic Ecology of Suspension-Feeding Rotifers. *Hydrobiologia* **2004**, *73*, 63–72. [CrossRef]

51. Starkweather, P.L.; Gilbert, J.J.; Frost, T.M. Bacterial Feeding by the Rotifer *Brachionus calyciflorus*: Clearance and Ingestion Rates, Behavior and Population Dynamics. *Oecologia* **1979**, *44*, 26–30. [CrossRef]

52. Kirk, K.L.; Gilbert, J.J. Suspended Clay and the Population Dynamics of Planktonic Rotifers and Cladocerans. *Ecology* **1990**, *71*, 1741–1755. [CrossRef]

53. Kirk, K.L. Inorganic Particles Alter Competition in Grazing Plankton: The Role of Selective Feeding. *Ecology* **1991**, *72*, 915–923. [CrossRef]

54. Schür, C.; Zipp, S.; Thalau, T.; Wagner, M. Microplastics but Not Natural Particles Induce Multigenerational Effects in *Daphnia Magna*. *Environ. Pollut.* **2020**, *260*, 113904. [CrossRef] [PubMed]

55. Yu, S.-P.; Cole, M.; Chan, B.K.K. Review: Effects of microplastic on zooplankton survival and sublethal responses. In *Oceanography and Marine Biology: An Annual Review*; Hawkins, S.J., Allcock, A.L., Bates, A.E., Evans, A.J., Firth, L.B., McQuaid, C.D., Russell, B.D., Smith, I.P., Swearer, S.E., Todd, P.A., Eds.; Editors Taylor and Francis; CRC Press: London, UK, 2020; Volume 58, pp. 351–393. ISBN 978-0-429-35149-5.

56. Klein, K.; Heß, S.; Nungeß, S.; Schulte-Oehlmann, U.; Oehlmann, J. Particle Shape Does Not Affect Ingestion and Egestion of Microplastics by the Freshwater Shrimp *Neocaridina Palmata*. *Env. Sci Pollut Res.* **2021**, *28*, 62246–62254. [CrossRef]

57. Botterell, Z.L.R.; Beaumont, N.; Cole, M.; Hopkins, F.E.; Steinke, M.; Thompson, R.C.; Lindeque, P.K. Bioavailability of Microplastics to Marine Zooplankton: Effect of Shape and Infochemicals. *Environ. Sci. Technol.* **2020**, *54*, 12024–12033. [CrossRef] [PubMed]

58. Desforges, J.-P.W.; Galbraith, M.; Ross, P.S. Ingestion of Microplastics by Zooplankton in the Northeast Pacific Ocean. *Arch Env. Contam Toxicol* **2015**, *69*, 320–330. [CrossRef] [PubMed]

59. Besseling, E.; Redondo-Hasselerharm, P.; Foekema, E.M.; Koelmans, A.A. Quantifying Ecological Risks of Aquatic Micro- and Nanoplastic. *Crit. Rev. Environ. Sci. Technol.* **2019**, *49*, 32–80. [CrossRef]

60. Lindeque, P.K.; Cole, M.; Coppock, R.L.; Lewis, C.N.; Miller, R.Z.; Watts, A.J.; Galloway, T.S. Are we underestimating microplastic abundance in the marine environment? A comparison of microplastic capture with nets of different mesh-size. *Environ. Pollut.* **2020**, *265*, 114721. [CrossRef]

61. Scherer, C.; Weber, A.; Stock, F.; Vurusic, S.; Egerci, H.; Kochleus, C.; Arendt, N.; Foeldi, C.; Dierkes, G.; Wagner, M.; et al. Comparative assessment of microplastics in water and sediment of a large European river. *Sci. Total. Environ.* **2020**, *738*, 139866. [CrossRef]

62. Weithoff, G.; Lorke, A.; Walz, N. Effects of Water-Column Mixing on Bacteria, Phytoplankton, and Rotifers under Different Levels of Herbivory in a Shallow Eutrophic Lake. *Oecologia* **2000**, *125*, 91–100. [CrossRef]

63. Zimmermann, L.; Göttlich, S.; Oehlmann, J.; Wagner, M.; Völker, C. What Are the Drivers of Microplastic Toxicity? Comparing the Toxicity of Plastic Chemicals and Particles to *Daphnia Magna*. *Environ. Pollut.* **2020**, *267*, 115392. [CrossRef]
64. Zimmermann, L.; Dierkes, G.; Ternes, T.A.; Völker, C.; Wagner, M. Benchmarking the in Vitro Toxicity and Chemical Composition of Plastic Consumer Products. *Environ. Sci. Technol.* **2019**, *53*, 11467–11477. [CrossRef]
65. Muncke, J. Exposure to Endocrine Disrupting Compounds via the Food Chain: Is Packaging a Relevant Source? *Sci. Total Environ.* **2009**, *407*, 4549–4559. [CrossRef] [PubMed]
66. Lithner, D.; Nordensvan, I.; Dave, G. Comparative Acute Toxicity of Leachates from Plastic Products Made of Polypropylene, Polyethylene, PVC, Acrylonitrile–Butadiene–Styrene, and Epoxy to *Daphnia Magna*. *Environ. Sci. Pollut. Res.* **2012**, *19*, 1763–1772. [CrossRef]
67. Schrank, I.; Trotter, B.; Dummert, J.; Scholz-Böttcher, B.M.; Löder, M.G.J.; Laforsch, C. Effects of Microplastic Particles and Leaching Additive on the Life History and Morphology of *Daphnia magna*. *Environ. Pollut.* **2019**, *255*, 113233. [CrossRef]
68. Weisse, T.; Laufenstein, N.; Weithoff, G. Multiple Environmental Stressors Confine the Ecological Niche of the Rotifer *Cephalodella Acidophila*. *Freshw. Biol.* **2013**, *58*, 1008–1015. [CrossRef] [PubMed]
69. Hiltunen, M.; Vehniäinen, E.-R.; Kukkonen, J.V.K. Interacting Effects of Simulated Eutrophication, Temperature Increase, and Microplastic Exposure on *Daphnia*. *Environ. Res.* **2021**, *192*, 110304. [CrossRef]
70. Aljaibachi, R.; Callaghan, A. Impact of Polystyrene Microplastics on *Daphnia magna* Mortality and Reproduction in Relation to Food Availability. *PeerJ* **2018**, *6*, e4601. [CrossRef] [PubMed]
71. Aljaibachi, R.; Laird, W.B.; Stevens, F.; Callaghan, A. Impacts of Polystyrene Microplastics on *Daphnia magna*: A Laboratory and a Mesocosm Study. *Sci. Total Environ.* **2020**, *705*, 135800. [CrossRef]

Article

Distribution and Seasonal Variation of Microplastics in Tallo River, Makassar, Eastern Indonesia

Ega Adhi Wicaksono [1], Shinta Werorilangi [2], Tamara S. Galloway [3] and Akbar Tahir [2,*]

1   Department of Fisheries, Universitas Hasanuddin, Jl. Perintis Kemerdekaan, KM 10 Tamalanrea, Makassar 90245, Indonesia; egaadhi@gmail.com

2   Department of Marine Science, Universitas Hasanuddin, Jl. Perintis Kemerdekaan, KM 10 Tamalanrea, Makassar 90245, Indonesia; shintakristanto@yahoo.com

3   College of Life and Environmental Sciences: Biosciences, University of Exeter, Geoffrey Pope Building, Stocker Road, Exeter EX4 4QD, UK; T.S.Galloway@exeter.ac.uk

*   Correspondence: akbar_tahir@mar-sci.unhas.ac.id

Abstract: Attention towards microplastic (MP) pollution in various environments is increasing, but relatively little attention has been given to the freshwater-riverine environment. As the biggest city in the eastern Indonesia region, Makassar can be a potential source of MP pollution to its riverine area. This study aimed to determine the spatial trends, seasonal variation, and characteristics of MPs in the water and sediment of Tallo River, as the main river in Makassar. Water samples were collected using a neuston net and sediment samples were collected using a sediment corer. The samples collected contained MPs with an abundance ranging from 0.74 ± 0.46 to 3.41 ± 0.13 item/m^3 and 16.67 ± 20.82 to 150 ± 36.06 item/kg for water and sediment samples, respectively. The microplastic abundance in the Tallo River was higher in the dry season and tended to increase towards the lower river segment. Fragments (47.80–86.03%) and lines (12.50–47.80%) were the predominant shapes, while blue (19.49–46.15%) and transparent (14.29–38.14%) were the most dominant color. Polyethylene and polypropylene were the common MP polymers found in the Tallo river. Actions to prevent MP pollution in the Makassar riverine area are needed before MP pollution becomes more severe in the future.

Keywords: plastics; riverine; coastal; estuary; characteristics; pollution

Citation: Wicaksono, E.A.; Werorilangi, S.; Galloway, T.S.; Tahir, A. Distribution and Seasonal Variation of Microplastics in Tallo River, Makassar, Eastern Indonesia. *Toxics* **2021**, *9*, 129. https://doi.org/10.3390/toxics9060129

Academic Editors: Costanza Scopetani, Tania Martellini and Diana Campos

Received: 3 May 2021
Accepted: 28 May 2021
Published: 1 June 2021

Publisher's Note: MDPI stays neutral with regard to jurisdictional claims in published maps and institutional affiliations.

1. Introduction

Plastic pollution is being reported everywhere and has become a major global problem. An increasing amount of plastic waste, primarily caused by anthropogenic activities in terrestrial locations, may eventually end up in the sea [1,2]. More than 190 coastal countries have been identified as contributors to an annual release of up to 12.7 million metric tons of plastic debris into the ocean [2]. Environmental stressors such as physical abrasion, elevated temperature, and UV-B exposure can all help plastic waste degrade into a smaller form of plastic in the environment [3,4]. These small-sized plastic particles that range from 1–5 mm eventually merge into a new form, called "microplastic" [5,6].

Microplastics (MPs) tend to receive a lot of attention from researchers, public communities, and governments worldwide due to their potential impacts on the ecosystem [7–9]. Microplastics are known to interact with other toxic compounds in the aquatic ecosystem [10–12]. Internal compounds in the MPs may also induce toxicity to the exposed organism [13]. The shape of MPs can resemble plankton, the primary food source in the aquatic environment, which makes it very easy to be consumed by aquatic organisms [14,15]. Reports on the incidence of MP ingestion by aquatic organisms have also been widely reported, as in plankton, fish, and shellfish [16–20]. This situation raises concerns about MPs' impact not only on the ecosystem but also on food security, which may have implications for human health [21,22].

Indonesia is branded as the world's second-largest contributor to ocean plastic pollution [2]. However, research regarding MP pollution in Indonesia is still in its early stages and needs further development. Currently, research on MP pollution in Indonesia focuses more on the marine environment. Microplastic is known to contaminate sediment [23,24], water [25,26] and biota [16,27,28] in Indonesia's marine environment. In contrast, research on MPs in the freshwater environment in Indonesia has received little attention. Only a few studies concerning MP pollution have been conducted in Indonesia's rivers [29,30]. According to these studies, MPs are reported to pollute rivers in the western Indonesia region, especially on Java Island [31].

To the best of our knowledge, even though studies regarding MPs have been conducted in western Indonesia's river, no MP pollution research has ever been performed in the riverine area in Indonesia's eastern region. Eastern Indonesia is an important location for plastic pollution research. This area is passed by the Indonesian throughflow (ITF) ocean current, which can carry plastic waste from the pacific ocean and its stream trajectory to the Indian Ocean [32,33]. The high input of plastic debris from the rivers in eastern Indonesia due to ITF ocean currents can further spread to other locations, posing risks to broader geographical areas.

As the biggest city in eastern Indonesia, Makassar needs more attention due to high anthropogenic pressure. Shuker and Cadman, in 2018 [34], reported that Makassar City produces more than 1200 tons of solid waste a day. The same report also stated that more than 44% of trash found in the Makassar coastal area is plastic waste. The coastal area of Makassar is already polluted by plastic waste in several colors and sizes [35,36]. The estuary areas in Makassar City also show MP contamination suspected from the river outflow [37]. Despite research into MPs in the marine environment of Makassar City being conducted at least five years earlier [16,33,35,38], information regarding MP pollution in Makassar's riverine environment is still lacking.

This study focuses on the MP pollution in Tallo River, as the main river trajectory in Makassar City. In general, Tallo riverbank is still covered by a mangrove ecosystem, as this river is utilized for recreational and fisheries purposes. The occurrence of MPs in Tallo River may pose threats to human health in Makassar City, considering that most of the freshwater fish and shrimp commodities in Makassar originate from this river. Tallo River is also directly feeding the Makassar Strait, the location of the ITF ocean current. This research aims to determine the abundance, spatial trend, and characteristics of MPs in the water and sediment of Tallo River during the wet and dry seasons. This research provides novel data on MP pollution in Makassar's riverine environment. It could be used as a baseline to evaluate and improve solid waste management in the east Indonesia region, particularly in Makassar City.

2. Materials and Methods

2.1. Study Sites and Sampling

The study was conducted in the section of the Tallo River that crosses Makassar City, Indonesia. Samples were taken in March and August 2019 to represent the wet and dry seasons, respectively. Six sampling points were distributed purposively based on their position from the upstream to the downstream part of the river section. Sampling points 1 and 2 were located on the upstream part of the river, where there is a thick *Nypa fruticans* green belt on the riverbank in this river segment. The mid-stream section was represented by sampling points 3 and 4, which are surrounded by a mangrove ecosystem and fisheries activities, such as a fish and shrimp pond. Between points 3 and 4, a flow of water enters from the Makassar industrial area. The downstream segment was represented by sampling points 5 and 6, which are surrounded by Makassar City's slum district. There is also a water flow that enters the Tallo river at point 5, originating from the Makassar urban area. Land use/cover area [39] and sampling points on Tallo River are described in Figure 1.

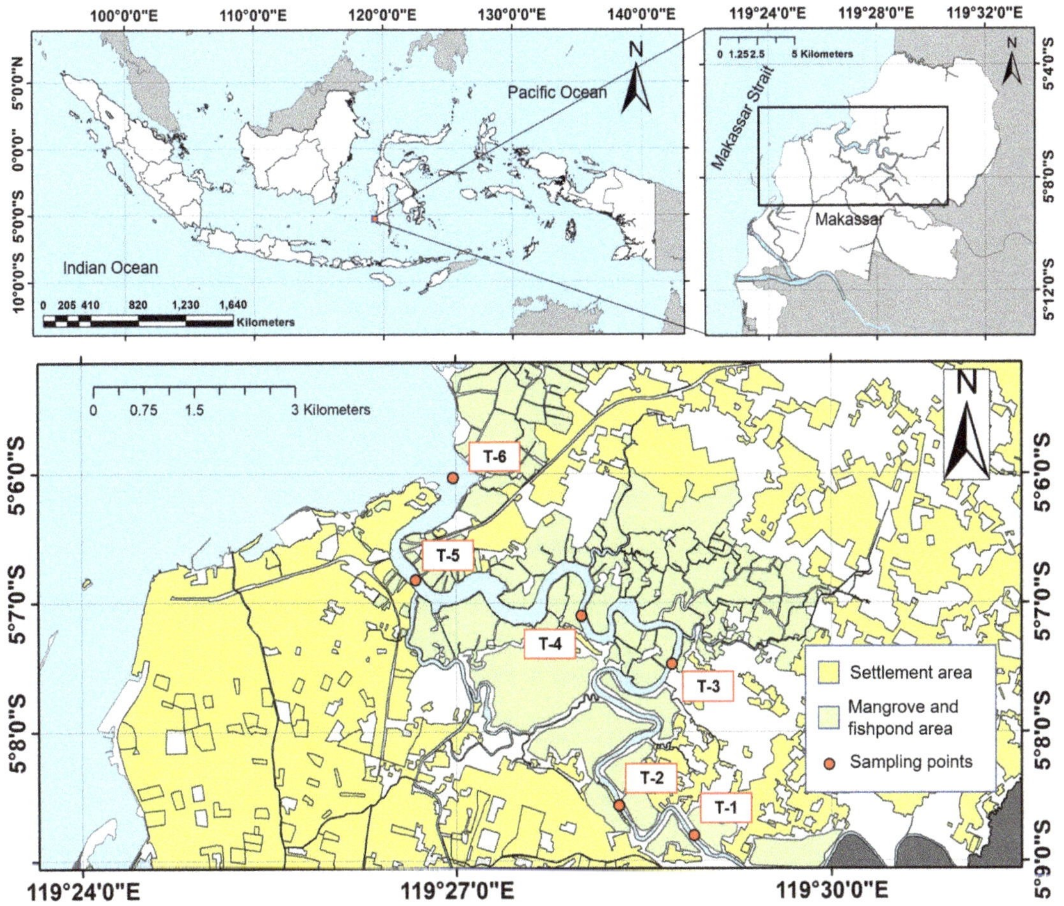

Figure 1. Sampling points on Tallo River.

Water samples were collected in triplicate from each sampling point using the neuston net method [26] with a slight modification to the net dimension. A custom rectangle-mouth neuston net (15 × 60 cm, 330 μm mesh size) was towed perpendicular to the river current at a constant speed (4 km/h) using a boat. Towing distance was measured using a GPS device (Garmin Montana 680, Schaffhausen, Switzerland). The amount of water filtered during towing was calculated by multiplying the net mouth area with the towing length. Water accrued in the cod-end was then transferred into a bottle sample and added to 30 mL of 10% KOH solution [40]. Following that, the samples were transported to the laboratory in a cool box. Water samples were preserved at 4 °C prior to further analysis. Samples were then filtered using a vacuum pump (Rocker 410, Kaohsiung, Taiwan) to a sterile 0.45 μm pore size cellulose filter (Whatman GE 7141-104, Buckinghamshire, UK). The filter paper was then placed into a clean glass Petri dish to be observed visually using a stereomicroscope.

Bulk sediment samples were taken in triplicate at every sampling point using a sediment corer (Ø 4.9 cm) in the river littoral zone (50 cm–1 m depth) [14]. Sediment was collected from the riverbed's top layer (5–7 cm). Sediment samples were then transferred to a Ziplock bag and preserved in the cool box for further analysis in the laboratory.

Sediment samples (400 g wet weight) were dried in an oven (60 °C for 48 h). For the density separator process, a total of 100 g of dry weight (DW) sediment was taken from the dried samples and subjected to 300 mL of a 30% NaCl solution (337 g analytical NaCl powder + 1 L distilled water, density $\approx 1.2 \, \text{g/cm}^3$) [41]. The samples were stirred at 1200 rpm for 2 min using a magnetic stirrer. Sediment samples were left at room temperature (27–28 °C) overnight to create a supernatant layer in the sample. The supernatant liquid was then filtered using the same method as that used in the water samples procedure described. The filter paper was then placed in a clean glass Petri dish for further visual analysis using a stereomicroscope.

Visual observations were performed using a stereomicroscope (Euromax SB-1902, Arnhem, Netherland; 45× magnification). The filter paper inspection was performed using a zigzag movement on filter paper until all of the areas on the filter were observed. Any MPs found in the filter paper were taken and placed into an object glass for preservation. The number, shape, size, and color of the MPs were then determined. The MPs' colors were classified according to Frias et al. [42] and the MPs' shape identification referred to GESAMP [43]. The MPs' size was determined using ImageJ (National Institute of Health, Bethesda, MD, USA, version 1.52a) software. Microplastic sizes were then classified into small MPs (SMPs, <1 mm) and large MPs (LMPs, 1–5 mm) [29,44]. The abundance of MPs in the samples was expressed in items/m^3 for water and items/kg DW for the sediment samples.

The polymer types of the representative MP samples were identified separately using the Fourier-transform infrared spectroscopy (FTIR) method. Microplastic was placed in the sample chamber and read using the FTIR machine (Bruker Tensor II, Ettlingen, Germany) with ATR accessories in a 500–4000 cm^{-1} spectral range and resolution of 4 cm^{-1}. The wave spectrum was then matched with the NICODOM spectra library to determine the polymer type.

2.2. Quality Controls

Several actions were taken to prevent contamination in the samples. All of the pieces of equipment were pre-cleaned with tap water and rinsed with distilled water. The MPs visual observation workspace was also cleaned using a dust roller prior to the MP identification process. All of the filter-filled Petri dishes were kept closed to prevent airborne contamination. During the visual observation process, Petri dish covers were opened for no longer than 30 s for every MP found, in order to move the MPs from the filter paper to an object glass.

Sample blanks and airborne controls were used as the negative control. A total of 12 sediment and 12 water sample blanks were created during this research. Water sample blanks were created by rinsing the clean neuston net from the net mouth with distilled water before towing. The flushed distilled water in the net cod-end was kept and analyzed as other water samples. For the sediment sample blanks, about 600 mL of the NaCl solution used in the density separator was filtered before use. The filter was then observed using the stereomicroscope.

Airborne controls were performed by placing three opened Petri dishes filled with distilled water next to the microscope during the visual observation process. Controls were placed 10 min before the sample observation and taken 10 min after the MPs visual analysis was complete. Controls were then observed visually using the same method that was used for the samples.

2.3. Data Analysis

The trends in MP abundance in water and sediment were analyzed using a one-way ANOVA with Tukey's post hoc analysis to determine the spatial MP abundance between the sampling points. The significant difference in MP abundance between the wet and dry seasons was determined using a parametric *t*-test. Microplastic color, shape, size and polymers were presented descriptively. Spatial distribution graphics and statistical analysis were conducted using GraphPad Prism (Graphpad Software, San Diego, CA, USA, version 9.0.2).

3. Results and Discussions

3.1. Contamination Control

Microplastic was not found in all water and sediment sample blanks. In the negative airborne control, from the 45 Petri dishes used during the MP identification process, only 1 MP (line, purple) was found with the average MP abundance found to be 0.02 items/Petri dish. Microplastic in the airborne blanks only had a proportion of about 0.28% of the MPs found in samples. Therefore, it is assumed that contamination does not affect the MPs' identification in water and sediment samples and can be ignored.

3.2. Microplastic Abundance on Water and Sediment

A total of 36 water and 36 sediment samples from the Tallo River were analyzed in this research. Microplastic was found in all of the samples. Microplastics are widespread in various environments, including the riverine system [14,45]. Mostly, the MPs found in the freshwater system come from anthropogenic pressures such as domestic, industry, wastewater treatment plants, and agrosystems [46,47]. All of the samples observed in this study contained MPs, which indicates that MPs have contaminated Tallo River.

The microplastic abundance found in water samples ranged from (mean $\pm$ SD) 0.74 ± 0.46 to 2.15 ± 0.68 items/m^3 in the wet season and 1.48 ± 0.26 to 3.41 ± 0.13 items/m^3 in the dry season (Figure 2). The microplastic abundance in water samples in this study is considered much lower than that which was reported in other river locations in Indonesia. Ciwalengke and Surabaya River in Indonesia were reported to have a MP abundance up to 600 items/m^3 and 21 items/m^3, respectively [29,30]. This result is understandable because the Ciwalengke and Surabaya Rivers flow directly through a densely populated district and an industrial area, which provide potential sources of MP pollution. In contrast, the Tallo River is mainly covered by mangrove areas on its riverbank and is not directly bordered by a resident/industrial area. The existence of mangrove areas could act as a MP trap. The muddy mangrove sediment could trap MPs and increase the magnitude of MP abundance up to eight times compared to non-mangrove sediment [48]. A mangrove ecosystem in the Tallo riverbank might prevent the run-off leakage of MPs entering the river. This condition could contribute to the lower MP abundance in the river water.

Microplastic abundance in water samples was significantly higher in the dry season (2.247 ± 0.688 items/m^3) compared to the wet season (1.457 ± 0.508 items/m^3) ($p < 0.05$) (Figure 3). In comparison, there was no significant difference in MP abundance in the sediment samples between the two seasons ($p > 0.05$). The tendency for a higher concentration of MP abundance in the dry season also happens in other rivers, such as the Maozhou and Yellow Rivers in China [49,50]. The difference in MP abundance in riverine water could happen because of the variation in topography, precipitation, and waste management in the sampling locations [49]. The Tallo River itself has a wide variety of water depths and velocities between the wet and dry seasons. Water depth in Tallo River during the wet season is due to high precipitation, and can be two times deeper than the depth during the dry season [51]. This difference could cause the river water volume:surface-water area ratio to be smaller in the dry season, which leads to a higher amount of MPs in the surface water [18].

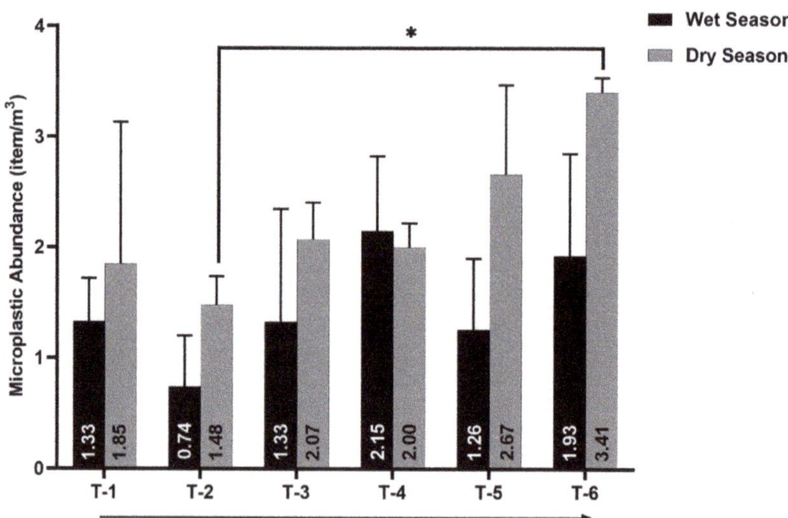

Figure 2. Microplastic abundance on the surface water of Tallo River. The arrows below the graph indicate the position of sampling points from the upstream to the downstream part of the river. The error bar indicates standard deviation ($n = 3$). The asterisk indicates the significant difference between sites based on a one-way ANOVA ($p < 0.05$).

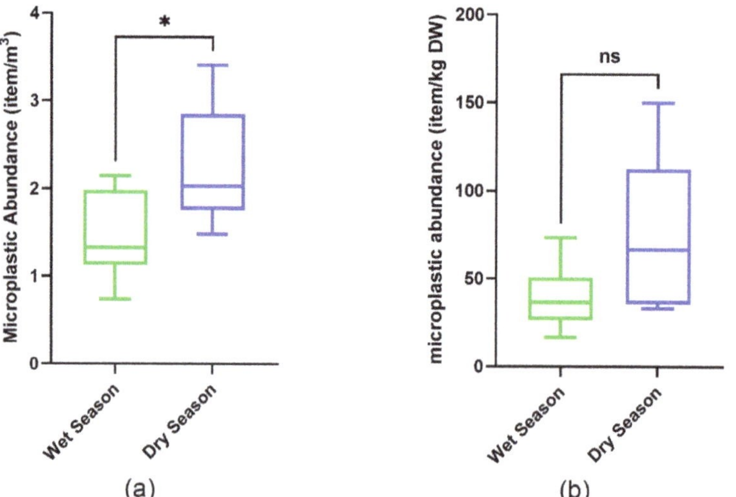

Figure 3. Boxplot diagram of microplastic abundance in water (**a**) and sediment (**b**) during the wet and dry seasons in Tallo River. The asterisk indicates the significant difference between the sites based on a *t*-test ($p < 0.05$). ns indicate no statistical difference between the sites based on a *t*-test ($p > 0.05$) explanation.

The microplastic abundance in sediment samples from Tallo River varied from 16.67 ± 20.82 to 73.33 ± 40.41 items/kg DW in the wet season and 33.33 ± 25.17 to 150 ± 36.06 items/kg DW in the dry season (Figure 4). The microplastic abundance in sediments from Tallo River was also considered lower compared to the other river sediments in Indonesia, such as in Ciwalengke River (≈300 items/kg DW), Jagir Estuary (90 to 590 items/kg DW), and Estuary in Jakarta Bay (up to 38,000 items/kg DW) [24,30,52]. This result suggests that MP abundance in Tallo sediment might not be as severe as that reported in riverine sediments from Java Island, the most populated island in western Indonesia. The higher anthropogenic pressures on the river catchment area will mostly lead to a higher MP abundance in its river environment. Jakarta City, where the MP abundance in riverine sediment exceeded 15,000 items/kg DW, for example, has a population of more than 10 million people [53], about 7.5 times higher than the population of Makassar City.

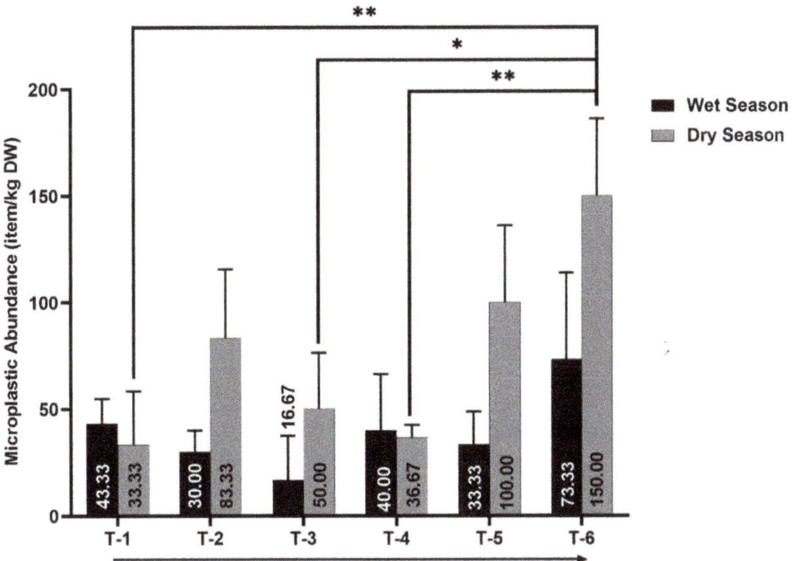

Figure 4. Microplastic abundance in sediment from Tallo River. The arrows below the graph indicate the position of every site from the upstream to the downstream part of the river. The error bar indicates standard deviation (n = 3). The asterisk (*) indicates the significant difference between the sites based on a one-way ANOVA ($p < 0.05$). The double asterisks (**) indicate the higher significant difference between the sites ($p < 0.01$).

The microplastic in water and sediment from the Tallo River has a similar spatial distribution. The microplastic abundance in the Tallo River tends to be higher in the river-mouth area compared to the upstream area. This pattern was more observable in the dry season. The microplastic abundance at site T-2 was significantly lower compared to site T-6, which was located at the river-mouth during the dry season ($p < 0.05$) (Figure 2). In the sediment samples, sites T-1, T-3, and T-4 were significantly lower compared to site T-6 ($p < 0.05$) (Figure 4). Even though there was no statistical difference in MPs' spatial distribution during the wet season, a similar trend to the dry season was observed, where the Tallo River's downstream segment had a greater MP abundance compared to the upstream section. An estuary location is more susceptible to MP contamination. The Tallo Estuary riverbank is directly located next to the slum settlement area of Makassar City, which potentially gives MPs input to the Tallo downstream area. Settlement area can provide various MP sources (e.g., laundry waste, beads from personal care products, and domestic trash) [14,54,55]. Estuaries with high anthropogenic pressure will generally have a higher MP abundance [47]. Water velocity in the estuary, in general, is lower than in the upstream river due to the more static marine water mass that influences this area. MPs' transport in the river is strongly affected by flow regime. The intense flow can cause the MPs' mobilization and transport, while the low stream velocity is causing the MP retention and deposition [56,57]. Low water velocity in Tallo Estuary can lengthen the MPs' residence time, leading to MPs' accumulation and increment in the estuary area.

3.3. Microplastic Characteristics

3.3.1. Microplastic Color

In general, there were six prominent MP colors found in the samples (Figure 5). Blue (19.49–46.15%) and transparent (14.29–38.14%) were the most dominant MP colors found in Tallo River, followed by white (10.17–20.59%), red (6.62–18.31%) and green (0.85–8.45%). Black MPs in Tallo River were only found in the water (3.30–12.71%) and were not present in the sediment compartment.

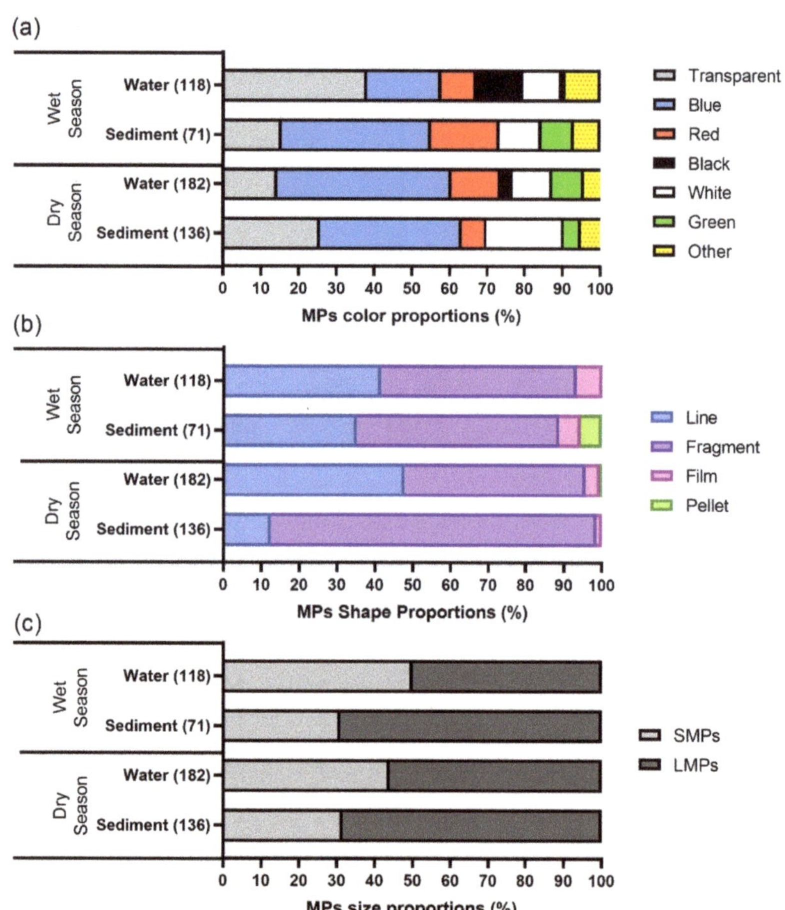

Figure 5. The proportions of the MPs' colors (**a**), shapes (**b**) and sizes (**c**) in the water and sediment samples from Tallo River. SMPs, small microplastics (<1 mm); LMPs, large microplastics (1–5 mm).

Microplastic color can provide information to predict the source and weathering process of MPs. For example, transparent color is often associated with polypropylene, commonly used as a food packaging material. The yellowish color of MPs can also indicate the photooxidation and weathering process of MPs [58]. In Tallo River, the most common MP colors found were blue and transparent. The pigmented MPs color may originate from textile and paint, which usually use various colors. The transparent MPs can be linked to a transparent food container that mainly consists of polyethylene and polypropylene polymer. The color of MPs may also influence a fish's preference to eat small plastic particles. Fish tend to prefer MPs with a similar color to their prey. For example, the scad fish collected from the South Pacific Gyre tend to ingest blue MPs due to their color similarity to the copepod species, which is scad's natural prey [59]. Some authors report that fish tend to prefer lighter colors of MPs, such as blue, white, and transparent, because it is easier to distinguish these colors compared to the brownish natural environment color [37,60]. The dominance of blue and transparent MPs might make these MPs more bioavailable for the aquatic organism in the river. In addition, a MPs' color usually comes from a synthetic colorant that can leach into the environment and pose additional risks to the aquatic organism [13].

3.3.2. Microplastic Shape

The microplastics in Tallo River were dominated by fragments (47.80–86.03%) and lines (12.50–47.80%) compared to other MP shapes, such as films (1.47–6.78) and pellets (0.55–5.63%) (Figures 5 and 6). A higher pellet proportion existed in Tallo sediment during the dry season (5.63%), while in the wet season, it only had a proportion of about 0.5% in the water. Tallo sediment during the dry season had a significant proportion of fragments.

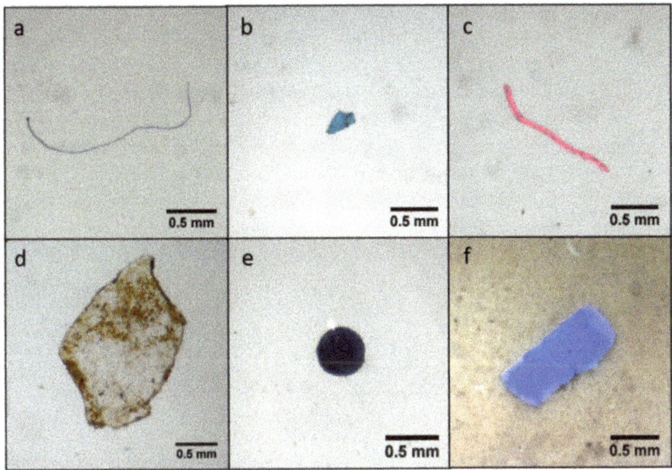

Figure 6. Representative of MPs found in the samples. Blue and red line (a,c), blue fragment (b), transparent fragment (d), blue pellet (e), and blue film (f) MPs.

The shape of MPs could mimic the natural prey of fish that exist in the environment [14,15]. For example, the line type of MPs has a similar shape to the filamentous algae in the aquatic environment, which is a fish's natural prey. The MPs' shape can also be an indicator of the MPs' origins. Fragments mainly originate from a secondary source of MPs (fragmentation of larger-sized plastic) [61]. The existence of pellets also shows the probability of primary MPs. Tallo River also receives water flow from the Makassar Industrial Area, where several plastics industries might be using the preproduction plastic pellet. Plastic pellets can leak into the environment due to production processes and raw pellet transportation [55]. However, the low proportion of pellets in this study suggests that MPs in the Tallo River do not primarily originate from primary MPs.

3.3.3. Microplastic Size

In general, there are a higher proportion of LMPs (50–69.01%) in the Tallo Riverine environment than SMPs (30.99–50%). Microplastic found in the water tends to be smaller compared to MP found in the sediment compartment. A more significant proportion of

LMPs in the Tallo River suggests that the MPs have not been further degraded. In a long trajectory river such as the Rhine River in Europe, SMP tend to dominate [62]. A large proportion of SMPs can indicate further plastic degradation due to physical and chemical stressors from the environment. The size of MPs can be gradually reduced because of degradation mechanisms in the river's trajectory. As MPs move towards river mouths, they can degrade to a smaller size. This condition leads to a higher proportion of SMPs in lower river segments [52].

Moreover, MPs' dimensions also affect their possible bioavailability. Microplastics with smaller sizes can be more easily ingested by zooplankton, making it easier for SMPs to enter the food web [20]. It is also easier for small-sized MPs to be transported into an organism's soft tissue, posing a greater risk to the organism [63].

3.3.4. Microplastic Polymer

A total of five polymers were identified in the study site (Figure 7). The most predominant polymers found in the water and sediment samples were polyethylene (43–50%) and polypropylene (30–36%). Poly(styrene:butadiene) was only found in the water samples (20%). Synthetic rayon and polyester were only found in the sediment samples (14% and 7%, respectively). Poly(styrene:butadiene) and polyethylene were mainly found in the shape of fragments, while polypropylene, rayon and polyester were found in the form of lines. As the highest-produced polymer globally, polypropylene and polyethylene are more available to reach the aquatic environment [64]. This condition means that polyethylene and polypropylene are commonly found in freshwater environments [65]. Poly(styrene:butadiene) is mainly used for anti-abrasion surfaces, such as in car tires and shoe soles, while rayon and polyester are commonly used as textile material [43,54,66,67]. A single wash of about 6 kg of polyester clothes can release nearly 500,000 polyester fibers in its waste effluent, leading to a higher polyester line in the environment [54]. The low density of polystyrene-butadiene (0.94 g/cm^3) means this polymer commonly accumulates in surface water. In contrast, rayon and polyester have a higher density than 1.35 g/cm^3, higher than the water density [43]. This condition means rayon and polyester tend to sink in the environment and end up in the sediment compartment.

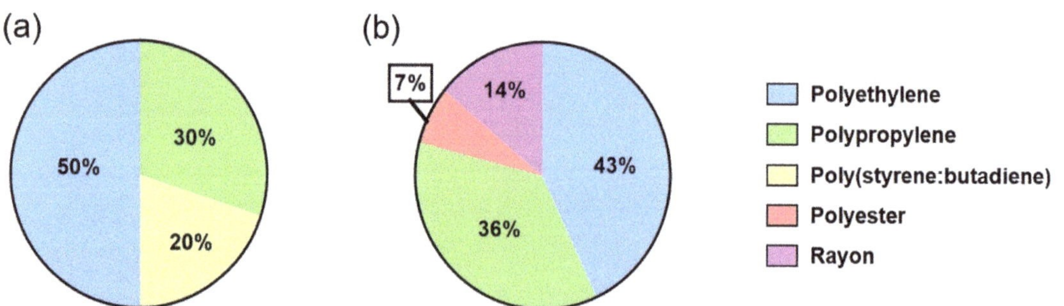

Figure 7. Microplastic polymer identified in water (**a**) and sediment (**b**) samples.

4. Conclusions

Tallo River has been contaminated by MPs, both in the water and sediment compartment. The MP abundance in the Tallo River is influenced by seasonal variations, where the MP abundance is higher in the dry season. The spatial trends suggest that MP abundance in the Tallo River tends to be higher in the lower river segment. Microplastics in the Tallo river mainly originate from secondary MPs, and polyethylene and polypropylene in the form of lines and fragments dominate. This is the first report of MP pollution in eastern Indonesia's river. The low MP abundance in water and sediment compared to that which is reported on the highly populated Java Island should be an incentive for early action to prevent MP contamination in Tallo River becoming more severe in the future.

Author Contributions: Conceptualization, E.A.W. and A.T.; methodology, E.A.W.; formal analysis, E.A.W.; writing—original draft preparation, E.A.W.; writing—review and editing E.A.W., S.W., T.S.G., and A.T.; supervision, S.W., T.S.G., and A.T.; funding acquisition, A.T. All authors have read and agreed to the published version of the manuscript.

Funding: This research was funded by The Indonesia Ministry of Research, Technology, and Higher Education on Higher Education through Master Program of Education Leading to Doctoral Degree for Excellent Graduates (PMDSU) Scholarship scheme, grant number 1739/UN4.21/PL.01.10/2019.

Institutional Review Board Statement: Not applicable.

Informed Consent Statement: Not applicable.

Data Availability Statement: Research data are available on request from the corresponding author.

Acknowledgments: Authors thank Amriana, Amri Yusuf, Kuasa Sari, Lulu A. L., and Anugrah Saputra for supporting the field sampling.

Conflicts of Interest: The authors declare no conflict of interest.

References

1. Lebreton, L.C.M.; van der Zwet, J.; Damsteeg, J.-W.; Slat, B.; Andrady, A.; Reisser, J. River plastic emissions to the world's oceans. *Nat. Commun.* **2017**, *8*, 1–10. [CrossRef] [PubMed]
2. Jambeck, J.R.; Geyer, R.; Wilcox, C.; Siegler, T.R.; Perryman, M.; Andrady, A.; Narayan, R.; Law, K.L. Plastic waste inputs from land into the ocean. *Science* **2015**, *347*, 768–771. [CrossRef] [PubMed]
3. Song, Y.K.; Hong, S.H.; Jang, M.; Han, G.M.; Jung, S.W.; Shim, W.J. Combined Effects of UV Exposure Duration and Mechanical Abrasion on Microplastic Fragmentation by Polymer Type. *Environ. Sci. Technol.* **2017**, *51*, 4368–4376. [CrossRef] [PubMed]
4. Corcoran, P.L.; Biesinger, M.C.; Grifi, M. Plastics and beaches: A degrading relationship. *Mar. Pollut. Bull.* **2009**, *58*, 80–84. [CrossRef]
5. Thompson, R.C.; Olson, Y.; Mitchell, R.P.; Davis, A.; Rowland, S.J.; John, A.W.G.; McGonigle, D.; Russell, A.E. Lost at Sea: Where Is All the Plastic? *Science* **2004**, *304*, 838. [CrossRef] [PubMed]
6. Frias, J.P.G.L.; Nash, R. Microplastics: Finding a consensus on the definition. *Mar. Pollut. Bull.* **2019**, *138*, 145–147. [CrossRef]
7. Andrady, A.L. Microplastics in the marine environment. *Mar. Pollut. Bull.* **2011**, *62*, 1596–1605. [CrossRef] [PubMed]
8. Jiang, J.-Q. Occurrence of microplastics and its pollution in the environment: A review. *Sustain. Prod. Consum.* **2018**, *13*, 16–23. [CrossRef]
9. Lestari, P.; Trihadiningrum, Y. The impact of improper solid waste management to plastic pollution in Indonesian coast and marine environment. *Mar. Pollut. Bull.* **2019**, *149*, 110505. [CrossRef]
10. Wang, F.; Wong, C.S.; Chen, D.; Lu, X.; Wang, F.; Zeng, E.Y. Interaction of toxic chemicals with microplastics: A critical review. *Water Res.* **2018**, *139*, 208–219. [CrossRef]
11. Klein, S.; Worch, E.; Knepper, T.P. Occurrence and Spatial Distribution of Microplastics in River Shore Sediments of the Rhine-Main Area in Germany. *Environ. Sci. Technol.* **2015**, *49*, 6070–6076. [CrossRef] [PubMed]
12. Mato, Y.; Isobe, T.; Takada, H.; Kanehiro, H.; Ohtake, C.; Kaminuma, T. Plastic Resin Pellets as a Transport Medium for Toxic Chemicals in the Marine Environment. *Environ. Sci. Technol.* **2001**, *35*, 318–324. [CrossRef] [PubMed]
13. Rochman, C.M. The Complex Mixture, Fate and Toxicity of Chemicals Associated with Plastic Debris in the Marine Environment. In *Marine Anthropogenic Litter*; Bergman, M., Gutow, L., Klages, M., Eds.; Springer International Publishing: Cham, Switzerland, 2015; pp. 117–140.
14. Dris, R.; Imhof, H.; Sanchez, W.; Gasperi, J.; Galgani, F.; Tassin, B.; Laforsch, C. Beyond the ocean: Contamination of freshwater ecosystems with (micro-)plastic particles. *Environ. Chem.* **2015**, *12*, 539–550. [CrossRef]
15. Cole, M.; Lindeque, P.; Halsband, C.; Galloway, T.S. Microplastics as contaminants in the marine environment: A review. *Mar. Pollut. Bull.* **2011**, *62*, 2588–2597. [CrossRef] [PubMed]
16. Rochman, C.M.; Tahir, A.; Williams, S.L.; Baxa, D.V.; Lam, R.; Miller, J.T.; Teh, F.-C.; Werorilangi, S.; Teh, S.J. Anthropogenic debris in seafood: Plastic debris and fibers from textiles in fish and bivalves sold for human consumption. *Sci. Rep.* **2015**, *5*, 1–10. [CrossRef]
17. Jabeen, K.; Su, L.; Li, J.; Yang, D.; Tong, C.; Mu, J.; Shi, H. Microplastics and mesoplastics in fish from coastal and fresh waters of China. *Environ. Pollut.* **2017**, *221*, 141–149. [CrossRef] [PubMed]
18. McNeish, R.E.; Kim, L.H.; Barrett, H.A.; Mason, S.A.; Kelly, J.J.; Hoellein, T.J. Microplastic in riverine fish is connected to species traits. *Sci. Rep.* **2018**, *8*, 1–12. [CrossRef] [PubMed]
19. Li, J.; Yang, D.; Li, L.; Jabeen, K.; Shi, H. Microplastics in commercial bivalves from China. *Environ. Pollut.* **2015**, *207*, 190–195. [CrossRef] [PubMed]

20. Cole, M.; Lindeque, P.; Fileman, E.; Halsband, C.; Goodhead, R.; Moger, J.; Galloway, T.S. Microplastic Ingestion by Zooplankton. *Environ. Sci. Technol.* **2013**, *47*, 6646–6655. [CrossRef]

21. Barboza, L.G.A.; Vethaak, A.D.; Lavorante, B.R.B.O.; Lundebye, A.; Guilhermino, L. Marine microplastic debris: An emerging issue for food security, food safety and human health. *Mar. Pollut. Bull.* **2018**, *133*, 336–348. [CrossRef]

22. Santillo, D.; Miller, K.; Johnston, P. Microplastics as contaminants in commercially important seafood species. *Integr. Environ. Assess. Manag.* **2017**, *13*, 516–521. [CrossRef]

23. Cordova, M.R.; Hadi, T.A.; Prayudha, B. Occurrence and abundance of microplastics in coral reef sediment: A case study in Sekotong, Lombok-Indonesia. *Adv. Environ. Sci. Bioflux* **2018**, *10*, 23–29.

24. Manalu, A.A.; Hariyadi, S.; Wardiatno, Y. Microplastics abundance in coastal sediments of Jakarta Bay, Indonesia. *AACL Bioflux* **2017**, *10*, 1164–1173.

25. Cordova, M.R.; Purwiyanto, A.I.S.; Suteja, Y. Abundance and characteristics of microplastics in the northern coastal waters of Surabaya, Indonesia. *Mar. Pollut. Bull.* **2019**, *142*, 183–188. [CrossRef]

26. Syakti, A.D.; Bouhroum, R.; Hidayati, N.V.; Koenawan, C.J.; Boulkamh, A.; Sulistyo, I.; Lebarillier, S.; Akhlus, S.; Doumenq, P.; Wong-Wah-Chung, P. Beach macro-litter monitoring and floating microplastic in a coastal area of Indonesia. *Mar. Pollut. Bull.* **2017**, *122*, 217–225. [CrossRef] [PubMed]

27. Tahir, A.; Rochman, C.M. Plastic Particles in Silverside (*Stolephorus heterolobus*) Collected at Paotere Fish Market, Makassar. *Int. J. Agric. Syst.* **2014**, *2*, 163–168.

28. Hastuti, A.R.; Lumbanbatu, D.T.F.; Wardiatno, Y. The presence of microplastics in the digestive tract of commercial fishes off pantai Indah Kapuk coast, Jakarta, Indonesia. *Biodiversitas* **2019**, *20*, 1233–1242. [CrossRef]

29. Lestari, P.; Trihadiningrum, Y.; Wijaya, B.A.; Yunus, K.A.; Firdaus, M. Distribution of microplastics in Surabaya River, Indonesia. *Sci. Total. Environ.* **2020**, *726*, 138560. [CrossRef]

30. Alam, F.C.; Sembiring, E.; Muntalif, B.S.; Suendo, V. Microplastic distribution in surface water and sediment river around slum and industrial area (case study: Ciwalengke River, Majalaya district, Indonesia). *Chemosphere* **2019**, *224*, 637–645. [CrossRef] [PubMed]

31. Alam, F.C.; Rachmawati, M. Development of Microplastic Research in Indonesia. *J. Presipitasi* **2020**, *17*, 344–352. (In Indonesian) [CrossRef]

32. Purba, N.P.; Handyman, D.I.W.; Pribadi, T.D.; Syakti, A.D.; Pranowo, W.S.; Harvey, A.; Ihsan, Y.N. Marine debris in Indonesia: A review of research and status. *Mar. Pollut. Bull.* **2019**, *146*, 134–144. [CrossRef]

33. Tahir, A.; Soeprapto, D.A.; Sari, K.; Wicaksono, E.A.; Werorilangi, S. Microplastic assessment in Seagrass ecosystem at Kodingareng Lompo Island of Makassar City. *IOP Conf. Ser. Earth Environ. Sci.* **2020**, *564*, 012032. [CrossRef]

34. Shuker, L.H.; Cadman, C.A. *Indonesia Marine Debris Hotspot Rapid Assessment: Synthesis Report*; World Bank: Washington, DC, USA, 2018; pp. 1–42.

35. Afdal, M.; Werorilangi, S.; Faizal, A.; Tahir, A. Studies on Microplastics Morphology Characteristics in the Coastal Water of Makassar City, South Sulawesi, Indonesia. *Int. J. Environ. Agric. Biotechnol.* **2019**, *4*, 1028–1033. [CrossRef]

36. Faizal, A.; Werorilangi, S.; Samad, W. Spectral characteristics of plastic debris in the beach: Case study of Makassar coastal area. *Indones. J. Geogr.* **2020**, *52*, 8–14. [CrossRef]

37. Wicaksono, E.A.; Tahir, A.; Werorilangi, S. Preliminary study on microplastic pollution in surface-water at Tallo and Jeneberang Estuary, Makassar, Indonesia. *AACL Bioflux* **2020**, *13*, 902–909.

38. Tahir, A.; Samawi, M.F.; Sari, K.; Hidayat, R.; Nimzet, R.; Wicaksono, E.A.; Asrul, L.; Werorilangi, S. Studies on microplastic contamination in seagrass beds at Spermonde Archipelago of Makassar Strait, Indonesia. *J. Phys. Conf. Ser.* **2019**, *1341*, 022008. [CrossRef]

39. Digital Topographic Map of Indonesia. Available online: http://tanahair.indonesia.go.id/portal-web/ (accessed on 24 May 2021).

40. Tahir, A.; Taba, P.; Samawi, M.F.; Werorilangi, S. Microplastics in water, sediment and salts from traditional salt producing ponds. *Glob. J. Environ. Sci. Manag.* **2019**, *5*, 431–440.

41. Coppock, R.L.; Cole, M.; Lindeque, P.K.; Queirós, A.M.; Galloway, T.S. A small-scale, portable method for extracting microplastics from marine sediments. *Environ. Pollut.* **2017**, *230*, 829–837. [CrossRef]

42. Frias, J.; Pagter, E.; Nash, R.; O'Connor, I.; Carretero, O.; Filgueiras, A.; Viñas, L.; Gago, J.; Antunes, J.; Bessa, F.; et al. *Standardised Protocol for Monitoring Microplastics in Sediments*; BASEMAN Project; JPI-Oceans: Brussels, Belgium, 2018. [CrossRef]

43. GESAMP. *Guidelines for the Monitoring and Assessment of Plastic Litter in the Ocean*; Kershaw, P.J., Turra, A., Galgani, F., Eds.; Rep. Stud. GESAMP No. 99; IMO/FAO/UNESCO-IOC/UNIDO/WMO/IAEA/UN/UNEP/UNDP/ISA Joint Group of Experts on the Scientific Aspects of Marine Environmental Protection: London, UK, 2019; 130p.

44. Hanke, G.; Galgani, F.; Werner, S.; Oosterbaan, L.; Nilsson, P.; Fleet, D.; Kinsey, S.; Thompson, R.; Palatinus, A.; Van Franeker, J.; et al. *Guidance on Monitoring of Marine Litter in European Seas*; EUR 26113; JRC83985; Publications Office of the European Union: Luxembourg, 2013.

45. Moore, C.J. Synthetic polymers in the marine environment: A rapidly increasing, long-term threat. *Environ. Res.* **2008**, *108*, 131–139. [CrossRef]

46. Dris, R.; Gasperi, J.; Rocher, V.; Tassin, B. Synthetic and non-synthetic anthropogenic fibers in a river under the impact of Paris Megacity: Sampling methodological aspects and flux estimations. *Sci. Total Environ.* **2018**, *618*, 157–164. [CrossRef]

47. Hitchcock, J.N.; Mitrovic, S.M. Microplastic pollution in estuaries across a gradient of human impact. *Environ. Pollut.* **2019**, *247*, 457–466. [CrossRef]
48. Zhou, Q.; Tu, C.; Fu, C.; Li, Y.; Zhang, H.; Xiong, K.; Zhao, X.; Li, L.; Waniek, J.J.; Luo, Y. Characteristics and distribution of microplastics in the coastal mangrove sediments of China. *Sci. Total Environ.* **2020**, *703*, 134807. [CrossRef]
49. Wu, P.; Tang, Y.; Dang, M.; Wang, S.; Jin, H.; Liu, Y.; Jing, H.; Zheng, C.; Yi, S.; Cai, Z. Spatial-temporal distribution of microplastics in surface water and sediments of Maozhou River within Guangdong-Hong Kong-Macao Greater Bay Area. *Sci. Total Environ.* **2020**, *717*, 135187. [CrossRef]
50. Han, M.; Niu, X.; Tang, M.; Zhang, B.; Wang, G.; Yue, W.; Kong, X.; Zhu, J. Distribution of microplastics in surface water of the lower Yellow River near estuary. *Sci. Total Environ.* **2020**, *707*, 135601. [CrossRef] [PubMed]
51. 51. Sutrisno. Study on Tallo River Potency as River Navigation. Master Thesis, Universitas Hasanuddin, Makassar, Indonesia, 2015. (In Indonesian).
52. Firdaus, M.; Trihadiningrum, Y.; Lestari, P. Microplastic pollution in the sediment of Jagir Estuary, Surabaya City, Indonesia. *Mar. Pollut. Bull.* **2020**, *150*, 110790. [CrossRef] [PubMed]
53. BPS-Statistic Indonesia. *Statistical Yearbook of Indonesia 2019*; Badan Pusat Statistik Indonesia: Jakarta, Indonesia, 2019; 738p.
54. Napper, I.E.; Thompson, R.C. Release of synthetic microplastic plastic fibres from domestic washing machines: Effects of fabric type and washing conditions. *Mar. Pollut. Bull.* **2016**, *112*, 39–45. [CrossRef] [PubMed]
55. Boucher, J.; Friot, D. *Primary Microplastics in the Oceans: A Global Evaluation of Sources*; IUCN: Gland, Switzerland, 2017; 43p.
56. Nizzetto, L.; Bussi, G.; Futter, M.N.; Butterfield, D.; Whitehead, P.G. A theoretical assessment of microplastic transport in river catchments and their retention by soils and river sediments. *Environ. Sci. Process. Impacts* **2016**, *18*, 1050–1059. [CrossRef]
57. Wicaksono, E.A.; Werorilangi, S.; Tahir, A. The influence of weirs on microplastic fate in the riverine environment (case study: Jeneberang River, Makassar City, Indonesia). *IOP Conf. Ser. Earth Environ. Sci.* **2021**, *763*, 1–7. [CrossRef]
58. Andrady, A.L. The plastic in microplastics: A review. *Mar. Pollut. Bull.* **2017**, *119*, 12–22. [CrossRef]
59. Ory, N.C.; Sobral, P.; Ferreira, J.L.; Thiel, M. Amberstripe scad Decapterus muroadsi (Carangidae) fish ingest blue microplastics resembling their copepod prey along the coast of Rapa Nui (Easter Island) in the South Pacific subtropical gyre. *Sci. Total Environ.* **2017**, *586*, 430–437. [CrossRef]
60. Crawford, C.B.; Quinn, B. *Microplastic Pollutants*; Elsevier: Amsterdam, The Netherlands, 2017; 315p.
61. da Costa, J.P.; Duarte, A.C.; Rocha-Santos, T.A.P. Microplastics—Occurrence, Fate and Behaviour in the Environment. In *Characterization and Analysis of Microplastics*; Rocha-Santos, T.A.P., Duarte, A.C., Eds.; Elsevier: Amsterdam, The Netherlands, 2017; pp. 1–24.
62. Mani, T.; Hauk, A.; Walter, U.; Burkhardt-Holm, P. Microplastics profile along the Rhine River. *Sci. Rep.* **2016**, *5*, 1–7. [CrossRef]
63. Triebskorn, R.; Braunbeck, T.; Grummt, T.; Hanslik, L.; Huppertsberg, S.; Jekel, M.; Knepper, T.P.; Krais, S.; Müller, Y.K.; Pittroff, M.; et al. Relevance of nano- and microplastics for freshwater ecosystems: A critical review. *TrAC-Trends Anal. Chem.* **2019**, *110*, 375–392. [CrossRef]
64. Geyer, R.; Jambeck, J.R.; Law, K.L. Production, use, and fate of all plastics ever made. *Sci. Adv.* **2017**, *3*, e1700782. [CrossRef]
65. Kukkola, A.; Krause, S.; Lynch, I.; Sambrook Smith, G.H.; Nel, H. Nano and microplastic interactions with freshwater biota—Current knowledge, challenges and future solutions. *Environ. Int.* **2021**, *152*, 106504. [CrossRef] [PubMed]
66. Wik, A.; Dave, G. Occurrence and effects of tire wear particles in the environment—A critical review and an initial risk assessment. *Environ. Pollut.* **2009**, *157*, 1–11. [CrossRef] [PubMed]
67. Leads, R.R.; Weinstein, J.E. Occurrence of tire wear particles and other microplastics within the tributaries of the Charleston Harbor Estuary, South Carolina, USA. *Mar. Pollut. Bull.* **2019**, *145*, 569–582. [CrossRef] [PubMed]

Article

Perfluoroalkylated Substances (PFAS) Associated with Microplastics in a Lake Environment

John W. Scott [1,*], Kathryn G. Gunderson [1], Lee A. Green [1], Richard R. Rediske [2] and Alan D. Steinman [2]

[1] Illinois Sustainable Technology Center, Prairie Research Institute, University of Illinois, Champaign, IL 61820, USA; kggunde2@illinois.edu (K.G.G.); leegreen@illinois.edu (L.A.G.)

[2] Annis Water Resources Institute, Grand Valley State University, Muskegon, MI 49441, USA; redisker@gvsu.edu (R.R.R.); steinmaa@gvsu.edu (A.D.S.)

* Correspondence: zhewang@illinois.edu; Tel.: +1-217-333-8407

Abstract: The presence of both microplastics and per- and polyfluoroalkyl substances (PFAS) is ubiquitous in the environment. The ecological impacts associated with their presence are still poorly understood, however, these contaminants are extremely persistent. Although plastic in the environment can concentrate pollutants, factors such as the type of plastic and duration of environmental exposure as it relates to the degree of adsorption have received far less attention. To address these knowledge gaps, experiments were carried out that examined the interactions of PFAS and microplastics in the field and in a controlled environment. For field experiments, we measured the abundance of PFAS on different polymer types of microplastics that were deployed in a lake for 1 month and 3 months. Based on these results, a controlled experiment was conducted to assess the adsorption properties of microplastics in the absence of associated inorganic and organic matter. The adsorption of PFAS was much greater on the field-incubated plastic than what was observed in the laboratory with plastic and water alone, 24 to 259 times versus one-seventh to one-fourth times background levels. These results suggest that adsorption of PFAS by microplastics is greatly enhanced by the presence of inorganic and/or organic matter associated with these materials in the environment, and could present an environmental hazard for aquatic biota.

Keywords: per- and polyfluoroalkyl substances; microplastics; Muskegon Lake

Citation: Scott, J.W.; Gunderson, K.G.; Green, L.A.; Rediske, R.R.; Steinman, A.D. Perfluoroalkylated Substances (PFAS) Associated with Microplastics in a Lake Environment. *Toxics* **2021**, *9*, 106. https://doi.org/10.3390/toxics9050106

Academic Editors: Tania Martellini, Costanza Scopetani and Diana Campos

Received: 30 March 2021
Accepted: 6 May 2021
Published: 11 May 2021

Publisher's Note: MDPI stays neutral with regard to jurisdictional claims in published maps and institutional affiliations.

1. Introduction

Per- and polyfluoroalkyl substances (PFAS) have received considerable attention from the scientific community and regulatory agencies. By nature of design, these compounds are thermally stable, oxidatively recalcitrant, and resist microbial degradation [1–3]. Bioaccumulation of legacy PFAS that was released into the environment has been observed in organisms at various trophic levels, such as phytoplankton, fish, porpoise, and polar bears [4–7]. Large knowledge gaps exist regarding bioavailability, bioaccumulation, and biotransformation of legacy and residual PFAS, particularly in lower-trophic level freshwater organisms, which may influence PFAS exposure to humans via fish-based consumption.

Plastic in the environment is also persistent, and rather than biodegrade, macroplastics ($\geq$5 mm) erode into microplastics (<5 mm) via physical and chemical processes and exposure to ultraviolet light [8]. Primary microplastics can also enter the environment through the loss of pre-production plastic pellets during manufacturing or transport, and more recently, wastewater effluent has been identified as a source of microbeads originating from cosmetic products and microfibers shed from clothing and textile laundering [9,10].

Certain persistent organic pollutants (POPs) are known for their carcinogenic, endocrine-disrupting, and reproductive effects [11]. In addition, POPs adsorb to plastics at concentrations greater than the surrounding environment and become biologically available for absorption after ingestion [12]. The bioaccumulation of plastic-borne POPs is prevalent in sea bird populations, for example, where the mass of plastic ingested by short-tailed shearwaters is correlated with polychlorinated biphenyl (PCB) body burden [13]. In the Great Lakes region, the bioaccumulation of polyaromatic hydrocarbons in salmonids was cited as a likely cause of thyroid deficiencies and goiter in wild herring gulls (*Larus argentatus*) and in lab rats sustained on a diet consisting of Great Lakes coho salmon (*Oncorhynchus kisutch*), suggesting that predation is a pathway for the bioaccumulation of POPs in the Great Lakes food web [14]. It is critical to identify routes of human exposure to PFAS because they have been detected in human blood and breast milk [15–17]. In addition to drinking water, diet may be a major exposure pathway for humans [17,18]. In the U.S., national fish monitoring studies suggested that fish consumption may be a source of human exposure to PFAS because these compounds have been frequently detected in fish tissues collected from the Great Lakes and urban rivers across the country [19,20]. In addition, PFAS have been found in shrimp and seafood [21,22]. In the aquatic environment, bioaccumulation from different media and organisms (i.e., water, sediment, phytoplankton, and fish) is well known as a major mechanism for PFAS transfer to the food chain [23]. PFAS is of special concern in Michigan, where some of the highest groundwater concentrations have been detected [24], and there are concerns about these plumes contaminating surface waters.

Like many of the chemicals known to sorb to plastics, PFAS have properties that can facilitate the potential of microplastics to serve as their carriers [25]. To the best of our knowledge, no previous studies have been conducted to investigate the nature and concentrations of PFAS adsorbed to microplastics in the environment. Another factor influencing the adsorption of chemicals to plastics is the role of biofilms, a consortium of algae, bacteria, and other microorganisms that can affect the fate and level of impact of adsorbed contaminants within freshwater systems [26]. Given the prevalence of PFAS and microplastics in natural waters, coupled with the extremely long persistence time of both classes of pollutants, these two groups of emerging contaminants may act synergistically in food webs to cause adverse effects in fish and wildlife, as well as humans.

Our study was designed to address this knowledge gap with experiments that examined the interactions of PFAS and microplastics in the field and in a controlled environment. For field experiments, we examined the abundance of seven common PFAS on three different polymer types of microplastics that were deployed in a lake over a time period of 1 and 3 months. Aqueous samples were also collected and analyzed at the time of deployment to serve as the background concentration of PFAS. Finally, based on the results of the field-based microplastic experiment, we conducted a controlled, lab-based experiment with the most abundant PFAS measured from the field experiment to assess the adsorption properties of microplastics in the absence of associated organic/inorganic matter and biofilm.

2. Materials and Methods

Microplastic Deployment (Field Study): Plastic materials were deployed at two sites located in Muskegon Lake, Michigan (Figure 1). The deployed materials included low-density polyethylene (LDPE), polypropylene (PP), and polyethylene terephthalate (PET), which were 2 to 4 mm in size, and incubated in separate containers (see below).

For lake deployment of the microplastics, incubation tubes were constructed and mounted to a deployment frame. Each tube contained approximately 42 g of each plastic type and each frame contained 3 polymer types with 4 replicates per frame. Therefore, a total of 12 tubes were randomly arranged on each frame. All frames were deployed on 4 June 2018. One of the sites was centrally located in mesotrophic Muskegon Lake (43.23834 N, 86.27923 W; depth = 12 m) and was placed at the water-sediment interface (Lake Bottom); this site was adjacent to the Muskegon Lake Observatory, which collects water quality data throughout the water column on a near-continuous basis (https://www.gvsu.edu/wri/buoy/, accessed on 10 May 2021). The other site chosen was near the sea wall at the more oligotrophic Lake Michigan–Muskegon Lake navigation channel (43.22769 N, 86.33911 W; depth = 2 m and 4 m). For the channel site, a frame was placed at a depth of 2 m and another at the sediment–water interface (channel water column and channel bottom, respectively). Incubation times were for 1 and 3 months and a total of 36 tubes were used. Aqueous samples were collected at the time of initial deployment and considered the background concentration of PFAS at these sites. In addition, water quality data including water temperature, conductivity, and dissolved oxygen were recorded during retrieval of the deployment racks at their respective timepoints (see Supplemental Table S1). Further details regarding sample deployment and treatment are published elsewhere [27].

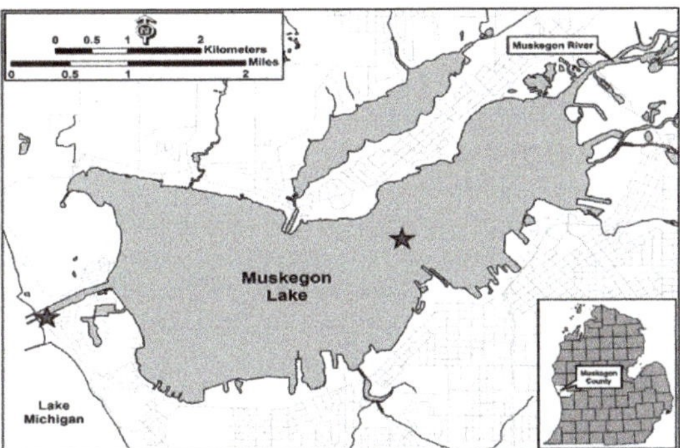

Figure 1. Locations (channel and lake) for Deployment (filled stars) of Microplastics in Muskegon Lake.

Controlled PFAS Exposure (Laboratory Study): The three most abundant PFASs from the field study (PFOA—perfluorooctanoic acid, PFHxA—perfluorohexanoic acid, and PFHpA—perfluoroheptanoic acid) were added to flasks containing 50 mL of deionized water. The exposure solution was prepared at a concentration of 5 µg/L for each PFAS. Ten grams each of fresh, non-incubated plastic type were added to the flasks. The solutions with microplastics were then placed in a laboratory incubator and shaken at 90 revolutions per minute (RPM) at room temperature for 1 month. After that time, the microplastics were collected by filtration (Whatman, Glass Microfibre (GF/F), pore size: 0.7 µm).

Sample Preparations and Analysis of PFAS: Sample preparation and analysis of PFAS was performed by US EPA Method 537 [28]. Isotopically enriched PFAS were spiked into all test materials to serve as surrogates for the native PFAS.

Pristine (non-incubated—laboratory study) and incubated (field study) microplastics were prepared by a solid-liquid extraction method utilizing a 10 g sample and methanol as an extraction solvent (3 × 20 mL). The pooled organic fractions were then concentrated to 1.0 mL before analysis.

Seven individual PFAS were targeted for field samples since they are the most abundant PFAS previously detected in the Great Lakes [29]. These PFAS compounds were perfluorobutanesulfonic acid (PFBSPFHxA, PFHpA, PFOA, perfluorononanoic acid (PFNA), perfluorodecanoic acid (PFDA), and perfluorooctanesulfonic acid (PFOS). This field study served as a "screening tool" for which PFASs were most relevant for a controlled experiment and based on these results, the laboratory study focused on PFOA, PFHxA, and PFHpA. PFAS compounds were analyzed by liquid chromatography tandem mass spectrometry (LC-MSMS) using a Waters Alliance 2695 coupled to a Quattro Micro tandem mass spectrometer (Waters Corporation, Milford, MA, USA).

Quality control parameters associated with the samples included reagent blanks, reagent blank spikes, and matrix spikes. Reagent blanks contained all the materials used for sample preparations and reagent blank spikes were similar yet contained the target PFAS. Matrix spikes were prepared by spiking a duplicate sample with PFAS.

All final PFAS results were calculated by the isotope dilution method, which utilizes the isotope surrogate and corrects the native PFAS concentrations based on their recoveries.

Reported results reflect the average of multiple sample preparation and analysis. The associated errors for these results were derived from either the relative percent difference (%RPD) or relative standard deviation (%RSD) of the multiple measurements. In situations where a target PFAS was detected in one replicate but not others, the value for the single result is reported.

Data Analysis—Field Study: Summed PFAS concentrations (when reported above minimum detection levels) were statistically analyzed separately for each deployed microplastic substrate using 2-way analysis of variance (ANOVA) to determine whether deployment site (channel water column, channel bottom, lake bottom), deployment duration (1 month, 3 months), or the interaction between site and duration had a significant effect on post-incubation PFAS concentrations. Each combination of site and duration factors had $n = 2$ tube replicates for each of the 1 month and 3 month sampling events. ANOVA assumptions of normality and equal variance were tested with Shapiro–Wilk and Brown–Forsythe tests, respectively. However, 2-way ANOVAs for each microplastic substrate violated assumptions of equal variance (i.e., Brown–Forsythe: $p > 0.05$), which were not improved by data transformation, and are presented herein using untransformed data. When 2-way ANOVAs detected significant differences, post hoc multiple comparisons were made using Holm–Sidak tests. A 1-way ANOVA was used to determine whether the plastic type (polypropylene, polyethylene, polyester) influenced final microplastic PFAS concentrations ($n = 3$ replicates per plastic type).

Data Analysis—Laboratory Study: Summed PFAS concentrations (PFHxA, PFHpA, and PFOA) were analyzed using 1-way ANOVA to determine whether plastic type (polypropylene, polyethylene, polyester) influenced final microplastic PFAS concentrations. Each microplastic type had $n = 3$ independent sample replicates. Assumptions of normality and variance were tested as described above and detected no violations and data were not transformed. Post hoc multiple comparison was completed using a Tukey test. All statistical analyses were completed using Sigma Plot (v14.0).

3. Results

Field Study: None of the seven target PFAS were detected above the detection limit for the trip blank, reagent blanks, and pristine (non-incubated) microplastics. This indicates that the sample collection, sample preparation techniques, and starting materials were free from PFAS contamination.

Unless otherwise stated, all PFAS concentrations are reported as a sum of the seven PFAS measured in the field study or the three PFAS in the laboratory study. The concentrations of PFASs measured from the field water samples were 2.8 ng/L (RPD = 16%) and 3.3 ng/L (RPD = 4.2%) in the channel and lake, respectively. PFOA, PFHpA, PFBS, and PFOS were detected in these samples, with PFOA at the greatest concentration. These results were considered the background concentration of PFAS to which the deployed microplastics were exposed.

PFAS concentrations associated with the plastics (including inorganic and organic matter associated with them) after incubation in Muskegon Lake ranged from 67 ng/kg to 730 ng/kg. These materials concentrated PFASs by factors ranging from 24 to 259 times the background aqueous concentration in the lake water within 1 to 3 months. Figure 2 presents the average PFAS by plastic type only, irrespective of location or exposure duration. The trend from lowest to highest concentrator is polypropylene < polyester < polyethylene. However, these differences were only marginally significant ($p < 0.10$) due to the high variance among plastics.

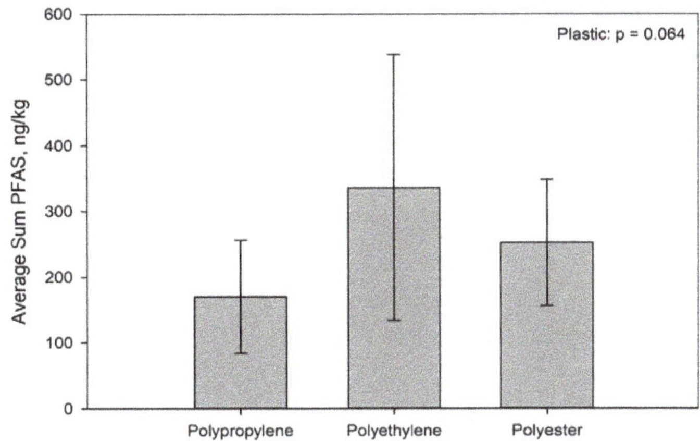

Figure 2. Average Sum of 7 PFAS (ng/kg) by Plastic Types for Materials Deployed in Muskegon Lake, MI for 1 Month and 3 Month Incubations in the Environment.

The concentrations of PFAS associated with the deployed microplastics by location and time are presented in Figures 3–5 for polyethylene, polypropylene, and polyester, respectively. On polyethylene (Figure 3), PFAS concentrations were not significantly different among sites at 1 month but were significantly different at 3 months due to concentrations on the plastics at the channel water column site exceeding those at both the channel bottom and lake bottom sites. On polypropylene (Figure 4), only time had a significant effect on PFAS concentration, with the 1 month concentrations greater than the 3 month concentrations; neither site nor the interaction term were statistically significant. Finally, on polyester (Figure 5), PFAS concentrations were not significantly affected by time or site.

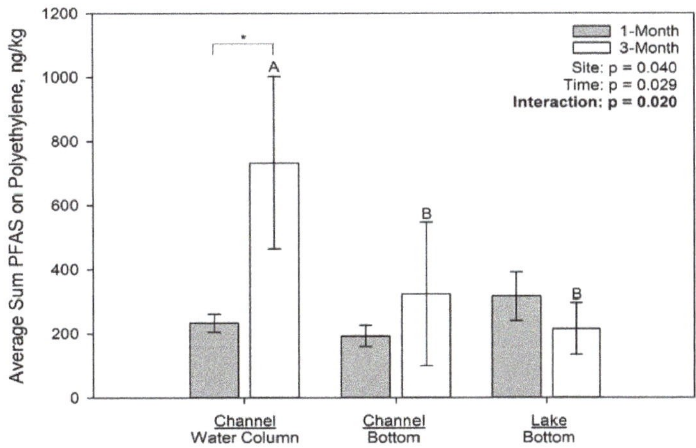

Figure 3. Average Sum of 7 PFAS (ng/kg) on Polyethylene Deployed at Different Locations in Muskegon Lake, MI. Different letters among bars indicate statistically significant differences among sites for either the 1 month or 3 month incubation period. Asterisks indicate statistically significant differences between the 1 month vs. 3 month incubation at a specific site.

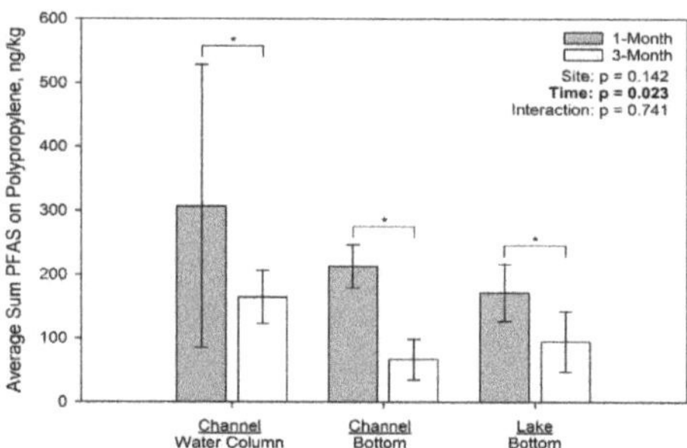

Figure 4. Average Sum of 7 PFAS (ng/kg) for Polypropylene Deployed in Muskegon Lake, MI. Asterisks indicate statistically significant differences between the 1 month vs. 3 month incubation at a specific site.

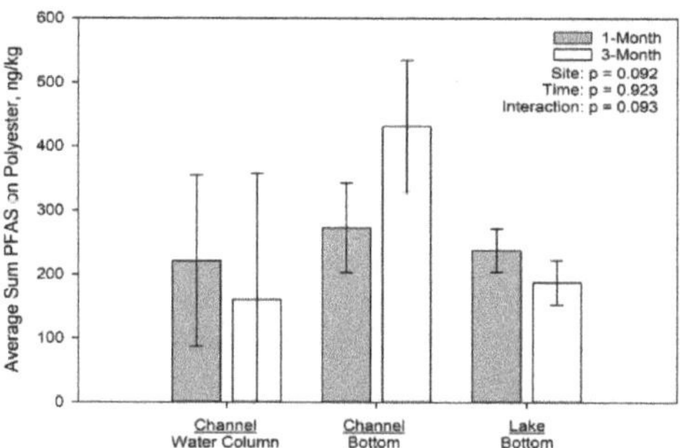

Figure 5. Average Sum of 7 PFAS (ng/kg) for Polyester Deployed in Muskegon Lake, MI.

Laboratory Study: PFHxA, PFHpA, and PFOA were the most abundant PFAS associated with the microplastics incubated at the lake sites, so these 3 were the focus of the laboratory experiments. Figure 6 presents the average PFAS concentration measured for each (non-incubated) plastic type and the average percent PFAS adsorbed for each plastic type in the absence of the associated inorganic and organic matter in relation to the total mass of PFAS spiked into the exposure solution. PFAS concentrations were significantly greater on polyester than polyethylene ($p < 0.01$), but there were no statistically significant differences between polyester and polypropylene or between polypropylene and polyethylene.

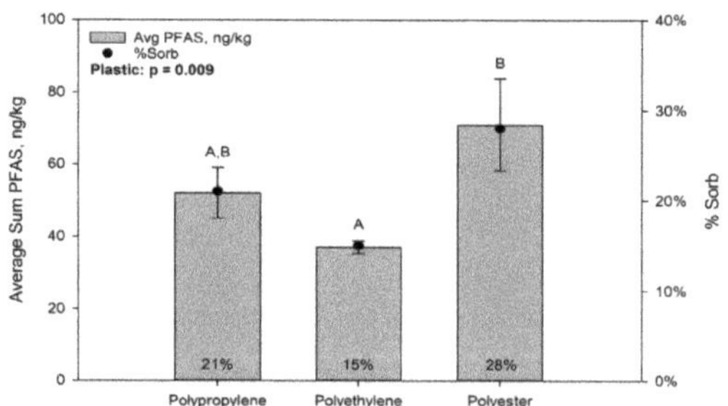

Figure 6. Average Summed 3 PFAS concentration (ng/kg) and Percent Adsorption (number above each bar) of PFAS on Plastic for Laboratory Study. Different letters among bars indicate statistically significant differences among sites for either the 1 month or 3 month incubation period.

All raw data tables are presented in the supplemental section.

4. Discussion

Environmental and health concerns over PFAS have increased dramatically in the past few years, although most of that attention has focused on groundwater and soil contamination [29]. In contrast, Remucal [30] measured PFAS concentrations in the open and nearshore Lake Michigan surface waters and found relatively low concentrations of 1.8 to 4.1 ng/L. Although these data are on the low-end of what has previously been reported for PFAS, their proximity to the shore could result in an increased ecosystem stressor [31]. Like a previous study that measured C6 to C10 perfluorocarboxylates and PFOS in Lake Michigan water samples, PFOA, PFHpA, and PFOS were the most commonly found PFAS [30]. PFHxA was not detected in the Muskegon Lake water samples. However, since it was detected on the incubated microplastics, it is likely this PFAS was present but at concentrations below the method detection limit. At the time of analysis for the background lake water samples in this study, the instrument detection limit for PFHxA was a factor of five greater than for other PFASs, such as PFHpA.

After the one month laboratory exposure to PFAS solutions, plastics adsorbed 11% to 36% of the PFAS contained in the exposure solution. A slight trend was observed with regards to the chain length and the amount adsorbed, with the longer chain (PFOA) being adsorbed more than the shorter chain (PFHx). This likely is a function of shorter chains being more water soluble and less adsorbent [32]. In a recent study of adsorption on filter membranes and centrifuge tubes, other researchers found that polypropylene tubes were able to adsorb 32% to 42% of the PFOA in solution that came in contact with this material [33]. Although the exposure time and surface areas were much different than this study, these results are similar.

All plastic types at all locations concentrated PFASs by factors ranging from 24 to 259 times the background lake water concentration. A great deal of variability was observed for PFAS concentrations for duplicate samples of the same type, same location, and same exposure duration. This degree of variability was not observed in the controlled laboratory experiments, analytical duplicate results, or in surrogate recoveries. This suggests that the PFAS distribution is very heterogeneous on the materials. The variability is likely associated with the heterogeneity of the biofilm (plastisphere) colonizing the plastic [34]. The observed variability makes definitive conclusions regarding the effect of plastic type, plastic location, and exposure duration on PFAS adsorption difficult to assess; however, polyethylene deployed in the channel water column drastically increased in PFAS from the 1 month to 3 month period, whereas polypropylene decreased from the 1 month to 3 month time period deployed at the channel bottom.

As part of this field study, adsorption of legacy persistent organic pollutants (POP), such as polycyclic aromatic hydrocarbons (PAH), PCB, and organochlorine pesticides, also were analyzed and the same plastic materials were found to concentrate POP up to 380 times background concentrations, similar in magnitude to what we measured for PFAS [27]. However, in that study, there were clear trends with regards to adsorption on material type (PE > PP > PET), location, and duration. In addition, the variability for samples obtained from the same material, duration, and location was much lower than what was observed for PFAS. The properties of legacy POP and PFAS are considerably different yet the degree of adsorption in the environment was quite similar.

The adsorption of PFAS was much greater in the field-incubated plastic than what was observed in the laboratory with plastic and water alone. Figure 7 displays images of polyethylene before and after field deployment (3-month). As shown, the deployed materials when retrieved had a great deal of organic matter and biofilm associated with them, particularly bacteria from the Burkholderiales, Rhodocyclaceae, Comamonadaceae, and Pseudomonadaceae [27]. Previous work has shown that PFASs prefer adsorption to lipids rather than being freely dissolved in water alone [35]. Furthermore, because the biofilm and organic matter accumulation on these materials is heterogenous, this is consistent with the large variability observed in the duplicate PFASs results associated with the same plastic types, locations, and durations reported in the present study. Therefore, the greater degree of PFAS adsorption observed in the field-deployed samples is most likely due to secondary adsorption of these compounds to the plastic-associated organic matter. This is consistent with the findings of Ateia et al. [36], who found that microplastics that were incubated with the natural organic matter had increased uptake of PFOA and PFOS compared to non-incubated microplastics, presumably due to an organic matter formation and/or co-sorption. The role of the biofilm, including the functional roles and adsorptive capacities of its taxonomic composition, is an area in need of additional research [34].

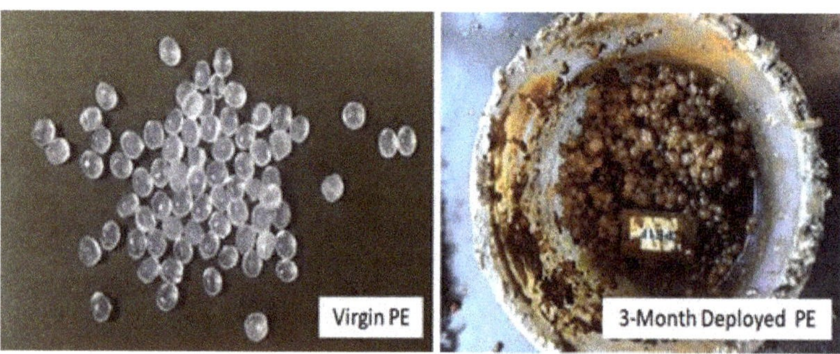

Figure 7. Low Density Polyethylene Before and After Deployment in Muskegon Lake, MI.

Although microplastics were found to significantly concentrate PFASs from background environmental concentrations, on a per mass basis they are relatively low. In the worst-case scenario found here (polyethylene/channel bottom/3 month duration), the highest concentration of microplastic-associated PFAS was 0.87 ng/g (lowest: 0.052 ng/g). Therefore, for every gram of plastic consumed there exists the potential for an organism to be exposed to an additional ~1 ng of common PFAS. However, it should be noted that several factors could influence the degree of PFAS adsorption. The exposure time of the plastic in Muskegon Lake was relatively short: 1 month and 3 month periods. Modeling studies have suggested that 50% of environmental plastics are 13 years or greater in age. Therefore, the degree of PFAS associated with actual microplastics in the environment may differ from those found in this study. Another factor that can impact the PFAS adsorption is related to the surface area of the microplastics. The size of microplastics in this study (2 mm to 5 mm) is much larger than most microplastics found in the environment. Smaller microplastics would have greater surface area per volume ratios per particle that could potentially provide more active sites of PFASs adsorption. To complicate this issue, PFAS adsorption in the environment appears to be related to secondary adsorption, and increased surface area could potentially facilitate more organic matter adsorption. In addition, over time biofilms can change in composition, which in turn can affect their adsorptive properties. The degree of influence these two parameters may have is unknown. However, it is suspected that they would increase PFASs adsorption, thereby making the results from this study biased low and conservative.

5. Conclusions

Three plastic materials (polyethylene, polypropylene, and polyethylene terephthalate) were shown to adsorb PFAS in aqueous environments. Materials deployed in the field (Muskegon Lake) demonstrated a much greater capacity for adsorption than those treated in the laboratory with PFAS and water alone. Concentrations of PFAS associated with plastic materials used in this study were relatively low and of themselves would not likely induce acute adverse effects to organisms exposed to them. However, given the short exposure times of these materials in the environment (3 months maximum) and large particle sizes (2 mm to 4 mm), these results are most likely a conservative estimate for microplastic adsorption of PFAS. These findings also demonstrate the need to consider not only the potential adverse effects of organisms exposed to microplastics alone but also the need to consider the biological and chemical materials associated with plastic materials in the environment.

Supplementary Materials: The following are available online at https://www.mdpi.com/article/10.3390/toxics9050106/s1, Table S1: Environmental data from the three MP deployment locations at the time of retrieval from Muskegon Lake, Table S2: Final Target Per- and Polyfluoroalkyl Substances (PFAS)—Channel Waters, Table S3: Final Target Per- and Polyfluoroalkyl Substances (PFAS)—Lake Waters, Table S4: Per- and Polyfluoroalkyl Substances (PFAS) Quality Control Summary—Channel and Lake Water Samples, Table S5: Final Target Per- and Polyfluoroalkyl Substances (PFAS)—Virgin Polyethylene, Table S6: Final Target Per- and Polyfluoroalkyl Substances (PFAS)—Virgin Polypropylene, Table S7: Final Target Per- and Polyfluoroalkyl Substances (PFAS)—Virgin Polyester, Table S8: Per- and Polyfluoroalkyl Substances (PFAS) Quality Control Summary—Virgin Plastics, Table S9: Final Target Per- and Polyfluoroalkyl Substances (PFAS)—Polyethylene Deployed for 1 Month—Channel Water, Table S10: Final Target Per- and Polyfluoroalkyl Substances (PFAS)—Polypropylene Deployed for 1 Month—Channel Water, Table S11: Final Target Per- and Polyfluoroalkyl Substances (PFAS)—Polyester Deployed for 1 Month—Channel Water, Table S12: Final Target Per- and Polyfluoroalkyl Substances (PFAS)—Polyethylene Deployed for 1 Month—Channel Bottom, Table S13: Final Target Per- and Polyfluoroalkyl Substances (PFAS)—Polypropylene Deployed for 1 Month—Channel Bottom, Table S14: Final Target Per- and Polyfluoroalkyl Substances (PFAS)—Polyester Deployed for 1 Month—Channel Bottom, Table S15: Final Target Per- and Polyfluoroalkyl Substances (PFAS)—Polyethylene Deployed for 1 Month—Lake Bottom, Table S16: Final Target Per- and Polyfluoroalkyl Substances (PFAS)—Polypropylene Deployed for 1 Month—Lake Bottom, Table S17: Final Target Per- and Polyfluoroalkyl Substances (PFAS)—Polyester Deployed for 1 Month—Lake Bottom, Table S18: Per- and Polyfluoroalkyl Substances (PFAS) Quality Control Summary—1-Month Deployed Plastics, Table S19: Final Target Per- and Polyfluoroalkyl Substances (PFAS)—Polyethylene Deployed for 3 Months—Channel Water, Table S20: Final Target Per- and Polyfluoroalkyl Substances (PFAS)—Polypropylene Deployed for 3 Months—Channel Water, Table S21: Final Target Per- and Polyfluoroalkyl Substances (PFAS)—Polyester Deployed for 3 Months—Channel Water, Table S22: Final Target Per- and Polyfluoroalkyl Substances (PFAS)—Polyethylene Deployed for 3 Months—Channel Bottom, Table S23: Final Target Per- and Polyfluoroalkyl Substances (PFAS)—Polypropylene Deployed for 3 Months—Channel Bottom, Table S24: Final Target Per- and Polyfluoroalkyl Substances (PFAS)—Polyester Deployed for 3 Months—Channel Bottom, Table S25: Final Target Per- and Polyfluoroalkyl Substances (PFAS)—Polyethylene Deployed for 3 Months—Lake Bottom, Table S26: Final Target Per- and Polyfluoroalkyl Substances (PFAS)—Polypropylene Deployed for 3 Months—Lake Bottom, Table S27: Final Target Per- and Polyfluoroalkyl Substances (PFAS)—Polyester Deployed for 3 Months—Lake Bottom, Table S28: Per- and Polyfluoroalkyl Substances (PFAS) Quality Control Summary—3 Month Deployed Plastics

Author Contributions: J.W.S., A.D.S., K.G.G., L.A.G. and R.R.R. contributed to conceptualization for this study and manuscript writing. Field experiment design and implementation were performed by A.D.S. and R.R.R., in addition to statistical analysis of the data. Sample preparations, analysis of PFAS, and experimental design and implementation of the controlled exposure study were performed by J.W.S., K.G.G., L.A.G. All authors have read and agreed to the published version of the manuscript.

Funding: This work was supported by the Illinois-Indiana Sea Grant (NA18OAR4170082), the Allen and Helen Hunting Research and Innovation Fund at Grand Valley State University (GVSU), and the Illinois Hazardous Waste Research Fund (HWR18-253).

Institutional Review Board Statement: Not applicable.

Informed Consent Statement: Not applicable.

Data Availability Statement: Data has been provided in supplementary section of this paper and can be made available on request from the corresponding author.

Acknowledgments: The authors would like to thank Mike Hassett, Maggie Oudsema, and Emily Kindervater for their assistance with sample deployment, sample collection, and sample preparations.

Conflicts of Interest: The authors declare no conflict of interest.

References

1. Kannan, K.; Hansen, S.; Franson, C.; Bowerman, W.; Hansen, K.; Jones, P.; Giesy, J. Perfluorochemical surfactants in the environment. *Environ. Sci. Technol.* **2001**, *35*, 3065–3070. [CrossRef] [PubMed]
2. Parsons, J.; Sáez, M.; Dolfing, J.; De Voogt, P. Biodegradation of perfluorinated compounds. *Rev. Environ. Contam. Toxicol.* **2008**, *196*, 53–71.
3. Babut, M.; Labadie, P.; Simonnet-Laprade, C.; Munoz, G.; Roger, M.; Ferrari, B.; Budzinski, H.; Sivade, E. Per-and poly-fluoroalkyl compounds in freshwater fish from the Rhône River: Influence of fish size, diet, prey contamination and biotransformation. *Sci. Total Environ.* **2017**, *605*, 38–47. [CrossRef] [PubMed]
4. Casal, P.; González-Gaya, B.; Zhang, Y.; JF Reardon, A.; Martin, J.; Jiménez, B.; Dachs, J. Accumulation of perfluoroalkylated substances in oceanic plankton. *Environ. Sci. Technol.* **2017**, *51*, 2766–2775. [CrossRef] [PubMed]
5. Fujii, Y.; Kato, Y.; Sakamoto, K.; Matsuishi, T.; Harada, K.; Koizumi, A.; Kimura, O.; Endo, T.; Haraguchi, K. Tissue-specific bioaccumulation of long-chain perfluorinated carboxylic acids and halogenated methylbipyrroles in Dall's porpoises (*Phocoenoides dalli*) and harbor porpoises (*Phocoena phocoena*) stranded in northern Japan. *Sci. Total Environ.* **2018**, *616*, 554–563. [CrossRef] [PubMed]
6. Smithwick, M.; Norstrom, R.; Mabury, S.; Solomon, K.; Evans, T.; Stirling, I.; Taylor, M.; Muir, D. Temporal trends of perfluoroalkyl contaminants in polar bears (*Ursus maritimus*) from two locations in the North American Arctic, 1972–2002. *Environ. Sci. Technol.* **2006**, *40*, 1139–1143. [CrossRef]
7. Shah, A.; Hasan, F.; Hameed, A.; Ahmed, S. Biological degradation of plastics: A comprehensive review. *Biotechnol. Adv.* **2008**, *26*, 246–265. [CrossRef]
8. Derraik, J.G. The pollution of the marine environment by plastic debris: A review. *Mar. Pollut. Bull.* **2002**, *44*, 842–852. [CrossRef]
9. Mason, S.; Garneau, D.; Sutton, R.; Yvonne Chu, Y.; Ehmann, K.; Barnes, J.; Fink, P.; Papazissimos, D.; Rogers, D. Microplastic pollution is widely detected in US municipal wastewater treatment plant effluent. *Environ. Pollut.* **2016**, *218*, 1045–1054. [CrossRef]
10. Athey, S.; Adams, J.; Erdle, L.; Jantunen, L.; Helm, P.; Finkelstein, S.; Diamond, M. The Widespread Environmental Footprint of Indigo Denim Microfibers from Blue Jeans. *Environ. Sci. Technol. Lett.* **2020**, *7*, 840–847. [CrossRef]
11. Hirai, H.; Takada, H.; Ogata, Y.; Yamashita, R.; Mizukawa, K.; Saha, M.; Charita Kwan, C. Organic micropollutants in marine plastics debris from the open ocean and remote and urban beaches. *Mar. Pollut. Bull.* **2011**, *62*, 1683–1692. [CrossRef] [PubMed]
12. Rios, L.; Moore, C.; Jones, P. Persistent organic pollutants carried by synthetic polymers in the ocean environment. *Mar. Pollut. Bull.* **2007**, *54*, 1230–1237. [CrossRef]
13. Yamashita, R.; Takada, H.; Fukuwaka, M.; Watanuki, Y. Physical and chemical effects of ingested plastic debris on short-tailed shearwaters, *Puffinus tenuirostris*, in the North Pacific Ocean. *Mar. Pollut. Bull.* **2011**, *62*, 2845–2849. [CrossRef] [PubMed]
14. Rolland, R. A review of chemically-induced alterations in thyroid and vitamin A status from field studies of wildlife and fish. *J. Wildl. Dis.* **2000**, *36*, 615–635. [CrossRef]
15. Cariou, R.; Veyrand, B.; Yamada, A.; Berrebi, A.; Zalko, D.; Durand, S.; Pollono, C. Perfluoroalkyl acid (PFAA) levels and profiles in breast milk, maternal and cord serum of French women and their newborns. *Environ. Int.* **2015**, *84*, 71–81. [CrossRef] [PubMed]
16. Glynn, A.; Berger, U.; Bignert, A.; Ullah, S.; Aune, M.; Lignell, S.; Darnerud, P. Perfluorinated alkyl acids in blood serum from primiparous women in Sweden: Serial sampling during pregnancy and nursing, and temporal trends 1996–2010. *Environ. Sci. Technol.* **2012**, *46*, 9071–9079. [CrossRef]
17. Kärrman, A.; Lindström, G. Trends, analytical methods and precision in the determination of perfluoroalkyl acids in human milk. *TrAC Trends Anal. Chem.* **2013**, *46*, 118–128. [CrossRef]
18. Ahrens, L.; Gashaw, H.; Sjöholm, M.; Gebrehiwot, S.; Getahun, A.; Derbe, E.; Bishop, K.; Åkerblom, S. Poly- and perfluoroalkylated substances (PFASs) in water, sediment and fish muscle tissue from Lake Tana, Ethiopia and implications for human exposure. *Chemosphere* **2016**, *165*, 352–357. [CrossRef] [PubMed]
19. De Silva, A.; Spencer, C.; Scott, B.; Backus, S.; Muir, D. Detection of a cyclic perfluorinated acid, perfluoroethylcyclohexane sulfonate, in the Great Lakes of North America. *Environ. Sci. Technol.* **2011**, *45*, 8060–8066. [CrossRef]
20. Stahl, L.; Snyder, B.; Olsen, A.; Kincaid, T.; Wathen, J.; McCarty, H. Perfluorinated compounds in fish from US urban rivers and the Great Lakes. *Sci. Total. Environ.* **2014**, *499*, 185–195. [CrossRef]
21. Kannan, K.; Corsolini, S.; Falandysz, J.; Oehme, G.; Focardi, S.; Giesy, J. Perfluorooctanesulfonate and related fluorinated hydrocarbons in marine mammals, fishes, and birds from coasts of the Baltic and the Mediterranean Seas. *Environ. Sci. Technol.* **2002**, *36*, 3210–3216. [CrossRef] [PubMed]
22. Loi, E.; Yeung, L.; Taniyasu, S.; Lam, P.; Kannan, K.; Yamashita, N. Trophic magnification of poly- and perfluorinated compounds in a subtropical food web. *Environ. Sci. Technol.* **2011**, *45*, 5506–5513. [CrossRef]
23. Kannan, K.; Tao, L.; Sinclair, E.; Pastva, S.; Jude, D.; Giesy, J. Perfluorinated compounds in aquatic organisms at various trophic levels in a Great Lakes food chain. *Arch. Environ. Contam. Toxicol.* **2005**, *48*, 559–566. [CrossRef] [PubMed]
24. Talpos, S. They persisted. *Science* **2019**, *364*, 622–626. [CrossRef] [PubMed]
25. Buck, R.; Franklin, J.; Berger, U.; Conder, J.; Cousins, I.; De Voogt, P.; Jensen, A.; Kannan, K.; Mabury, S.; van Leeuwen, S. Perfluoroalkyl and polyfluoroalkyl substances in the environment: Terminology, classification, and origins. *Integr. Environ. Assess. Manag.* **2011**, *7*, 513–541. [CrossRef]

26. Lagarde, F.; Olivier, O.; Zanella, M.; Daniel, P.; Hiard, S.; Aurore Caruso, A. Microplastic interactions with freshwater microalgae: Hetero-aggregation and changes in plastic density appear strongly dependent on polymer type. *Environ. Pollut.* **2016**, *215*, 331–339. [CrossRef] [PubMed]
27. Steinman, A.; Scott, J.; Green, L.; Partridge, C.; Oudsema, M.; Hassett, M.; Kindervater, E.; Rediske, R. Persistent organic pollutants, metals, and the bacterial community composition associated with microplastics in Muskegon Lake (MI). *J. Great Lakes Res.* **2020**, *46*, 1444–1458. [CrossRef]
28. Shoemaker, J. Determination of selected perfluorinated alkyl acids in drinking water by solid phase extraction and liquid chromatography/tandem mass spectrometry (LC/MS/MS). In *United States Environmental Protection Agency Method 537*; U.S. Environmental Protection Agency: Washington, DC, USA, 2009.
29. Mahinroosta, R.; Senevirathna, L. A review of the emerging treatment technologies for PFAS contaminated soils. *J. Environ. Manag.* **2020**, *255*, 109896. [CrossRef] [PubMed]
30. Remucal, C. Spatial and temporal variability of perfluoroalkyl substances in the Laurentian Great Lakes. *Environ. Sci. Process. Impacts* **2019**, *21*, 1816–1834. [CrossRef] [PubMed]
31. Allan, J.D.; McIntyre, P.B.; Smith, S.D.; Halpern, B.S.; Boyer, G.L.; Buchsbaum, A.; Burton, G.A.; Campbell, L.M.; Chadderton, W.L.; Ciborowski, J.J.; et al. Joint analysis of stressors and ecosystem services to enhance restoration effectiveness. *Proc. Natl. Acad. Sci. USA* **2013**, *110*, 372–377. [CrossRef]
32. Pereira, H.C.; Ullberg, M.; Berggren Kleja, D.; Gustafsson, J.; Ahrens, L. Sorption of perfluoroalkyl substances (PFASs) to an organic soil horizon—Effect of cation composition and pH. *Chemosphere* **2018**, *207*, 183–191. [CrossRef]
33. Lath, S.; Knight, E.; Navarro, D.; Kookana, R.; McLaughlin, M. Sorption of PFOA onto different laboratory materials: Filter membranes and centrifuge tubes. *Chemosphere* **2019**, *222*, 671–678. [CrossRef] [PubMed]
34. Amaral-Zettler, L.; Zettler, E.; Mincer, T. Ecology of the plastisphere. *Nat. Rev. Microbiol.* **2020**, *18*, 139–151. [CrossRef] [PubMed]
35. Higgins, C.; Luthy, R. Modeling sorption of anionic surfactants onto sediment materials: An a priori approach for perfluoroalkyl surfactants and linear alkylbenzene sulfonates. *Environ. Sci. Technol.* **2007**, *41*, 3254–3261. [CrossRef] [PubMed]
36. Ateia, M.; Zheng, T.; Calace, S.; Tharayil, N.; Pilla, S.; Karanfil, T. Sorption behavior of real microplastics (MPs): Insights for organic micropollutants adsorption on a large set of well-characterized MPs. *Sci. Total Environ.* **2020**, *720*, 137634. [CrossRef] [PubMed]

Article

Interaction between Styrofoam and Microalgae *Spirulina platensis* in Brackish Water System

Hadiyanto Hadiyanto [1,2,*], Amnan Haris [1], Fuad Muhammad [1,3], Norma Afiati [4] and Adian Khoironi [5]

[1] Departments of Environmental Science, School of Postgraduate Studies, Diponegoro University, Semarang 50275, Indonesia; amnan.haris43@gmail.com (A.H.); fuad.muh@gmail.com (F.M.)
[2] Center of Biomass and Renewable Energy (CBIORE), Chemical Engineering Department, Faculty of Engineering, Diponegoro University, Semarang 50271, Indonesia
[3] Biology Department, Faculty of Sciences and Mathematic, Diponegoro University, Semarang 50271, Indonesia
[4] Fisheries Science, Faculty of Fisheries and Marine Science, Diponegoro University, Semarang 50271, Indonesia; normaafiati.na@gmail.com
[5] Department of Environmental Health, Faculty of Public Health, Dian Nuswantoro University, Semarang 50131, Indonesia; batik.ayacantik@gmail.com
* Correspondence: hadiyanto@live.undip.ac.id

Abstract: Styrofoam is a thermoplastic with special characteristics; it is an efficient insulator, is extremely lightweight, absorbs trauma, is bacteria resistant, and is an ideal packaging material, compared to other thermoplastics. The aim of this study was to analyze the interaction between Styrofoam and *S. platensis*. The study examined the growth of *S. platensis* under Styrofoam stress, changes in Styrofoam functional groups, and their interactions. The research method was culture carried out in brackish water (12 mg/L salinity) for 30 days. *S. platensis* yields were tested by FTIR and SEM-EDX and Styrofoam samples by FTIR. The results showed the highest growth rate of *S. platensis* in cultures treated with 150 mg Styrofoam that is 0.0401 day^{-1}. FTIR analysis shows that there has been a change in the functional group on Styrofoam. At a wavelength of 3400–3200 cm^{-1} corresponds to the alcohol group and there was an open cyclic chain shown by the appearance of a wavelength at 1680–1600 cm^{-1} assignment to alkene. SEM-EDX test results show that Styrofoam can be a resource of nutrition, especially carbon for *S. platensis* to photosynthesize. Increased carbon content of 24.56% occurred in culture, meanwhile, Styrofoam is able to damage *S. platensis* cells.

Keywords: microplastic pollutant; polystyrene; biodegradation; microalgae

Citation: Hadiyanto, H.; Haris, A.; Muhammad, F.; Afiati, N.; Khoironi, A. Interaction between Styrofoam and Microalgae *Spirulina platensis* in Brackish Water System. *Toxics* **2021**, *9*, 43. https://doi.org/10.3390/toxics9030043

Academic Editors: Costanza Scopetani, Tania Martellini and Diana Campos

Received: 15 January 2021
Accepted: 23 February 2021
Published: 26 February 2021

Publisher's Note: MDPI stays neutral with regard to jurisdictional claims in published maps and institutional affiliations.

1. Introduction

The increasing human population causes an increase in the amount of plastic waste. Plastic pollution has become a major issue in Sustainable Development Goals (SDGs) and is stated in point number 12, under the header "Responsible Consumption and Production". Plastics are a material that degrade very slowly and may stay in the environment for a long period [1]. Plastics are available in environment in a wide range of size and forms with different chemical composition, density and color [2,3]. Plastics with microscopic sizes are called microplastics and have a diameter between 1 μm to 5 mm [4,5]. Furthermore, European Chemical Agency (ECHA) [6] defined microplastics as a solid polymer material and their additives or other substances, most of which have particle dimensions of 1 nm to 5 mm, and for fiber form, the size is mostly in the length of 3 nm to 15 mm with length to diameter ratio greater than 3. Auta et al. [7] and Frias and Nash [8] categorized microplastics in aquatic environment into two types: primary microplastics and secondary microplastics. In the first type, they include plastic based products for daily domestic and industrial usages, i.e., personal care products, facial scrubs, insect repellents [3,4,9] as well as products from the ship-breaking industry and air-blasting technology [1,4]. The second type includes smaller fragments of plastic from breaking of larger plastic items in aquatic systems through biological degradation, photo-degradation, chemical deposition, and physical fragmentation [1,3,4,10]. The common microplastics found in aquatic environment are polypropylene, polyethylene, polystyrene, polyvinylchloride, and polyethylene terephthalate [10]. Both types of microplastics present in aquatic environments are reaching certain concentrations and may have effects on aquatic organism including microalgae.

Polystyrene or Styrofoam is a type of plastic with light properties, heat resistance, and low production costs. Until now, Styrofoam is sold freely in shops, stalls, and even supermarkets. Styrofoam is widely used as a food and beverage container. After use, this Styrofoam container would be discarded, though it is still in good condition and can be reused. Most of the consumers lack the knowledge that Styrofoam needs a long time to be completely degraded. Styrofoam can be recycled. However, the high cost and complicated process make producers prefer to produce new Styrofoam, rather than recycle it [11]. Styrofoam is light because 95% of it is air, making it unsinkable [12]. Styrofoam waste is easily caught in dams and aquatic plants. The nontransparent color of Styrofoam can reduce the amount of sunlight entering the water, which makes algal photosynthesis less than optimal [13]. Environmental factors such as weather changes and water microorganism cause plastic to degrade into microplastics [14]. Microplastics are plastic particles < 5 mm [15] and through the degradation process, the polymer chains in plastics turn into monomers. Frequently, new chemical bonds will also be formed as a byproduct of this process [16]. Microplastics can be found in all parts of the aquatic system [17]. Due to their very small size, microplastics can be ingested by aquatic biota and cause disease [18]. Microplastics also spread through the food chain [19].

Styrofoam consists of long hydrocarbon chains, providing an opportunity for microalgae to use the chemical content in Styrofoam as nutrients. The carbon content in Styrofoam can spur the growth of microalgae. According to Li et al., [14] though polystyrene could inhibit *C. reindhartii* growth, they are still able to adapt because they obtain organic carbon sources from polystyrene and use it for growth. However, there are additive substances in plastics such as Bisphenol-A (BPA), phthalates, trace elements, and refractory substances, which make plastic durable and dangerous, especially for microalgae. One of them is *S. platensis*, which is often used in the food, cosmetic, and medicinal industries. These components of Styrofoam damage *S. platensis* cells, as a result of which photosynthetic activity is decreased and cell growth is inhibited [9].

Microalgae is a photosynthetic microorganism that utilizes carbon source and sunlight for the photosynthesis process lead to biomass production. Microalgae biomass can be extracted for value added products mostly containing protein, lipids, and carbohydrate. Because of their importance, the potential effect of microplastic on their growth must be studied. Microalgae cells of *Chlorella* sp. and *Scenedesmus* sp, are able to absorb nanoplastic beads (0.02 µm) and resulting inhibition of photosynthesis and induction of oxidative stress [20]. Moreover, Khoironi et al. [10] showed that there was an interaction between *Spirulina* sp. cells with microplastics. Microplastic can be absorbed by the *Spirulina* sp. cell and it utilizes them as a source of carbon for photosynthesis. Marquez et al. [21] stated that *S. platensis* is capable of growth on glucose heterotrophically under aerobic-dark conditions and that the photosynthetic activity and oxidative assimilation of glucose can independently operate mixotrophically under light conditions. These phenomena are mainly caused by physical and chemical properties of the microplastics and the morphological and biochemical properties of the algae. Furthermore, Bhattacharya et al. [20] reported that algae and microplastic has a great affinity in which microplastic particles have positive charges.

Microalgae can also produce Extracellular Polymeric Substances (EPS), which stimulates formation of biofilms on the microplastic surface, which is the main indicator of damage to microplastic material. Since biofilms contain nutrients, they can be a suitable living environment for other micro-organisms such as bacteria, fungi, and protozoa. The presence of these micro-organisms will form a protein structure such as enzyme that acts as a metabolic catalyst and breaks down chemical elements in the polymer into other elements. The chemical elements of polymers can form nutrition for micro-organisms, so that the latter obtain two resources of nutrition simultaneously, viz., *S. platensis* biofilms and chemical compounds of microplastic. The ability of micro-organisms to utilize chemical elements from polymers as nutrients is called biodegradation [22], because it will have an impact on changes in the chemical compounds in polymers [11].

Brackish water is found in estuary areas, has its own diversity, and is usually used for aquaculture such as milkfish. Microalgae serving as major producers of aquatic ecosystems are also found here [23]. However, microalgae might also be affected by the presence of microplastics in water bodies [24]. It has been proved that microplastic particles and doses can cause toxic effects on microalgae, including inhibition of growth, decreased photosynthetic efficiency, etc. [10,14,18]. However, the opposite results were also found by some researchers. Sjollema et al. [25] emphasized the impact of microplastic on growth rate, but not on photosynthetic efficiency for marine flagellates *Dunaliella tertiolecta* under a high exposure concentration of 250 mg/L with a particle size of 0.05 mm. Canniff and Hoang [26] showed that plastic microbeads could serve as a substrate for *Raphidocelissubcapitata*, thus, benefiting microalgae growth. Further, high concentrations of microplastics with a size of N400 μm had no deleterious effect on freshwater microalgae *Chlamydomasreinhardtii* [15]. Considering the contradictory discoveries and the limited number of microalgae species tested, more investigation is needed.

This research aims to investigate the inhibitory effects of different dosages of PS microplastics on the growth and photosynthetic efficiency of *S. platensis* and the effect of microalgae on the physical morphology of PS. The results of this study are expected to provide information useful for updating knowledge relating to the toxicity of PS with different dosages in the aquatic environment.

2. Materials and Methods

2.1. Styrofoam Preparation

The microplastics used in this study were Styrofoam granules obtained from CV. Mitra Sejati Foamindo, Genuk, Semarang City, Indonesia. The Styrofoam was weighed carefully with mass concentrations of 150, 250, and 400 mg in 500 mL culture volume, washed with ethanol and dried at room temperature for 24 h.

2.2. Culture Preparation of S. platensis

Microalgae *S. platensis* was obtained from Neoalgae, Sukoharjo, Central Java, Indonesia. Microalgae cultivation, testing, and result analysis were carried out at the UPT C-BIORE Laboratory, Diponegoro University, Semarang, Indonesia. Culturing was performed in 500 mL Erlenmeyer glasses, each equipped with an aerator (BS-410, Amara, Shanghai, China) (Figure 1). The cultures were placed into an illumination incubator under an 8W Philips tube lamp with light intensity of 1500 lux (light/dark ratio was 24 h/0 h). The cultivation temperature was controlled at about $23 \pm 2\,^\circ\text{C}$. Styrofoam was put into an Erlenmeyer, which already contained the culture of *S. platensis*. The experiment was set up for four different Styrofoam concentrations (*Spirulina* A = *Spirulina* culture without Styrofoam or as a control, *Spirulina* B = *Spirulina* culture with 150 mg Styrofoam, *Spirulina* C = *Spirulina* culture with 250 mg Styrofoam, and *Spirulina* D = *Spirulina* culture with 400 mg Styrofoam). Each culture was conducted in triplicate experiments while the Optical Density (OD) was measured for 30 days. Nutrient was given every two days in the form of a mixture of 15 ppm TSP, 70 ppm Urea, and 1 g/L $NaHCO_3$, to maintain the growth of *S. platensis*. OD was measured using a spectrophotometer (OPTIMA SP-300, Osaka, Japan) to determine the density of cells in *S. platensis* under the wavelength of 680 nm. Growth rate (μ) was measured using the formula [27]:

$$u = \frac{lnXn - lnXo}{tn - to}$$

where *ln X* is the natural logarithm of optical density and *t* is the time observed for *S. platensis*.

Figure 1. Microplastic Styrofoam with a diameter of 2 mm (**left**) and implementation of Styrofoam in microalgae culture (**right**).

2.3. Harvesting of S. platensis

After a 30-day exposure under the toxicity test (PS microplastics), *S. platensis* was harvested. Before harvesting, Styrofoam was separated by filtering *Spirulina* sp. containing micro plastic with a Whatman filter diameter of 1 mm to obtain *Spirulina* sp. without microplastic. Harvesting of *S. platensis* was carried out on the 30th day of culture by the filtration method. Filtrate obtained was in the form of wet biomass, which was dried in the oven at 35–40 °C temperature. Dry *S. platensis* samples were taken randomly for SEM-EDX analysis and Styrofoam samples for FTIR analysis.

2.4. FTIR and SEM Analysis

FTIR is a common technique used to determine any changes in the functional group of Styrofoam and was adopted for investigation of plastic degradation as stated in ISO 4582 and ISO 4892 for UV exposure, and for microorganism's surface colonization in ISO 846 and ISO 11266 [14]. The Styrofoam plastics that were applied in *Spirulina* sp. were taken every two days for about 30 days. Prior to the FTIR test, plastics were rinsed with distilled water and left to dry for 24 h, then, the Styrofoam was cut at a size of 2 mm. A FTIR apparatus Perkin Elmer Type Frontier (USA) was used to collect spectra from 4000–200 cm^{-1} (SNI 19-4370-2004 method) and ASTM D6288-89. FTIR test was also conducted in *Spirulina* sp, which had interacted with microplastic treatment for 30 days.

The morphology of microplastic Styrofoam was observed using scanning electron microscope (SEM) and combination with Energy Dispersive X-ray spectroscopy (EDX) to determine the inorganic elements contained in the material [14]. The analysis was conducted at room temperature and metalized using Au.A Jeol (model JSM-6510 LA, Tokyo, Japan) at 3000× magnification.

2.5. Statistical Analysis

Triplicates were applied and results were presented as means ± standard error of the mean. *S. platensis* growth rate data were statistically analyzed using the IBM SPSS application version 25, using the one-way ANOVA test followed Post-Hoc analysis with a confidence level of 95%. A value of $p < 0.05$ was used to reveal a significant difference.

3. Results

3.1. Spirulina platensis Growth under Styrofoam Pressure

The brackish water cultivation was imbued with 12 mg/L of NaCl, for maintaining the consistency of the culture in a brackish condition until harvest. According to Astuti, Jamali, and Amin [28], brackish water has a salinity of 0.5–17 mg/L. For 30 days, the salinity of the media fluctuates, but still in the brackish water range. *S. platensis* prefers higher salinity conditions. According to Hadiyanto dan Azim [29], *S. platensis* is able to grow in environments of high salinity, because in these conditions, some contaminants such as microbes are not able to survive. The graph of *S. platensis* growth in brackish water culture with Styrofoam treatment can be seen in Figure 2. In Figure 2, there is a point that shows an extreme increase in optical density. Culture A on day 28 from 1.42 to 1.52; culture B on day 29 from 1.52 to 1.62; C culture on day 28 from 1.05 to 1.09 and culture D on day 27 from 0.71 to 0.76. This extreme increase in optical density value shows the *S. platensis* culture experiencing an exponential phase [30].

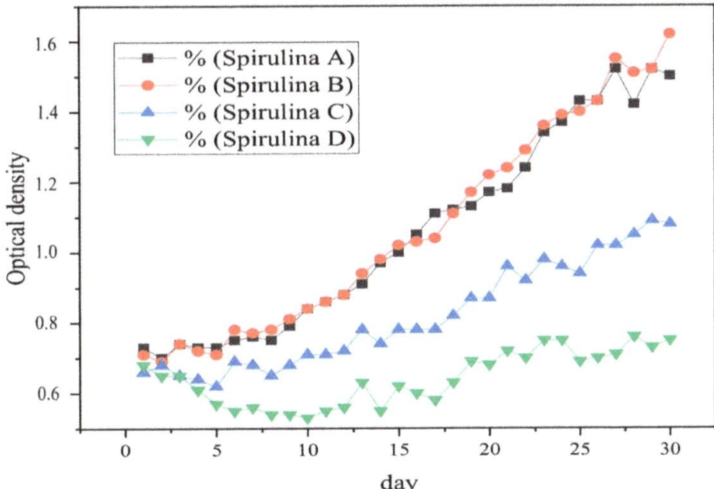

Figure 2. Brackish water culture *S. platensis* growth in each treatment (*Spirulina* A is a control (without Styrofoam), *Spirulina* B = 150 mg Styrofoam/500 mL culture, *Spirulina* C = 250 mg Styrofoam/500 mL culture, *Spirulina* D = 400mg Styrofoam/500 mL culture).

In order to evaluate the significance difference between experiments, One-way ANOVA followed by Post-Hoc Tukey HSD (honestly significant difference) was used in this research. Based on Figure 3 and calculation of the means of growth rate constant (μ) of each experiment (Table 1), it was revealed that the growth rate of *S. platensis* A (control) is 0.035925 day^{-1}. *S. platensis* B with 150 mg/500 mL Styrofoam treatment was 0.03525 day^{-1}. *S. platensis* C treated with Styrofoam 250 mg/500 mL was 0.02675 day^{-1}. *S. platensis* D treated with Styrofoam 400 mg/500 mL was 0.020425 day^{-1}. Furthermore, Table 2 also shows that the p-value (2.295×10^{-10}) between group corresponding to the F-statistic of one-way ANOVA is lower than 0.05, hence, H0 (null hypothesis 0 is rejected and H1 is accepted [31], indicating a difference in the *S. platensis* growth in brackish water, treated with different levels of Styrofoam.

The Tukey HSD test (Table 3) was then used to identify which pairs of these experiments are significantly different from each other. Comparing experiment A (control) and B (150 mg Styrofoam/500 mL *Spirulina*) revealed that they are insignificantly different of their growth rate as its p-value (0.7948595) is higher than 0.01. Moreover, the pairs of experiments A–C, A–D, B–C, B–D, and C–D show significant differences since all the Tukey HSD p-value are lower than 0.01 (Table 3).

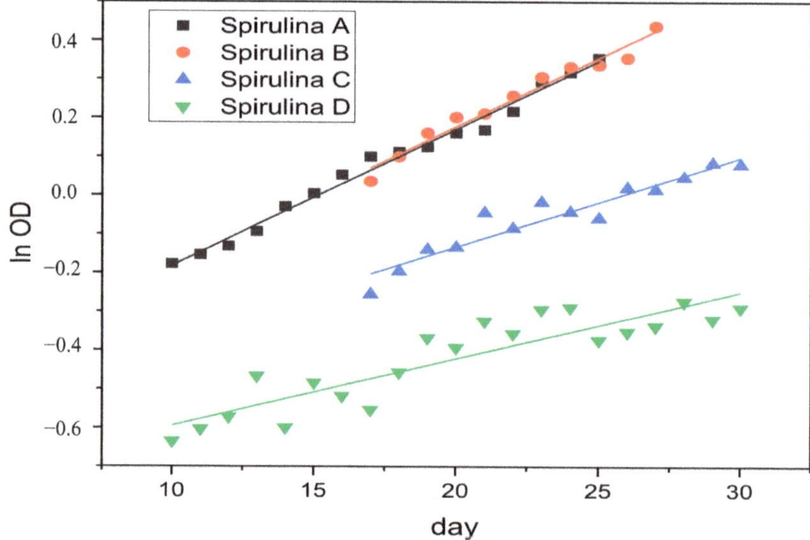

Figure 3. The logarithmic of optical density of *S. platensis* at the exponential phase in brackish water in various concentrations of microplastic treatment (**A**) control, (**B**) 150 mg, (**C**) 250 mg, and (**D**) 400 mg.

Table 1. Means value and their variances of each experiment.

Treatments	Sum	Average μ (day^{-1})	Variance
Control (A)	0.1437	0.035925	1.2425×10^{-6}
150 mg/500 mL (B)	0.141	0.03525	0.00000259
250 mg/500 mL (C)	0.107	0.02675	0.00000055
400 mg/500 mL (D)	0.0817	0.020425	2.49167×10^{-7}

Table 2. Analysis of variance (ANOVA) of F and p values between experiments.

Source of Variation	df	MS	F	p-Value	F Crit
Between Groups	3	0.000218974	189.1104714	2.2956×10^{-10}	3.49029482
Within Groups	12	1.15792×10^{-6}			
Total	15				

df, degree of freedom; MS, Mean Square is just the Sum of Squares divided by its degrees of freedom, and the F value is the ratio of the mean squares.

Table 3. The post-hoc Tukey HSD analysis of four group experiment.

Treatment Pair	Tukey HSD Q Statistic	Tukey HSD p-Value	Tukey HSD Interference
A–B	1.2546	0.7948595	insignificant
A–C	17.0529	0.0010053	** $p < 0.01$
A–D	28.8087	0.0010053	** $p < 0.01$
B–C	15.7983	0.0010053	** $p < 0.01$
B–D	27.5541	0.0010053	** $p < 0.01$
C–D	11.7558	0.0010053	** $p < 0.01$

**, significant.

3.2. Styrofoam Degradation

Fourier Transform Infrared (FTIR) is a tool for determining the functional groups and molecular bonds of a chemical compound in a specimen. Its working principle is the interaction between spectrum originating from the source and the test sample material. The sample will generate vibrations, which will be captured by the detector and finally translated into a transmittance curve that has certain peaks with a spectrum of 4000–400 cm^{-1} [32]. In this research, FTIR was employed detect degradation in plastic by considering changes in functional groups [14].

Figure 4 shows the effect of presence of microplastics with different concentration in microalgae *Spirulina* sp. culture. According to Dmytryk et al. [33], the wavelength of 3800–3200 cm^{-1} indicates the amine functional group (NH$_3$) in the protein. The following peak, 1750–1600 cm^{-1} represents the primary amide and carbonyl (C=O) groups in the protein. The stretching vibrations observed in the frequency range of peaks 1450 cm^{-1} and peaks at 1400–1300 cm^{-1} represent carboxyl (COO-) and alkyl groups, respectively. Then at a wavelength of 1050–1000 cm^{-1} stretching of CO, CC, and OH in the presence of ether, ester, and hydroxyl of polysaccharides are observed.

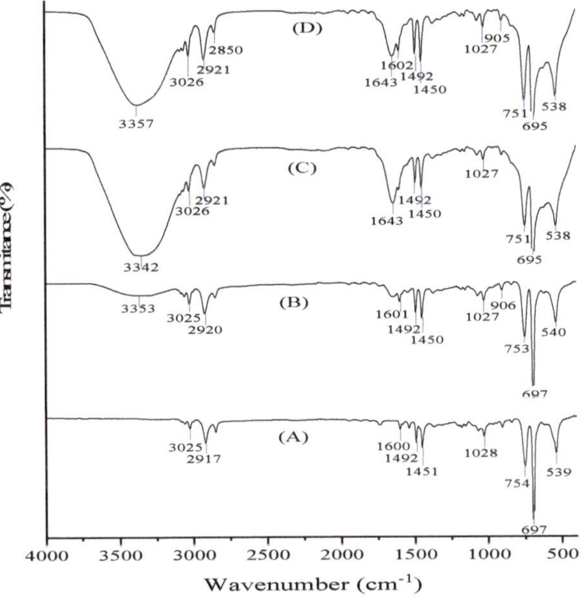

Figure 4. FTIR results of the ratio of Styrofoam (**A**) before treatment, (**B**) 150 mg, (**C**) 250 mg, and (**D**) 400 mg; after 30-day treatment with *S. platensis* in brackish water culture.

Furthermore, Figure 4 shows that no O-H groups in Styrofoam, which was found also in brackish water Styrofoam, where peaks (3353 cm^{-1}) began to form with low intensity. O-H groups were clearly visible in Styrofoam C, D-brackish water. The peak read was in the range of 3378–3345 cm^{-1} with an intensity of 59.18–67.65%. The presence of an O-H group also has been confirmed with a C-O group (1300–1000 cm^{-1}). Which can be seen in brackish water B, C, D-Styrofoam. This shows a change in the functional group on Styrofoam, with evidence of the formation of an alcohol group (-COOH) [11].

3.3. Interaction of S. platensis with Styrofoam

Scanning Electron Microscopy (SEM) is a tool for determining the surface morphology of a specimen, including changes caused by micro-organisms [34]. SEM performance using a magnification of 3000× is supported by EDX, which is able to determine the content of inorganic elements in a specimen using X rays [35].

SEM analysis results on brackish water *S. platensis* showed that around the *S. platensis* A, B, C, and D, cells produced EPS in the form of small spheres and large nuggets, thought to be salt or urea given during culture (Figure 5). Further, the morphology of *S. platensis* A was still normal, while *S. platensis* B, C, and D were seen to be damaged. According to Li et al., [4] the presence of microplastics can damage microalgae cell membranes, thus inhibiting the photosynthesis process.

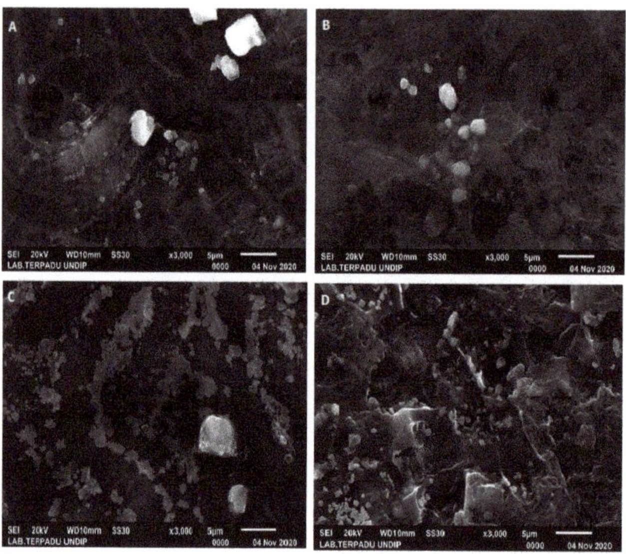

Figure 5. SEM analysis results of brackish water culture *S. platensis* for 30 days. (**A**) *S. platensis* without Styrofoam treatment. (**B**) *S. platensis* treated with Styrofoam 150 mg/500 mL. (**C**) *S. platensis* treated with Styrofoam 250 mg/500 mL. (**D**) *S. platensis* treated with Styrofoam 400 mg/500 mL.

The results of EDX analysis (Table 4) on brackish water *S. platensis* showed that in culture B and C, there was an increase in carbon content, namely 24.56% and 4.24%, compared to *S. platensis* A culture, whereas in D culture, there was a decrease in carbon content by 2.14%.

Table 4. Energy Dispersive X-ray spectroscopy (EDX) analysis results for the chemical constituents of *S. platensis* cultured in brackish water for 30 days.

S. platensis Content	Styrofoam Levels			
	S. platensis A (Control)	*S. platensis* B + 150 mg	*S. platensis* C + 250 mg	*S. platensis* D + 400 mg
Carbon, C	64.3	85.23	67.15	62.92
Nitrogen, N	18.59	-	16.5	23.69
Natrium Oxide, Na_2O	4.14	5.39	5.14	3.75
Magnesium Oxide, MgO	0.51	0.2	0.27	0.43
Alumina, Al_2O_3	-	-	-	-
Silica Dioxide, SiO_2	-	-	-	0.31
Phosphor Pentoxide, P_2O_5	2.29	1.67	1.75	2.55
Sulfide, SO_3	1.95	2.15	2.3	1.76
Chloride, Cl	4.56	3.58	4.9	2.8
Kalium Oxide, K_2O	3.67	1.78	1.99	1.78
Calcium Oxide CaO	-	-	-	-
Cuprum (II) Oxide, CuO	-	-	-	-
Zinc Oxide, ZnO	-	-	-	-

4. Discussions

Our research reported an interaction between microalgae and Styrofoam microplastic. Infusion of Styrofoam had an impact on the S. platensis growth rate, because Styrofoam gave a shading effect on the culture surface, thereby reducing the light intensity used by *S. platensis* for photosynthesis [15]. Imposing Styrofoam 150 mg in 500 mL *Spirulina* culture did not significantly affect the growth rate as compared to control (Figure 2), which means that at this concentration the Styrofoam did not give a shading effect and eventually microalgae cell could use carbon from the Styrofoam (Tables 1 and 2). However, increasing Styrofoam concentration (250 mg/500 mL and 400 mg/500 mL) the growth of algae cell was significantly retarded by the Styrofoam particles concentration (Figure 2). Moreover, the decrease in the growth rate of S. platensis may be also influenced by the formation of excess Extracellular Polymeric Substances (EPS), which is toxic to *S. platensis* itself. The presence of EPS will be a place for other micro-organisms to compete with algae cells in the absorption of nutrients, both from the culture and from the breakdown of carbon chains from Styrofoam [14].

The growth rate of *S. platensis* B culture (given Styrofoam 150 mg/500 mL) in brackish waters was the highest as compared to 250 mg/500 mL and 400 mg/500 mL. This is presumably because *S. platensis* obtains additional nutrients from the degradation of Styrofoam (Table 1 and Figure 3). In addition, the Styrofoam in culture B did not cover the entire surface of the culture, so that the light could still enter and be used properly by *S. platensis*. Increased levels of Styrofoam resulted in a decrease in the growth rate of the *S. platensis* culture as evidenced by culture D, which has a lower growth rate than culture C, due to *S. platensis* being under pressure from the environment in the form of Styrofoam. The number of Styrofoam floating on the surface is also able to block light from entering the culture, thus, inhibiting the photosynthesis process [15].

The FTIR analysis (Figure 4) depicts that no carboxyl groups (C=O, at a wavelength of $1810-1630$ cm^{-1}) are formed, indicating the absence of oxidation reaction to Styrofoam. The structure of Styrofoam showed the presence of an aromatic C=C group and no aliphatic C=C group was formed, indicating that the initial structure of Styrofoam in the form of styrene has a closed chain (cyclic) shape. However, all the FTIR test results on Styrofoam that were included in the brackish water *S. platensis* culture, showed the presence of aromatic C=C groups and aliphatic C=C groups, proving that there is an open cyclic chain [11]. Mohamed et al. [32] stated that Styrofoam is stable because its constituent structure is a cyclic chain with a very long arrangement. The opening of the cyclic chain proves the occurrence of degradation, although such degradation has not yet reached physical fragmentation and changes into simpler chemical monomers [14]. Another phenomena showed that all FTIR in Styrofoam showed a peak at a wavelength of 754–538 cm^{-1} with a sharp peak at 697–695 cm^{-1}. According to Nandiyanto, Oktiani, and Ragadhita [31], the peak of 750 cm^{-1} is a characteristic of aromatic compounds. These FTIR test data results on concluded that Styrofoam has interaction with *S. platensis* cells in the culture. According to Chentir et al. [36], increasing the concentration of NaCl can reduce the availability of nutrients such as nitrogen, thereby triggering the incorporation of carbon both from *S. platensis* and from Styrofoam into EPS. The decrease in carbon content in algae culture indicates damage to the cell membrane of *S. platensis*, which affects the ability of photosynthesis. Li et al. [4] stated that although microalgae are able to absorb carbon from plastics, these plastics are at risk of damaging cell membranes; hence, plastic is not a good source of nutrition for microalgae.

Styrofoam is composed of styrene chains, which are a source of carbon for micro-organisms in the waters. This causes the nutrients needed for photosynthesis of *S. platensis* especially from the element carbon supplied by Styrofoam, which is available in the medium. The availability of this carbon can support the growth of *S. platensis*, which will have an impact on increasing the production of Extracellular Polymeric Substances (EPS), which in turn plays a role in producing a biofilm on the Styrofoam surface [37,38]. Biofilms are a suitable abode for other micro-organisms such as bacteria, fungi, protozoa etc., which play a role in the degradation of the Styrofoam surface. During this microbial activity, micro-organisms will form protein structures in the form of enzymes that play a role in changing the chemical content in Styrofoam into other forms. The presence of other inorganic elements in the EDX analysis proved that *S. platensis* was able to absorb contaminants, which can come from the release of additives from Styrofoam, such as Mg, Al, Si, S, Ca, K, Cl, Cr, Zn, Cu etc., as well as from the nutrients given such as C, N, P, Na, Cl etc. [14].

5. Conclusions

This interaction between Styrofoam and microalgae *Spirulina* sp. has been investigated in this research. The growth of microalgae, the change of morphological structure of Styrofoam and chemical functional groups were measured and used in determining the effect of interactions. The results of the variations of Styrofoam concentration from 300 g/L to 800 g/L in microalgae culture showed significant inhibitory effects on *Spirulina* sp. growth. There was a change in the functional group on Styrofoam as an indicator of biodegradation, with evidence of the formation of an alcohol group (-COOH) at a wavelength of 3400–3200 cm^{-1} and an open cyclic chain (peaks appearing at a wavelength of 1680–1600 cm^{-1}). SEM-EDX test results show that Styrofoam can be a source of nutrients, especially carbon, needed by *S. platensis* for photosynthesis. However, the presence of microplastic Styrofoam also gives a deterioration effect to the microalgae cell, which cause photosynthetic inhibition. The findings of this work essentially improve understanding of the interaction between microplastics and microalgae cell in aquatic environments. The continuous influence of different sizes of microplastics on microalgae or other organisms should be further investigated. Nevertheless, this study only showed the preliminary findings on the interaction between Styrofoam with microalgae and further investigation and detail analysis should be done in more replications experiments to obtain a statistical significance of the results.

Author Contributions: H.H.: principle investigator, main concept formulation, data analysis, writing; A.H.: performed experiment, data analysis; F.M.: supervising, data interpretation; N.A.: data interpretation, writing; A.K.: data analysis, writing, analysis. All authors have read and agreed to the published version of the manuscript.

Funding: This research was funded by Ministry of Culture and Education under Riset Penugasan (World Class Research) grant with a contract number 201-07/UN7.6.1/PP/2020.

Institutional Review Board Statement: Not applicable.

Informed Consent Statement: Not applicable.

Data Availability Statement: The data presented in this study are available on request from the corresponding author.

Acknowledgments: The authors greatly appreciate the Biomass and Renewable Energy (C-BIORE) Laboratory, Diponegoro University for providing funds to carry out this research. The authors also would like to thank the anonymous reviewers for their helpful comments and suggestions that greatly improved the manuscript.

Conflicts of Interest: The authors declare no conflict of interest.

References

1. Andrady, A.L. Microplastics in the marine environment. *Mar. Pollut. Bull.* **2011**, *62*, 1596–1605. [CrossRef]
2. Hidalgo-Ruz, V.; Gutow, L.; Thompson, R.C.; Thiel, M. Microplastics in the marine environment: A review of the methods used for identification and quantification. *Environ. Sci. Technol.* **2012**, *46*, 3060–3075. [CrossRef] [PubMed]
3. Duis, K.; Coors, A. Microplastics in the aquatic and terrestrial environment: Sources (with a specific focus on personal care products), fate and effects. *Environ. Sci. Eur.* **2016**, *28*, 1–25. [CrossRef] [PubMed]
4. Cole, M.; Lindeque, P.; Halsband, C.; Galloway, T.S. Microplastics as contaminants in the marine environment: A review. *Mar. Pollut. Bull.* **2011**, *62*, 2588–2597. [CrossRef]

5. Thompson, R.C.; Olsen, Y.; Mitchell, R.P.; Davis, A.; Rowland, S.J.; John, A.W.G.; McGonigle, D.; Russell, A.E. Lost at sea: Where is all the plastic? *Science* **2004**, *304*, 838. [CrossRef]

6. European Chemical Agency (ECHA). Available online: https://echa.europa.eu/documents/10162/23665416/rest_microplastics_qa_v1.0_16524_en.pdf/c9849410-c360-d95b-e287-ae635b0b7b3f (accessed on 17 September 2020).

7. Auta, H.S.; Emenike, C.U.; Fauziah, S.H. Distribution and importance of microplastics in the marine environment: A review of the source, fate, effect, and potential solution. *Environ. Int.* **2017**, *102*, 165–176. [CrossRef]

8. Frias, J.P.G.L.; Nash, R. Microplastics: Finding a consensus on the definition. *Mar. Pollut. Bull.* **2019**, *138*, 145–147. [CrossRef] [PubMed]

9. Castañeda, R.A.; Avlijas, S.; Simard, M.A.; Ricciardi, A. Microplastic pollution in St. Lawrence River sediments. *Can. J. Fish. Aquat. Sci.* **2014**, *70*, 1767–1771.

10. Khoironi, A.; Anggoro, S.; Sudarno, S. Evaluation of the Interaction among Micoalgae Spirulina sp, Plastics Polyethylene Terephtalate and Polypropylene in Freshwater Environment. *J. Ecol. Eng.* **2019**, *20*, 161–173. [CrossRef]

11. Ho, B.T.; Roberts, T.K.; Lucas, S. An overview on biodegradation of polystyrene and modified polystyrene: The microbial approach. *Crit. Rev. Biotechnol.* **2018**, *38*, 308–320. [CrossRef] [PubMed]

12. Chandra, M.; Kohn, C.; Pawlitz, J.; Powell, G. *Real Cost of Styrofoam*; Saint Luis University: St. Louis, MO, USA, 2016; Available online: https://greendiningalliance.org/wp-content/uploads/2016/12/real-cost-of-styrofoam_written-report.pdf (accessed on 8 November 2020).

13. Rummel, C.D.; Jahnke, A.; Gorokhova, E.; Kühnel, D.; Schmitt-Jansen, M. Impacts of biofilm formation on the fate and potential effects of microplastic in the aquatic environment. *Environ. Sci. Technol. Lett.* **2017**, *4*, 258–267. [CrossRef]

14. Li, S.; Wang, P.; Zhang, C.; Zhou, X.; Yin, Z.; Hu, T.; Hu, D.; Liu, C.; Zhu, L. Influence of polystyrene microplastics on the growth, photosynthetic efficiency and aggregation of freshwater microalgae *Chlamydomonas reinhardtii*. *Sci. Total Environ.* **2020**, *714*, 136767. [CrossRef] [PubMed]

15. Lagarde, F.; Olivier, O.; Zanella, M.; Daniel, P.; Hiard, S.; Caruso, A. Microplastic interactions with freshwater microalgae: Hetero-aggregation and changes in plastic density appear strongly dependent on polymer type. *Environ. Pollut.* **2016**, *215*, 331–339. [CrossRef] [PubMed]

16. Fachrul, M.F.; Rinanti, A. Bioremediasi Pencemar Mikroplastik di Ekosistim Perairan Menggunakan Bakteri Indigenous (Bioremediation of Microplastic Pollutant in Aquatic Ecosystem by Indigenous Bacteria). *Semin. Nas. Kota Berkelanjutan* **2018**, *1*, 302. [CrossRef]

17. Fahrenfeld, N.L.; Arbuckle-Keil, G.; Beni, N.N.; Bartelt-Hunt, S.L. Source tracking microplastics in the freshwater environment. *TrAC Trends Anal. Chem.* **2019**, *112*, 248–254. [CrossRef]

18. Prata, J.C.; Lavorante, B.; BS MMontenegro, M.; Guilhermino, L. Influence of microplastics on the toxicity of the pharmaceuticals procainamide and doxycycline on the marine microalgae Tetraselmis chuii. *Aquat. Toxicol.* **2018**, *197*, 143–152. [CrossRef]

19. Harding, S. Marine Debris: Understanding, Preventing and Mitigating the Significant Adverse Impacts on Marine and Coastal Biodiversity, CBD Technical Series. *Biodiversity* **2016**. [CrossRef]

20. Bhattacharya, P.; Lin, S.; Turner, J.P.; Ke, P.C. Physical adsorption of charged plastic nanoparticles affects algal photosynthesis. *J. Phys. Chem.* **2010**, *C 114*, 16556–16561. [CrossRef]

21. Marquez, F.J.; Sasaki, K.; Kakizono, T.; Nishio, N.; Nagai, S. Growth characteristics of Spirulina platensis in mixotrophic and heterotrophic conditions. *J. Ferment. Bioeng.* **1993**, *76*, 408–410. [CrossRef]

22. Song, Y.; Qiu, R.; Hu, J.; Li, X.; Zhang, X.; Chen, Y.; Wu, W.M.; He, D. Biodegradation and disintegration of expanded polystyrene by land snails Achatina fulica. *Sci. Total Environ.* **2020**, *746*, 141289. [CrossRef]

23. Troell, M. Integrated Marine and Brackishwater Aquaculture in Tropical Regions', Integrated Mariculture-A Global Review-FAO Fisheries and Aquaculture Technical Paper N0. 529, (October 2013). 2009, pp. 47–132. Available online: http://linkinghub.elsevier.com/retrieve/pii/S0044848603004691 (accessed on 16 October 2020).

24. Besseling, E.; Wang, B.; Lurling, M.; Koelmans, A.A. Nanoplastic affects growth of S. obliquus and reproduction of D. magna. *Environ. Sci. Technol.* **2014**, *48*, 12336–12343. [CrossRef]

25. Sjollema, S.B.; Redondo-Hasselerharm, P.; Leslie, H.A.; Kraak, M.H.S.; Vethaak, A.D. Do plastic particles affect microalgal photosynthesis and growth? *Aquat. Toxicol.* **2016**, *170*, 259–261. [CrossRef] [PubMed]

26. Canniff, P.M.; Hoang, T.C. Microplastic ingestion by Daphnia magna and its enhancement on algal growth. *Sci. Total Environ.* **2018**, *633*, 500–507. [CrossRef]

27. Fakhri, M.; Antika, P.W.; Ekawati, A.W.; Arifin, N.B. Pertumbuhan, Kandungan Pigmen, dan Protein Spirulina platensis yang Dikultur Pada Ca(NO$_3$)$_2$ Dengan Dosis yang Berbeda. *J. Aquac. Fish Health* **2020**, *9*, 38–47. [CrossRef]

28. Astuti, W.; Jamali, A.; Amin, M. Desalinasi Air Payau Menggunakan Surfactant Modified Zeolite (SMZ). *J. Zeolit Indones.* **2007**, *6*, 32–37.

29. Nur, M.A.; Hadiyanto, H. Enhancement of chlorella vulgaris biomass cultivated in pome medium as biofuel feedstock under mixotrophic conditions. *J. Eng. Technol. Sci.* **2015**, *47*, 487–497. [CrossRef]

30. Islam, M.T. *Learning SPSS without Pain: A Comprehensive Manual for Data Analysis and Interpretation of Outputs*, 1st ed.; ASA Publications: Dhaka, Bangladesh, 2020. [CrossRef]

31. Nandiyanto, A.B.D.; Oktiani, R.; Ragadhita, R. How to read and interpret ftir spectroscope of organic material. *Indones. J. Sci. Technol.* **2019**, *4*, 97–118. [CrossRef]

32. Mohamed, M.A.; Jaafar, J.; Ismail, A.F.; Othman, M.H.D.; Rahman, M.A. *Fourier Transform Infrared (FTIR) Spectroscopy, Membrane Characterization*; Elsevier B.V.: Amsterdam, The Netherlands, 2017. [CrossRef]
33. Dmytryk, A.; Saeid, A.; Chojnacka, K. Biosorption of microelements by spirulina: Towards technology of mineral feed supplements. *Sci. World J.* **2014**, *2014*, 1–15. [CrossRef] [PubMed]
34. Sujatno, A.; Salam, R.; Bandriyana, B.; Dimyati, A. Studi Scanning Electron Microscopy(SEM) untuk Karakterisasi Proses Oxidasi Paduan Zirkonium. *J. Forum Nukl.* **2015**, *9*, 44–50. [CrossRef]
35. Abd Mutalib, M.; Rahman, M.A.; Othman, M.H.D.; Ismail, A.F.; Jaafar, J. *Scanning Electron Microscopy (SEM) and Energy-Dispersive X-ray (EDX) Spectroscopy, Membrane Characterization*; Elsevier B.V.: Amsterdam, The Netherlands, 2017. [CrossRef]
36. Chentir, I.; Hamdi, M.; Doumandji, A.; HadjSadok, A.; Ouada, H.B.; Nasri, M.; Jridi, M. Enhancement of extracellular polymeric substances (EPS) production in *Spirulina* (*Arthrospira* sp.) by two-step cultivation process and partial characterization of their polysaccharidic moiety. *Int. J. Biol. Macromol.* **2017**, *105*, 1412–1420. [CrossRef]
37. Khoironi, A.; Hadiyanto, H.; Anggoro, S.; Sudarno, S. Evaluation of polypropylene plastic degradation and microplastic identification in sediments at Tambak Lorok coastal area, Semarang, Indonesia. *Mar. Pollut. Bull.* **2020**, *151*, 110868. [CrossRef] [PubMed]
38. Dianratri, I.; Hadiyanto, H.; Khoironi, A.; Pratiwi, W.Z. The influence of polypropylene and polyethylene microplastics on the quality of spirulina sp. Harvests. *Food Res.* **2020**, *4*, 1739–1743. [CrossRef]

Review

Occurrence of Natural and Synthetic Micro-Fibers in the Mediterranean Sea: A Review

Saul Santini [1], Eleonora De Beni [1], Tania Martellini [1,2,*], Chiara Sarti [1], Demetrio Randazzo [1], Roberto Ciraolo [1], Costanza Scopetani [3] and Alessandra Cincinelli [1,2,*]

[1] Department of Chemistry "Ugo Schiff", University of Florence, Via della Lastruccia 3, Sesto Fiorentino, 50019 Florence, Italy; saul.santini@unifi.it (S.S.); eleonoradebeni@msn.com (E.D.B.); chiara.sarti@unifi.it (C.S.); demetrio.randazzo@unifi.it (D.R.); roberto.ciraolo@unifi.it (R.C.)

[2] Department of Chemistry "Ugo Schiff", University of Florence and Consorzio Interuniversitario per lo Sviluppo dei Sistemi a Grande Interfase (CSGI), Via della Lastruccia 3, Sesto Fiorentino, 50019 Florence, Italy

[3] Faculty of Biological and Environmental Sciences Ecosystems and Environment Research Programme, University of Helsinki, Niemenkatu 73, FI-15140 Lahti, Finland; costanza.scopetani@helsinki.fi

*   Correspondence: tania.martellini@unifi.it (T.M.); alessandra.cincinelli@unifi.it (A.C.)

Abstract: Among microplastics (MPs), fibers are one of the most abundant shapes encountered in the aquatic environment. Growing attention is being focused on this typology of particles since they are considered an important form of marine contamination. Information about microfibers distribution in the Mediterranean Sea is still limited and the increasing evidence of the high amount of fibers in the aquatic environment should lead to a different classification from MPs which, by definition, are composed only of synthetic materials and not natural. In the past, cellulosic fibers (natural and regenerated) have been likely included in the synthetic realm by hundreds of studies, inflating "micro-plastic" counts in both environmental matrices and organisms. Comparisons are often hampered because many of the available studies have explicitly excluded the micro-fibers (MFs) content due, for example, to methodological problems. Considering the abundance of micro-fibers in the environment, a chemical composition analysis is fundamental for toxicological assessments. Overall, the results of this review work provide the basis to monitor and mitigate the impacts of microfiber pollution on the sea ecosystems in the Mediterranean Sea, which can be used to investigate other basins of the world for future risk assessment.

Keywords: microplastics; fibers; cellulose; Mediterranean Sea; pollution; chemical characterization; environmental pollution; biota contamination

Citation: Santini, S.; De Beni, E.; Martellini, T.; Sarti, C.; Randazzo, D.; Ciraolo, R.; Scopetani, C.; Cincinelli, A. Occurrence of Natural and Synthetic Micro-Fibers in the Mediterranean Sea: A Review. *Toxics* **2022**, *10*, 391. https://doi.org/10.3390/toxics10070391

Academic Editor: Roberto Rosal

Received: 17 June 2022
Accepted: 8 July 2022
Published: 13 July 2022

Publisher's Note: MDPI stays neutral with regard to jurisdictional claims in published maps and institutional affiliations.

1. Introduction

Plastic is considered a persistent and ubiquitous pollutant, and it is considered among the top environmental concerns of the Anthropocene [1,2]. Microplastics (MPs) are small plastic fragments ranging from 1 µm to 5 mm in size that can be found in different environmental compartments [3]. MPs accumulate in the environment and increase stress on the marine, freshwater and terrestrial ecosystems [4]. Several studies have evidenced their presence in the marine environment [5–7], aquatic sediments [8], freshwaters [9], soils [10] and the atmosphere [11,12]. MPs can act as a carrier of hydrophobic organic contaminants, transporting the pollutants inside the organisms through ingestion and subsequent chemical release. However, it has been shown that sometimes, ingested MPs can adsorb the pollutants already present in the organisms and remove them once they are excreted [13]. Plastics themselves contain toxic chemical additives (such as plasticizers, antistatic agents, flame retardants, heat stabilizers, acid scavengers, colorants, etc.) that can be released into the environment [14]. Moreover, chemical additives in plastics can adsorb organic contaminants from other matrices and increase the exposure of several contaminants to the environment [15,16]. These chemicals, if present in the food chain and absorbed by humans, could cause many diseases linked with hormonal disruption, reproductive problems, nervous tissue, liver and kidney damage, etc. [17]. Although the effects of plastic litter on the marine environment and organisms have been recently investigated in several oceanic areas, more information is needed for the Mediterranean Sea [18], which is an enclosed sea with limited exchange with the ocean basins and high diversity of sensitive ecosystems. This particularity, together with other factors such as the high-density population in the coastal areas, intense navigation traffic, and industrial and fishing activities, makes the Mediterranean basin one of the most affected seas by plastic accumulation all over the world [19]. The determination and characterization of MPs for shape, color, size and type is fundamental to better understand their impact on the environment. Among MPs, fibers are the predominant shape in the aquatic environment, often accounting for more than 80% of the total items [11,20–33]. For this reason, increasing attention is being paid to micro-fibers and their potential toxicological and environmental effects, as evidenced by the growing number of studies on microfiber pollution over the past decade (Figure 1). According to the general definition proposed by Liu et al. (2019), microfibers (MFs) are any natural or artificial fibrous materials of threadlike structure with a diameter less than 50 μm, length ranging from 1 μm to 5 mm, and length to diameter ratio greater than 100 [34].

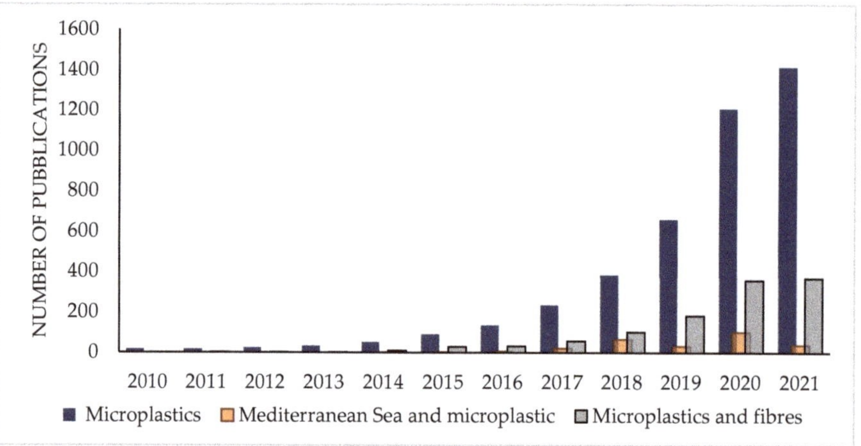

Figure 1. Number of publications per year studying MPs in the environment, MPs in the Mediterranean Sea and MPs/fibers. Source: Web of Science Database.

Microplastics, especially MFs, contaminate and affect many aquatic organisms or species of birds or mammals that feed on aquatic species since they are often mistaken for food and ingested by prey species, which, in turn, are eaten by predators, allowing MPs to move up the trophic chain [35,36].

However, information about the microfiber distribution in the Mediterranean Sea is still limited and filling this knowledge gap would be the first step to take to tackle the microfiber pollution issue. The second important step is to characterize the nature of the fibers because they are not always plastic but rather dyed cellulose. In the last decade, cellulosic fibers (natural and regenerated) have been likely included in the synthetic realm by hundreds of studies, inflating "microplastic" counts in both environmental matrices and organisms; this error has resulted from the assumption that all colored fibers are synthetic [37]. The separation of textile MFs from other MPs does not necessarily add complexity but, conversely, might bring consistency to the comparison across different investigations [38]. A recent study by Pedrotti et al., 2021, shows that fibers analyzed from textiles considered 100% synthetic constituted 17.4% of natural or derived from the transformation of natural polymers. In the seawater samples, 14–50% of the fibers analyzed were synthetic, 35–72% were of natural origin (cotton, wool) or made by processing natural polymers (especially cellulose), and the rest were a mixture of different materials or could not be identified (14–21%) [39]. Most microfibers of natural origin come from anthropogenic sources; however, a very small percentage can be released into the environment from "natural" sources such as bast fibers, leaf fibers, seed fibers, grass and all other types such as roots and wood [40]. As shown in a study by Athey et al., 2021, many of the methods used to investigate the occurrence of MPs do not provide data on the nature of synthetic or non-synthetic. Moreover, some steps of the methods, such as chemical digestion, could generate mistakes [41–44]. Comparisons between different studies are often hampered because many of them highlight the predominance of fibers in environmental samples without including a chemical characterization of the fibers. Thus, to ensure that studies of the presence of microplastics in the environment, and particularly in the marine environment, provide information to understand the ecological damage from these pollutants, it is essential to use appropriate instrumentation. While a stereomicroscope is sufficient to separate MFs from MPs, more complex instrumentation is required to identify the nature of the MPs and specifically whether an MF is natural or synthetic, cellulose or not. To this aim, chemical analysis of the polymeric composition using, for example, Fourier Transform Infrared Spectroscopy (FTIR), μ-Raman and scanning electron microscope (SEM) [5] need to be performed. The present review aims at examining the current literature on the occurrence of cellulose and cellulose-based fibers in the Mediterranean Sea, providing a picture of MF contamination in coastal marine environments.

Non-Synthetic MFs Toxicity

In the industry of non-synthetic textiles, a similar cocktail of dyes and chemicals as in synthetic textiles is used, and many of these substances are toxic and can accumulate in the environment [45]. The toxic chemicals released by MPs into the tissue of fishes and marine animals are several and include, e.g., colorants, plasticizers, elasticizers, and together with the microfiber particles, can physically damage various organs, the digestive tract, stomach lining, immune function and stymie growth, and thus, affect the entire ecosystem [44,46]. The textile industry, a source of pollution of MPs, including MFs, in the environment, involves the use of many dyes that can be toxic to organisms [47,48]. Several dyes such as: Acid Red 26, Basic Red 9, Basic Violet 14, Direct Black38, Direct Blue 6, Direct Red 28, Disperse Blue 1, Disperse Orange11 and Disperse Yellow 3 are classified as carcinogenic in the European standard of textile ecology [49]. The effect of the carcinogenic dyes in rats is included in the IARC monographs [50]. Moreover, experiments were conducted to observe the toxic effects of these dyes if dispersed in the environment and absorbed by marine organisms. Shen et al. (2015) studied the toxic effects of Basic Violet 14, Direct Red 28 and Acid Red 26 on zebrafish larvae, observing acute effects: cardiovascular toxicity and molecular mechanism by Acid Red 26 and hepatotoxicity effects by Basic Violet 14 [51]. In a study by Remy et al. (2015), the presence of non-synthetic fibers was identified in the invertebrate community that live in Neptune grass, Posidonia Oceanica (L.) Delile, a species heavily predated by fishes, in the Mediterranean coastal zone [25]. The dyes of these fibers were two: Direct Red 28 and Direct Blue 22, and they are used in the textile industry for natural and artificial fibers. Direct Blue 22 is not considered harmful to humans, but Direct Red 28 is classified as carcinogenic, mutagenic or toxic to reproduction. Direct Red 28 can be reduced by the intestine bacteria and generate carcinogenic molecules in humans [52]. Non-synthetic and semi-synthetic microfibers and their additives or dyes may interact negatively with biota in aquatic environments similar to plastic microfibers, but ingestion, chemical leaching and degradation rates in marine environments are poorly understood [25]. Natural fibers, although considered environmentally friendly by their faster environmental degradation, pose a global threat comparable to synthetic polymers. In fact, due to the processing of textiles, they can be mixed with flame retardants and/or resins, and this not only represents a problem related to the release of toxic compounds but also has an effect on degradation times, which become longer [37]. Moreover, since they constitute a major component of litter in water bodies and aquatic animals, they could become important vectors not only of contaminants but also of bacteria [53]. Espinosa et al. (2016) have associated the presence of MFs in fish with a mixture of several polybrominated diphenyl ethers at concentrations that can cause effects on the endocrine system [54]. The presence of these substances in the environment can hamper reproduction, in particular, for fish. This is due to the high sensitivity of juvenile and adult fishes to endocrine disruptors [55,56]. The adverse effects caused to the aforementioned organisms by fibers might be relevant also for humans since MPs and their associated chemicals can be transferred through the food chain and reach us [57]. Another way through which the human organism is exposed to MPs is airborne contamination. The MPs get deposited in our lung tissues and lead to lung inflammation [58]. These fibers are known to have adverse impacts on terrestrial and marine ecosystems [59]. Unfortunately, MPs are present in all environmental compartments and rayon, and polyester fibers are commonly present in marine animal species [60]; they can be absorbed through herds and cause problems to the respiratory and gastrointestinal systems. The aim of this review is to report the current state of research on the environmental impacts of microfibers and to identify gaps in knowledge. In light of the findings, it appears essential that future research should focus on the characterization of microfibers, the chemical and physical properties of various fabrics, both synthetic and natural, and the ability of microfibers to become carriers of toxic substances.

2. Discussion

We summarize the 2015–2021 literature data on the abundance of fibers in the Mediterranean Sea, including the abundance of synthetic or non-synthetic fibers, colors and size. Based on published literature from the Web of Science, SCOPUS, Google Scholar, Science Direct, Pubmed and Sci-Finder, we obtained studies by searching for "microfibers and microplastics", "microplastics and fibers", "filaments and plastic pollution", "plastic and microfibers", "microplastics and filaments", "microplastic fibers", "synthetic fibers and microplastics", "Textile fibers and microplastics", "fragments and microfibers", Microplastics and Mediterranean sea", Microplastics and biota", Sediment and microplastics", "Microfibers and source and fate", "Microfibers and toxic effects". Then, we eliminated irrelevant studies by reading the title and abstract and supplemented our literature database by reading all references of the selected papers. Moreover, only available data on fiber abundance in the Mediterranean Sea over the 2015–2021 timeframe for biota, sediment and seawater were selected, and they are summarized in Tables 1 and 2, respectively. Finally, we selected 49 studies.

Table 1. Literature review about percentages of the predominant type of microplastic (fibers, fragments) in the Mediterranean Sea, region and year of sampling and instrumental method for the characterization of MPs in biota (invertebrates, fishes and sea turtles).

Area	Year of Sampling	Predominant Type (%)	Instrumental Method	References
Calvi Bay (Corsica)	2011–2012	All fibrous in shape	Raman	[25]
Southern Adriatic Sea	2013	78.5% fragments	FTIR	[61]
Central and North Adriatic Sea	2014	57% fragments	FTIR	[62]
Gulf of Lions (France)	2013	37.1%. fibers	Raman	[63]
Spanish Mediterranean coast	2014	71% fibers	n.a.	[26]
Mediterranean coast of Turkey	2015	70% fibers	FTIR	[27]
Mallorca Island (Balearic Islands, Western Mediterranean)	2014–2015	97% fibers	FTIR	[64]
Mallorca Island (Balearic Islands, Western Mediterranean)	n.a.	86.4% fibers	FTIR	[65]

Table 1. *Cont.*

Area	Year of Sampling	Predominant Type (%)	Instrumental Method	References
Giglio Island	2014	60% fragments	FTIR	[66]
Western Spanish Mediterranean coast	2015	83% fibers	FTIR	[67]
Northern Ionian Sea (*M. galloprovincialis; S. pilchardus, P. erithrinus, M. barbatus*)	2015	77.8% fragments 80% fragments 73.3% fragments 83.3% fragments	FTIR	[68]
Northern Cyprus	n.a.	85.3% fibers	FTIR	[69]
Adriatic and NE Ionian Sea (Croatian Sea; Slovenian Sea, NE Ionian sea)	2014–2015	75.6% fibers 97.7% fibers 79% fragments	n.a.	[70]
Spanish Catalan coast	2018	~60% fragments	FTIR	[71]
Tyrrhenian Sea (Northern coasts of Sicily, Gulf of Patti)	2019	93.3% fibers	μ-Raman, XPS and SEM-EDX	[72]
Mediterranean Sea (European hake, Red mullet)	n.a.	81% fibers 44% fibers	n.a.	[73]
Anzio coast (south of Rome, Tyrrhenian Sea)	2018	85.7% fibers	FTIR	[74]
Tyrrhenian Sea (northern coasts of Sicily, Gulf of Patti	2017	97.1% fibers	ATR-FTIR and μ-Raman	[75]
Iberian Peninsula coast and Balearic Islands (Western Mediterranean Sea)	2015	92.9% fibers	n.a.	[30]
Ligurian Sea	2011–2014	n.a. fibers	FTIR	[76]
Silba Island and Telašćica (Croatia, Adriatic Sea)	2007 and 2018	39.4–43.3% fibers 35.7–57.5% fibers	μ-FTIR	[77]
NW Mediterranean (Catalan coast)	n.a.	97% fibers	Raman	[78]
Northern, Central and Southern Adriatic Sea (Pelagic, benthopelagic, demerdal and benthic organism)	2016	38% fragments 50% fragments 53% fragments 61% fragments	μ-FTIR	[38]
Gulf of Patti (Southern Tyrrhenian Sea)	2019	93.3% fibers	FTIR and Raman	[79]
Catalan coast (NW Mediterranean Sea)	2007, 2017 and 2018	84.6% fibers	FTIR	[80]
Southeast Spain	2018, 2019	71.7% fibers	FTIR	[81]
Turkey, Izmir bay	2020	87.2% fibers	n.a.	[82]
Egypt cost (Mars Mtruh, Port Said, Alexandria, Damietta)	2020	100% fibers 50% fragments 96.2% fragments 85.2% fragments	ATR-FTIR	[83]

n.a: not available.

Table 2. Literature review about percentages of the predominant type of microplastic (fibers, fragments) in the Mediterranean Sea, region and year of sampling and instrumental method for characterization of MPs in sediments and seawaters.

Area	Year of Sampling	Predominant Type (%)	Instrumental Method	References
Mediterranean Sea	2001–2012	All fibrous in shape	FTIR	[22]
Gulf of Lion, the BalearicIslands, Sardinia and Corsica	2012	72% fragments	n.a.	[84]
Southern Adriatic Sea	2013	78.5% fragments	FTIR	[61]
Mediterranean Sea	2013	n.a.	ATR-FTIR	[85]
Aeolian Archipelago (central Mediterranean and Tyrrhenian sea)	n.a.	>85% fibers	n.a.	[86]
Mediterranean cost of Turkey	2015	70% fibers	FTIR	[27]
Balearic Islands, Adriatic and Ionian Sea	2011 and 2013	87.3% fragments	n.a.	[87]
Israeli Mediterranean coast	2013–2015	96.2% fragments	n.a.	[88]
Tyrrhenian Sea	2012	>88% fibers	n.a.	[89]
Central Adriatic Sea	2015	69.3% fibers	FTIR	[90]
Northern Tunisian coast(South Lake of Tunis, North Lake of Tunis, Carthage, Goulette)	2017	66.8% fibers 87.3% fibers 71% fibers 98.8% fibers	FTIR	[91]
Alboran, Catalan, Cretan and Levantine Sea	2009–2015	All fibrous in shape	FTIR	[92]
Ebro River Delta (Catalonia, Spain, Northwestern Mediterranean) (Sand, benthic sediment, surface water)	2017	89.5% fibers 75.1% fibers 46.1% fibers	μ-FTIR	[93]
Spanish Mediterranean Coast	2014–2015	82.9% fibers	n.a.	[19]
Silba Island and Telašćica (Croatia, Adriatic Sea)	2007 and 2018	33.1–76.9% fibers 82.7–97.3% fibers	μ-FTIR	[77]
Central-western Mediterranean Sea	2017	All fibrous in shape	FTIR	[94]
Mediterranean Sea	2017	All fibrous in shape	μ-FTIR	[37]
Northwestern Mediterranean Sea (Naples, Corsica north and south-east cost of France)	2014	All fibrous in shape	FTIR	[39]
Danube delta	2018	74.6% fibers	ATR-FTIR	[95]
Montenegrin cost	2019	55.5% fibers	ATR-FTIR	[96]
Ligurian Sea coastal	2018	80% fibers	n.a.	[97]

2.1. Most Abundance Shapes

The available literature data on the abundance of fibers in the Mediterranean Sea in the time frame 2015–2021 for biota, sediment and seawater are summarized in Tables 1 and 2, respectively.

Figure 2 summarizes all data presented in Tables 1 and 2, providing a global view of the occurrence of fibers, fragments, films and other shapes (i.e., spheres, pellets, sheets) in the Mediterranean Sea. The uniformly high presence of MFs in the water environment and biota samples of the Mediterranean area reflect a wider distribution of sources of textile fibers along the coastlines of the Mediterranean Sea, but also, the potential for atmospheric transport is much higher for MFs than for MPs [38].

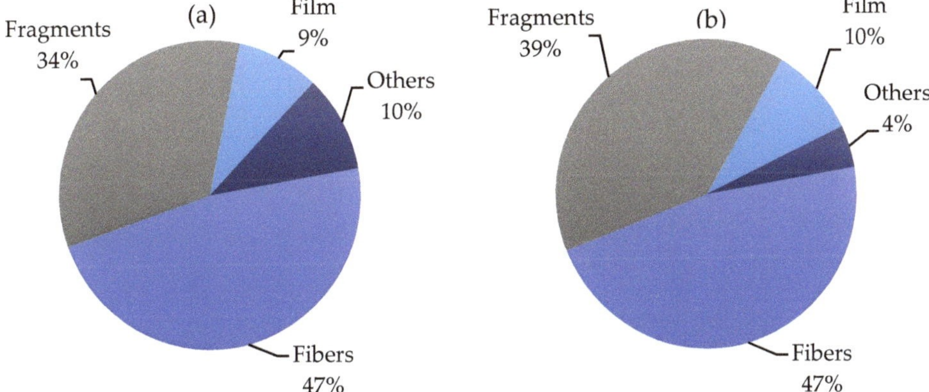

Figure 2. Pie charts showing the relative abundance (%) of fibers, fragments, films and other shapes (i.e., spheres, pellets, sheets) in the literature data globally in biota (**a**) and water (**b**) from the Mediterranean Sea.

In the Mediterranean Sea, MFs account for approximately 40% (range 1.6–85.9%) of fragments of micrometric size in the seawater and seabed, followed by fragments (mean 34.5%, range 1.6–72.7%), films (mean 17.3%, range 1.5–14.1%) and other shapes, such as spheres, pellets and sheets (mean 8.2%, range 1.6–24.1%). When considering MPs occurrence in marine organisms (invertebrates, fishes and sea turtles) collected from the Mediterranean Sea, we found 39.1% fragments, 37.8% fibers, 14.5% films and 8.7% other shapes. The matrices containing the higher amounts of fibers were sediments and seawater, where they reached 43.9%. The remaining part was formed by fragments (26.8%), films (22%) and others (7.3%).

Microfiber pollution has also been documented in all major ocean basins [21–23,28,37,98] as well as within the entire trophic web [20,24,29,32,33,59,99–104]. Natural microfibers are infrequently documented and not typically included in marine environment impact analyses, resulting in the underestimation of a potentially ubiquitous and harmful pollutant [28]. The literature data on the abundance of non-synthetic materials, including natural (i.e., cellulose), artificial (i.e., cellulose-based) and other (i.e., wool, silk and natural rubber) MFs, found in the Mediterranean Sea, are shown in Tables 3 and 4 for biota and sediment and seawater samples, respectively. The number of investigated individuals, the total amount of fibers, and the sub-sample analyzed are also reported. Table 3 focuses on the literature data on the abundance of natural (i.e., cellulose), artificial (i.e., cellulose-based), other non-synthetic (i.e., wool, silk) and plastic microfibers in biota (invertebrates, fish and sea turtles) of the Mediterranean Sea, together with the number of specimens sampled and the relative number of fibers found and analyzed.

Table 3. The literature data on the abundance of natural (i.e., cellulose), artificial (i.e., cellulose-based), other non-synthetic (i.e., wool, silk) and plastic microfibers in biota (invertebrates, fish and sea turtles) of the Mediterranean Sea, together with the number of specimens sampled and the relative number of fibers found and analyzed. Polyvinyl chloride (PVC), polyethylene terephthalate (PET), polypropylene (PP), polyethylene (PE), polyamide (PA), polyester (PEST), polystyrene (PS).

Species	No of Individuals	The total Amount of MFs	No of Identified MFs	Plastic Materials (%)									Non-Synthetic Materials (%)			Reference
				PVC	PET	PP	PE	PA	Nylon	PEST	PS	Others	Cellulose	Cellulose-Based	Others	
Macroinvertebrates *	235	91	11											100		[25]
Holothuria tubulosa (Gmelin, 1788) (Telaščica, Silba)	170	n.a.	n.a.	13.2		21.6	27.5	1.2	9.8			13.9	12.7			[77] **
				17.6	12.9	12.8	12.7		11.6		5.3	25.8	1.4			
Invertebrates and fishes *	<500	2079	100					1		10			74	8	7	[38]
Boobs boops (Linnaeus, 1758)	30	80	16										n.a.			[72]
Teleosts * / Elasmobranchs *	125	18	n.a.			31.2	6.2		12.5			31.2		18.7		[75] **
Mullus barbatus (Linnaeus, 1758)	118	167	39		31.1								56.8			[78]
Mullus surmuletus (Linnaeus, 1758)	417	n.a.	n.a.		36.3							33.3		30.3		[64]
Galeus melastomus (Rafinesque, 1810)	125	n.a.	n.a.		27.3	12.1	4.5	3				19.7		33.3		[65]
Caretta caretta (Linnaeus, 1758) / *Chelonia mydas* (Linnaeus, 1758)	102	811	169				20.7	4.9				61.2		5.8	7.4	[69]
Engraulis encrasicolus (Linnaeus, 1758)	9	35	19			45.7							54.3			[63] **
Plankton	29	1140	n.a.			10	41	3		12	5	22	7			[61] **
Sardina pilchardus (Walbaum, 1792) / *Engraulis encrasicolus* (Linnaeus, 1758)	105	41	24		12.5		8.3	4.2				8.3	54.1	8.3	4.2	[67] **
Sardina pilchardus (Walbaum, 1758) / *Engraulis encrasicolus* (Linnaeus, 1758)	264	n.a.	n.a.			n.a.	n.a.	n.a.						n.a.	—	[79] **
Sparus aurata (Linnaeus, 1758)	17	279	n.a.	2.2		2.9	21.5	2.2		4.4	2.2	71.3				[81]
Caranx crysos (Mitchill, 1815), *Liza aurata* (Risso, 1810), *Siganus rivulatus* (Rüppell, 1828), and *Epinephelus caninus* (Valenciennes, 1843)	3	480	n.a.		35.0		6.7		5			16.6		36.7		[83]
	3	383	n.a.		23.3				6.7			15.7		53.3		
	3	526	n.a.		18.8		8.4		8.4			6.7		56.7		
	3	648	n.a.		18.8		8.4		8.4			6.7		56.7		

n.a. not available. * Macroinvertebrates: *Gammarella fucicola* (Leach, 1814), *Gammarus aequicauda* (Martynov, 1931), *Melita hergensis* (Reid, 1939), *Nototropis guttatus* (Costa, 1853), *Nebalia strausi* (Risso, 1826), *Palaemon xiphias* (Risso, 1816), *Liocarcinus navigator* (Herbst, 1794), *Athanas nitescens* (Leach, 1813), *Galathea intermedia* (Liljeborg, 1851); invertebrates: *Mytilus galloprovincialis* (Lamarck, 1819), *Ostrea edulis* (Linnaeus, 1758), *Sabella spallanzanii* (Gmelin, 1805), *Actinia* sp., *Squilla mantis* (Linnaeus, 1758), *Penaeus kerathurus* (Forskål, 1775), *Nephrops norvegicus* (Linnaeus, 1758), *Palaemon* sp., *Paracentrotus lividus* (Lamarck, 1816), *Mnemiopsis leydi* (Agassiz, 1865), *Rhizostoma pulmo* (Macri, 1778); fishes: *Sardina pilchardus* (Walbaum, 1792), *Scomber scombrus* (Linnaeus, 1758), *Trachurus trachurus* (Linnaeus, 1758), *Merluccius merluccius* (Linnaeus, 1758), *Mullus barbatus* (Linnaeus, 1758), *Chelidonichthys lucernus* (Linnaeus, 1758), *Solea solea* (Linnaeus, 1758), *Sardinella aurita* (Valenciennes, 1847), *Diplodus vulgaris* (Geoffroy Saint-Hilaire, 1817), *Pagellus erythrinus* (Linnaeus, 1758), *Spondilosoma cantharus* (Linnaeus, 1758), *Tracinus draco* (Linnaeus, 1758), *Lithognathu mormyrus* (Linnaeus, 1758); Teleosts: *M. barbatus* and *Trigla lyra* (Linnaeus, 1758); Elasmobranchs: *Galeus melastomus* (Rafinesque, 1810), *Scyliorhinus canicula* (Linnaeus, 1758) and *Raja miraletus* (Linnaeus, 1758); *Sparus aurata* (Linnaeus, 1758); *Caranx crysos* (Mitchill, 1815), *Liza aurata* (Risso, 1810), *Siganus rivulatus* (Rüppell, 1828) and *Epinephelus caninus* (Valenciennes, 1843). ** in this study, percentages refer not only to fibers composition but to MPs generally.

Table 4. The literature data on the abundance of natural (i.e., cellulose), artificial (i.e., cellulose-based), other non-synthetic (i.e., wool, silk) and plastic microfibers in the sediment and water column from the Mediterranean Sea, together with the number of specimens sampled and the relative number of fibers found and analyzed.

Sample	No of Samples	The Total Amount of MFs	Subset of MFs for Analysis	Plastic Materials (%)										Non-Synthetic Materials (%)			Reference
				PVC	PET	PP	PE	PA	Nylon	PEST	PAN	PS	Others	Cellulose	Cellulose-Based	Others	
Sediment	12	n.a.	n.a.				23	14.7					5.3		56.9		[22]
	29	202	all		12.9	1	1	1					4.5	79.7			[92]
Sediment (Telaščica)	51	n.a.	n.a.	23					26.9				22.6	9.7	3.6		[77] *
Sediment (Silba)				18	16.2	14	12.2		17.2			1.2	6.8	13.7	0.8		
Beaches	5	197	25														[93] *
Sediment	n.a.	229				8	16	24		12			16	12			
Surface water	n.a.	293															
	29	1140	n.a.			10	41	3		12		5	22	7			[61] *
Seawater	916	23,593	2134			0.4		0.3	0.7	6.2			0.7		79.5	12.3	[37] **
	108	5466	336			0.9		0.6	0.9	4.2			0.3	47.3	39.6	5.4	[94]
Seawater (Haliotis outfall)	9	65	27			9		30		22	13	17			35		[39] ***
Seawater (Point B)	9	23	15					17		33	17		33		72		
Seawater (Bastia)	9	32	24					60		40					58		
Seawater (Dyfamed)	9	178	38					38		62					47		
Surface water	12	3289	93			33.3	30.1	1.1		1.1	4.3	4.3	3.3				[95]
Sediment	10	688	103			54.5	9.7	2.0					22.2	5.1		6.4	[96]

n.a. not available. * percentages refer not only to fiber composition but to MPs generally. ** percentages refer not only to the Mediterranean Sea but also include ocean basins. *** plastic material percentages refer only to 14–50% of synthetic material and the non-synthetic material percentages to total microfibers.

2.2. Non-Synthetic Composition of MFs in the Mediterranean Sea

Studies are increasingly documenting the ingestion of cellulose fibers by fishes and other organisms. A large portion of MFs found in biota from the Mediterranean Sea is cellulose-based, which consists of both dyed natural cellulose and manufactured fibers composed of regenerated cellulose. Natural fibers originating from plants are grouped into seed (e.g., cotton), bast (e.g., flax, hemp, kenaf, ramie), leaf (e.g., sisal) as well as tree fibers (e.g., wood), which have been extensively used for clothing, domestic woven fabrics and ropes for thousands of years [105]. Over the last years, and due to their wide availability, low cost, good recyclability, low density and high-specific mechanical strength, natural fibers have aroused interest in several applications as reinforcements in, e.g., the automotive and construction industries [106]. Wood pulp is the most important resource for producing cellulose-based human-made fibers, which can be manufactured through derivative and direct methods [107]. Human-made cellulosic materials represent a good compromise as the fiber-forming processes currently in use can lead to innovative fiber materials that combine the advantages of natural fibers and the possibility of tailor-made properties and chemical modifications [108]. In Europe, fibers and fabrics produced from regenerated cellulose are known as "viscose" whereas in the U.S., they are called rayon. Rayon makes up a significant proportion of synthetic microparticles found in the marine environment [20]. Rayon is used in cigarette filters, personal hygiene products and clothing and is introduced to the marine environment through sewage (e.g., washing of clothes) [23].

As reported above, Remy et al. (2015) identified the presence of artificial fibers in invertebrate communities; the artificial fibers were made of viscose, and the chemical characterization was confirmed by Raman spectroscopy. In addition, the colors of these fibers were two: Direct Red 28 and Ingrain Blue. These colors are used in the textile industry both for natural and artificial fibers. This shows that specific dyes cannot be linked to natural only or artificial only fibers, and thus, dyes cannot be used as reliable indicators for identifying synthetic or natural MFs or MPs [25].

Similar levels of non-synthetic fibers were detected in sea cucumbers, *Holothuria Tubulosa* (Gmelin, 1788), from Croatia, in which cellulose and cellulose acetate in stomach contents reached 13.3% and 14.8% in samples collected from Silba Island and Telašćica, respectively (ranging within 0–33.3% of total items). In the same study proposed by Renzi and Blašković (2020), fibers represented the larger number of recorded MPs in sediments from both Silba and Telašćica (ranging within 0–67.9% of total items). Among benthic species, sea cucumbers were selected as a target because they are widely representative of marine benthic species and are considered a key benthic taxonomic group to preserve marine ecosystem integrity (they are listed as protected species in some EU countries). Moreover, they play a crucial role in the food web through predation by stars, crustaceans, gastropods and fishes [77]. The presence of anthropogenic fibers both in *H. Tubulosa* and sediments (see Table 3) shows the large diffusion of these pollutants, supporting the hypothesis of active ingestion by these organisms from the surrounding environment. Similar results were obtained from Boskovic et al., 2021, where cellulose fibers in nine out of ten sediment samples of the Montenegrin coast were detected, which highlighted the predominance of fibers among all other MPs [96]. PP was detected in all the different sampling locations, while PE was in seven out of ten. The results showed the highest concentrations of MPs were in locations near highly populated centers, municipal effluent discharge restaurants, fishing and tourist activities, such as cruises.

The semipelagic fish bogue *Boops boops* (Linnaeus, 1758) is a commonly agreed-upon bioindicator in the Mediterranean Sea [18]. Italy is one of the European countries required to implement the Marine Strategy Framework Directive (MSFD), and the use of bioindicator species is strongly recommended by MSFD and other monitoring programs (e.g., UNEP/MAP) to increase the knowledge on the extent of marine litter pollution and its impacts on marine species [109]. Since *B. boops* is an omnivorous species, which feeds both benthic and pelagic preys, living on diverse types of the sea bottom (sandy, muddy, rocky and seagrass beds) [100], it has been proposed to act as a sentinel for microplastic pollution in the Mediterranean small-scale pelagic environment (https://plasticbustersmpas.interreg-med.eu, accessed on 15 June 2022). In the study conducted by Savoca et al. (2019) in the Gulf of Patti [72], the authors reported, for the first time in the Mediterranean Sea, the ingestion of human-made cellulose fibers in bogue specimens, assuming that the high presence of fibers found in their stomach might depend on the habitat and its extension. As a matter of fact, the urban wastewater treatment of the area is not powerful enough to retain all the fibers, especially during the summer when many tourists populate the area [110]. Their data complied with the studies of Fastelli et al. (2016) and Cannas et al. (2017) carried out in the same area of the Mediterranean Sea [86,89]. Similar results were also obtained by Rios-Fuster et al. (2019), who evaluated the ingestion of anthropogenic particles in four species of fish, including *B. boops* [30], and found a percentage of 92.86% of fibers and 7.14% of fragments. Previous studies carried out using the same species as a bioindicator detected similar MFs occurrence levels in the Balearic Islands [100]. In this study, a total of 731 items were observed in 195 full gastrointestinal tracts of bogue. The fibers were only detected and characterized by different colors. Similarly, Neves et al. (2015) recorded a total of 73 MPs in the 32 bogues sampled in the North Atlantic, off the Portuguese coast, 48 of which (65.8%) were fibers and 25 (34.2%) were particles [24]. On the contrary, Garcia-Garin et al. (2019) found a prevalence of fragments (60%) in bogues samples collected from the Spanish Catalan coast near Barcelona. The authors suggested that the high amount of fragments found in the organisms was due to the severe MPs pollution present in the sampling area [71].

Avio et al. (2020) provided a comprehensive characterization of the ingestion of microplastics in several fish and invertebrate species from the Adriatic Sea, which is considered a preferential area of plastic accumulation in the Mediterranean. Almost 500 organisms, including benthic and pelagic invertebrates and benthopelagic, pelagic and demersal fish species, were collected (see Table 3). Textile MFs were abundant in Adriatic food webs occurring in all the analyzed species with frequencies (ranging between 40% and 70%) higher than those reported for MPs; an elevated percentage of MFs was of natural (74% cotton, 8% wool) and non-synthetic origin (8%) [38]. One of the species studied by Avio et al. (2020) was the European hake, *Merluccius Merluccius* (Linnaeus, 1758), which is an important predatory species inhabiting a wide range of depths (20–1000 m) throughout the Mediterranean Sea and the north-eastern Atlantic region. It is one of the main commercial and most exploited species of fish in all northern Mediterranean countries [111]. Bellas et al. (2016) and Giani et al. (2019) investigated the occurrence of MPs in *M. Merluccius*, and their results were comparable to those of Avio et al. (2020): the detected MPs were mostly constituted by fibers (71% and 81%, respectively). In both studies, however, no chemical characterization of the fibers was provided [26,73]. Interestingly, previous studies conducted by Suaria et al. (2015) and Avio et al. (2015) in the same area of the Mediterranean Sea reported a predominance of fragments over fibers in plankton and *M. Merluccius* specimens (78.5% and 57%, respectively).

The usual hake diet consists mainly of Crustacea (especially Decapoda) and teleost fishes (i.e., *Engraulis Encrasicolus* and *Cepola microphthalmia*). European anchovy *E. Encrasicolus*, together with *Sardina pilchardus* (Walbaum, 1792), are some of the most captured fish species in the Mediterranean Sea and are thus of economic importance. Moreover, they are directly subjected to MP's pollution because they are planktivorous and are mainly filter-feeding. Both of the species have been used in MP studies, and natural and plastic microparticles have been found in both of the organisms with a predominance of MFs (83%) [61,62]. Natural fibers (such as cotton) accounted for 54.1% and other cellulose-based fibers for 12.5%. Plastic materials, especially PET, PE and PA, accounted for 33.3%. A study conducted by Collard et al. (2015) showed that the majority of "non-plastic" particles found in *E. Encrasicolus* collected from the Gulf of Lions were made of cellulose (54.3%) [67]. Similar results to those presented by Collard et al. (2015) and Compa et al. (2018) in the same Mediterranean area were confirmed by Sanchez-Vidal et al. (2018) [67]. Sanchez-Vidal et al. (2018) reported the predominance of cellulosic fibers (79.7%) over other synthetic polymers (see Table 4) in the sediment on the Spanish Mediterranean coast [92]. Moreover, a recent study carried out in the Southern Tyrrhenian Sea by Savoca et al. (2020) confirmed the presence of polymers, such as PP, PA, Nylon and PE, and human-made cellulose, such as rayon, in *E. Encrasicolus*, and *S. Pilchardus*. Instead, Neves et al. (2015) noted the presence of MPs in fish from the coast of Portugal, highlighting the presence of rayon fibers through μ-FTIR, one of the techniques more suitable for distinguishing and determining the chemical composition of fibers [24].

Red mullet, *Mullus Barbatus* (Linnaeus, 1758), and striped red mullet, *Mullus surmuletus* (Linnaeus, 1758), are demersal fish species widely spread in the Mediterranean Sea and the NE Atlantic [78], and are considered important resources for coastal Mediterranean fisheries [112]. Due to its dietary habits, *M. Barbatus* is in constant contact with sediment and, therefore, it is exposed to the pollutants present in this matrix. Thus, it has been widely proposed as a sentinel species for several pollutants. Fiber ingestion by the red mullet has been widely reported in *M. Barbatus* samples collected from several areas of the Mediterranean Sea, including the Turkish shore, Adriatic and Tyrrhenian Seas and the Mediterranean Spanish Coast [26,27,61,73,75,78]. It is interesting to note that some of these studies showed that 56.79% of the fibers found in the fishes were cellulose-based, almost twice as many as PET (31.14%) [78].

M. Surmuletus is sensitive to marine debris contamination and microplastic ingestion [112]. In the study carried out by Alomar et al. (2017), the vast majority of identified microplastics in *M. Surmuletus* samples were filaments (30% of which were non-plastic material) [64].

Capillo et al. (2020) investigated five demersal fish species from the Southern Tyrrhenian Sea, including the red mullet *M. Barbatus*, the piper gurnard *Trigla Lyra* (Linnaeus, 1758) and the blackmouth catshark *Galeus Melastomus* (Rafinesque, 1810). A total of 97.1% of the microparticles found in all the samples were fibers. Specifically, the red mullet presented high values of plastic material (mainly PTFE, 75%), while the items found in specimens of *T. Lyra* were all composed of cellulose (100%). The feeding behavior of *T. Lyra* is the same of *M. Barbatus*, i.e., the fish swallows sediment (together with the prey) and then expels them through the gills.

G. Melastomus has a different feeding behavior compared to *T. Lyra*; it is a benthopelagic predator that feeds mainly on demersal invertebrates (shrimps and cephalopods) and mesopelagic fish. It could ingest MPs during predation, biomagnifying them along the food chain. The estimation of the percentage of MFs ingestion in *G. Melastomus* in this study (especially nylon) is different from those reported in other areas of the Mediterranean Sea [75]. Indeed, a high proportion of cellulosic-based fibers in this species was found in samples from the Balearic Islands (western Mediterranean Sea) area, where Alomar and Deudero (2016) reported the dominance of cellophane over other synthetic polymers. In the stomachs of this elasmobranch species, the authors showed that 86.36% of the identified particles were filaments, while the rests were fragments and films. Woodall et al. (2014), Sanchez-Vidal et al. (2018), Filgueiras et al. (2019) and Suaria et al. (2020) suggested that *G. Melastomus* ingests fibers directly from the seafloor and water column [19,22,37,65,92]. Similar results were achieved by Valente et al. (2019), who identified the presence of 221 synthetic fibers (85.7% of the particles) in *G. Melastomus* collected from the Tyrrhenian Sea. These data comply with the results reported in a study conducted by Cannas et al. (2017) in the same part of the Mediterranean [74,89]. Anastasopoulou et al. (2013) have also recorded MP ingestion by *G. Melastomus* in the Ionian Sea, but unlike the results obtained from the previous studies, the percentage of fibers reached only 3% [113]. In agreement with these results, Ruiz-Orejòn et al. (2016) reported 87.3% of hard plastic fragments as the majority of the material observed in the Ionian Sea, demonstrating how the marine environment can affect biota microparticles ingestion [87].

Finally, in a recent study by Sayed et al. (2021) along Egypt's coast, the presence of cellulose-based fibers was observed while analyzing the level of MPs in the digestive tracts of *Caranx Crysos*, *Liza Aurata*, *Siganus Rivulatus* and *Epinephelus Caninus* from the Eastern

Harbor. Plastic particles were evident in all fish samples, including seven thermoplastic polymers. Rayon and polyethylene terephthalate were the most dominant types of polymers in fish [83].

Due to the concentrations of plastic in the Mediterranean Sea, loggerhead sea turtles, *Caretta Caretta* (Linneaus, 1758), were confirmed by Matiddi et al. (2017) as the main target species for monitoring MP ingestion by marine organisms. The turtles tend to ingest marine litter, confusing it with natural prey [114]. The study conducted by Duncan et al. (2018) provides an overview of the presence of microplastics in various marine turtle specimens. The analysis of marine turtles' specimens reveals a high abundance of fibers unanimous in the three basins (Atlantic 77.1%, Mediterranean Sea 85.3%, Pacific 64.8%). Of these, a subsample of the isolated particles was tested using FTIR to determine the polymeric composition, revealing the presence of both synthetics (mainly PE, ethylene propylene, PEST and polyacrylamide) and cellulose-based materials (rayon, natural rubber and plant protein) [69].

2.3. Color of MFs

From the available literature data, four different colors in MFs were found to be more abundant in the Mediterranean Sea, both in biota (fishes, invertebrates and sea turtles) and in the seabed and seawater samples. As indicated in Figure 3, the dominant color was black (ranged between 12.1–100%), followed by transparent and clear colors (2.5–50.3%) and blue (10.1–45.8%). Red (3.8–27%) and others (2.2–20%) were less abundant.

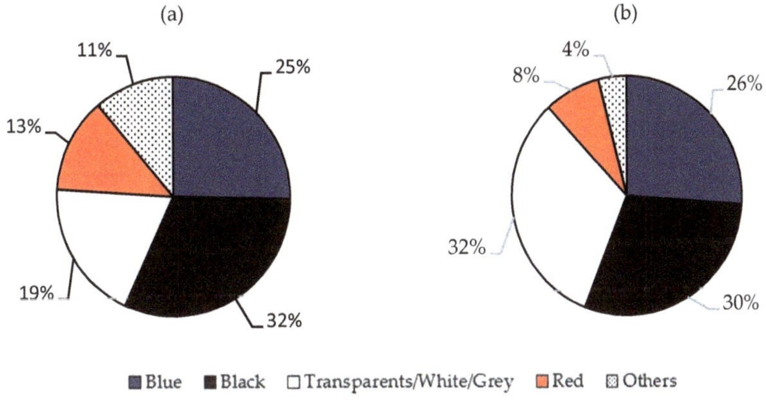

Figure 3. Most abundant colors in MFs present in the literature data from the Mediterranean Sea, both in the biota (**a**), and in seabed and seawater samples (**b**).

Instead, in open basins (Atlantic, Pacific, Indian Arctic and Southern Oceans) the following order was observed: blue (10.1–88%) > black (8.8–57.1%) > transparent (2.5–47%) > red (5.2–42%) > others (1–9%). The MF's color could potentially increase their bioavailability due to their resemblance to prey objects. There is evidence of visual confusion between prey and anthropogenic particles [30]. Predatory fish show a preference for ingesting blue fibers, while transparent fibers may be confused due to their resemblance to gelatinous prey or can be ingested accidentally via filtration [76]. Furthermore, studies noticed, without providing any explanation, that planktivorous fish seem to ingest whiter, lighter and bluer fiber colors [115]. The only speculation that was made to explain this observation was that these colors are the most abundant found in the fibers collected from the Mediterranean Sea. Another aspect that has to be taken into account is that some chemical treatments used during the extraction procedure of the fibers can cause physical damage and discoloration of the microplastics, as shown by Cole et al. (2014) [116].

2.4. Self-Contamination

During the analysis of MFs, one of the biggest problems is the contamination of the sample by those who carry out the sampling, treatment and analysis of the samples. Contaminations can occur through the use of instrumentation that releases particles into the environment or from the researcher's clothing [117]. The procedures attempt to control contaminants entering samples from analyst clothing, airborne sources, laboratory surfaces, equipment and consumables used, but there is not yet a standardized method to prevent contaminations. Over the years, more and more precautions have been taken for the treatments of the samples, in fact, initially, the procedures did not take into account the possible self-contamination [118], while techniques have recently been adopted to avoid this problem [119–121]. For example, Gaylarde et al. (2021) cleaned all materials used with ethanol and filtered deionized water, put on colored suits and performed the fish dissection and digestion protocols in a clean airflow cabin [122]; instead, Barrows et al. (2018) tested microplastic contamination during the treatment of the sample: cleaning all laboratory surfaces, analyzing laboratory water and laboratory air and analyzing blanks of the filtrate used to rinse the sample bottle and filtration apparatus. The results showed average contamination of 0.005 pieces per 0.010 L of water and 0.154 pieces per 8 min of exposure to air from synthetic and non-synthetic MPs [28]. As highlighted by Prata et al. (2021), less than 50% of studies on MPs do not collect and analyze controls and blanks during the sampling phase and processing step of the sample [119]. Moreover, only some studies involve taking "control" samples of possible sources of contamination from MPs and MFs and the use of colored cotton clothes [96,121–125]). Finally, as highlighted in a study by Scopetani et al. (2020), the level of self-contamination in MPs studies is not negligible, highlighting the importance of finding a standardized method to avoid the overestimation of MPs and MFs in environmental matrices [126].

2.5. Size

From the data available in the literature, we can notice that most of the studies conducted in the Mediterranean Sea that focused on microfibers pollution investigated microfibers with a length ranging between 1 and 2 mm (Figure 4).

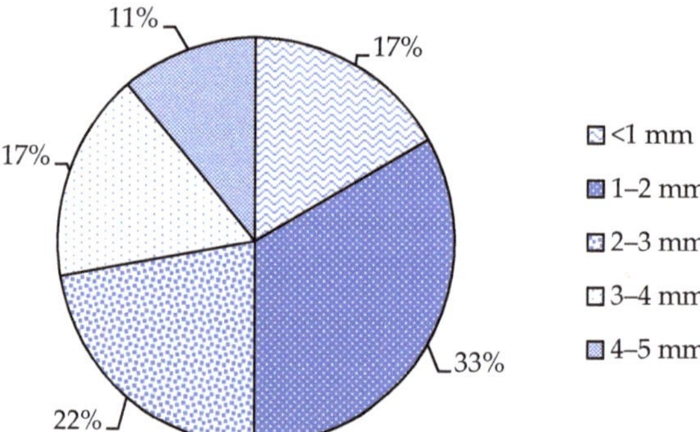

Figure 4. A comparison of the literature data of percentages frequency of different fiber lengths in biota and water samples from the Mediterranean Sea.

We can hypothesize that this may probably represent the optimal size to carry out investigations regarding the chemical composition of the fibers, but further research is needed to deepen this aspect. The small size of the MFs is relevant as it determines the potential impact of these contaminants on the ecosystem and the bioaccumulation/biomagnification in biota from ingestion. If the fibers are ingested by marine organisms, they can damage them, block and affect the physical performance of the digestive tract of fish [20]. The effect caused by the volume occupied in the digestive tracts does not depend on the size of the individual fibers because these can tangle and form larger agglomerates. Indeed, fibers longer than 5 mm (usually not considered in studies on microplastics) can tangle with themselves and with other fibers and occupy large volumes in the stomach, volumes similar to those of agglomerates of shorter fibers [20,80]. Therefore, it is difficult to find a correlation between fiber size and the effects on the organism, but if they do not tangle, as shown in a study by Grigorakis et al. (2017), they can cross the entire digestive tract and be expelled from the body without causing damage [127]. Not all the studies agree on the possibility of detecting the presence of microplastics up to 0.6 mm in organs not belonging to the digestive system, as detected by Avio et al. (2015) in fish mullet liver [62]. Instead, many authors believe that probably only MPs and MFs smaller than 100 μm or their additives can come into contact with organs not belonging to the digestive system and cross the intestinal barrier [78].

3. Conclusions

As described above, the investigations in the Mediterranean Sea provide insight into the level of microfiber pollution and underline the necessity to use specific analytical techniques to explore and confirm MFs composition to avoid overestimation when assessing the level of MP occurrence in the marine environment. This review underlines the need to distinguish natural fibers from plastic ones, given the high number of fibers found in the marine environment and biota. Additionally, future studies should better investigate the impact of fibers on biota since synthetic fibers tangle easily and can originate bundles of fibers causing obstruction in organs and hindering or preventing feeding. The same consideration is applied to cellulosic fibers, even if they do not constitute an environmental problem in themselves, but any additives or dyes within them could potentially be carcinogenic and harmful to sea organisms and, consequently, to humans. Overall, the results of this review provide the basis to monitor the impacts of microfiber pollution on the sea ecosystems in the Mediterranean Sea, which can be used to investigate other basins of the world for future risk frameworks.

Author Contributions: S.S., Conceptualization, Data Discussion, Writing—review and editing; E.D.B., Investigation, Data discussion, Writing—review and editing; T.M., Conceptualization, Data Curation, Writing—review and editing; C.S. (Chiara Sarti), Data Curation, Review the text; R.C., Data Discussion, Review the text; D.R., Conceptualization, Supervision, Review the text; C.S. (Costanza Scopetani), Conceptualization, Supervision, Review the text, Data Discussion; A.C., Conceptualization, Supervision, Investigation, Data Discussion, Review the text. All authors have read and agreed to the published version of the manuscript.

Funding: This research received no external funding.

Institutional Review Board Statement: Not applicable.

Informed Consent Statement: Not applicable.

Conflicts of Interest: The authors declare no conflict of interest.

References

1. Abeynayaka, A.; Kojima, F.; Miwa, Y.; Ito, N.; Nihei, Y.; Fukunaga, Y.; Yashima, Y.; Itsubo, N. Rapid Sampling of Suspended and Floating Microplastics in Challenging Riverine and Coastal Water Environments in Japan. *Water* **2020**, *12*, 1903. [CrossRef]
2. Cozzolino, L.; Nicastro, K.R.; Zardi, G.I.; de los Santos, C.B. Species-specific plastic accumulation in the sediment and canopy of coastal vegetated habitats. *Sci. Total Environ.* **2020**, *723*, 138018. [CrossRef] [PubMed]
3. Hartmann, N.B.; Hüffer, T.; Thompson, R.C.; Hassellöv, M.; Verschoor, A.; Daugaard, A.E.; Rist, S.; Karlsson, T.; Brennholt, N.; Cole, M.; et al. Are We Speaking the Same Language? Recommendations for a Definition and Categorization Framework for Plastic Debris. *Environ. Sci. Technol.* **2019**, *53*, 1039–1047. [CrossRef] [PubMed]
4. Antunes, J.; Frias, J.; Sobral, P. Microplastics on the Portuguese coast. *Mar. Pollut. Bull.* **2018**, *131*, 294–302. [CrossRef] [PubMed]
5. Cincinelli, A.; Martellini, T.; Guerranti, C.; Scopetani, C.; Chelazzi, D.; Giarrizzo, T. A potpourri of microplastics in the sea surface and water column of the Mediterranean Sea. *TrAC Trends Anal. Chem.* **2019**, *110*, 321–326. [CrossRef]

6. Cincinelli, A.; Scopetani, C.; Chelazzi, D.; Lombardini, E.; Martellini, T.; Katsoyiannis, A.; Fossi, M.C.; Corsolini, S. Microplastic in the surface waters of the Ross Sea (Antarctica): Occurrence, distribution and characterization by FTIR. *Chemosphere* **2017**, *175*, 391–400. [CrossRef] [PubMed]
7. Wu, P.; Huang, J.; Zheng, Y.; Yang, Y.; Zhang, Y.; He, F.; Chen, H.; Quan, G.; Yan, J.; Li, T.; et al. Environmental occurrences, fate, and impacts of microplastics. *Ecotoxicol. Environ. Saf.* **2019**, *184*, 109612. [CrossRef]
8. Hurley, R.; Woodward, J.; Rothwell, J.J. Microplastic contamination of river beds significantly reduced by catchment-wide flooding. *Nat. Geosci.* **2018**, *11*, 251–257. [CrossRef]
9. Klein, S.; Worch, E.; Knepper, T.P. Occurrence and spatial distribution of microplastics in river shore sediments of the rhine-main area in Germany. *Environ. Sci. Technol.* **2015**, *49*, 6070–6076. [CrossRef]
10. Nizzetto, L.; Bussi, G.; Futter, M.N.; Butterfield, D.; Whitehead, P.G. A theoretical assessment of microplastic transport in river catchments and their retention by soils and river sediments. *Environ. Sci. Process. Impacts* **2016**, *18*, 1050–1059. [CrossRef]
11. Dris, R.; Gasperi, J.; Rocher, V.; Tassin, B. Synthetic and non-synthetic anthropogenic fibers in a river under the impact of Paris Megacity: Sampling methodological aspects and flux estimations. *Sci. Total Environ.* **2018**, *618*, 157–164. [CrossRef] [PubMed]
12. Gasperi, J.; Wright, S.L.; Dris, R.; Collard, F.; Mandin, C.; Guerrouache, M.; Langlois, V.; Kelly, F.J.; Tassin, B. Microplastics in air: Are we breathing it in? *Curr. Opin. Environ. Sci. Health* **2018**, *1*, 1–5. [CrossRef]
13. Koelmans, A.A.; Bakir, A.; Burton, G.A.; Janssen, C.R. Microplastic as a Vector for Chemicals in the Aquatic Environment: Critical Review and Model-Supported Reinterpretation of Empirical Studies. *Environ. Sci. Technol.* **2016**, *50*, 3315–3326. [CrossRef] [PubMed]
14. Hahladakis, J.N.; Velis, C.A.; Weber, R.; Iacovidou, E.; Purnell, P. An overview of chemical additives present in plastics: Migration, release, fate and environmental impact during their use, disposal and recycling. *J. Hazard. Mater.* **2018**, *344*, 179–199. [CrossRef] [PubMed]
15. Wagner, M.; Scherer, C.; Alvarez-Muñoz, D.; Brennholt, N.; Bourrain, X.; Buchinger, S.; Fries, E.; Grosbois, C.; Klasmeier, J.; Marti, T.; et al. Microplastics in freshwater ecosystems: What we know and what we need to know. *Environ. Sci. Eur.* **2014**, *26*, 12. [CrossRef]
16. Kärrman, A.; Schönlau, C.; Engwall, M. Exposure and Effects of Microplastics on Wildlife. A review of existing data. *DiVA.* 2016, p. 39. Available online: https://www.diva-portal.org/smash/get/diva2:921211/FULLTEXT01.pdf (accessed on 14 March 2022).
17. Anbumani, S.; Kakkar, P. Ecotoxicological effects of microplastics on biota: A review. *Environ. Sci. Pollut. Res.* **1999**, *25*, 14373–14396. [CrossRef] [PubMed]
18. Fossi, M.C.; Pedà, C.; Compa, M.; Tsangaris, C.; Alomar, C.; Claro, F.; Ioakeimidis, C.; Galgani, F.; Hema, T.; Deudero, S.; et al. Bioindicators for monitoring marine litter ingestion and its impacts on Mediterranean biodiversity. *Environ. Pollut.* **2018**, *237*, 1023–1040. [CrossRef]
19. Filgueiras, A.V.; Gago, J.; Campillo, J.A.; León, V.M. Microplastic distribution in surface sediments along the Spanish Mediterranean continental shelf. *Environ. Sci. Pollut. Res.* **2019**, *26*, 21264–21273. [CrossRef]
20. Lusher, A.L.; McHugh, M.; Thompson, R.C. Occurrence of microplastics in the gastrointestinal tract of pelagic and demersal fish from the English Channel. *Mar. Pollut. Bull.* **2013**, *67*, 94–99. [CrossRef]
21. Lusher, A.L.; Tirelli, V.; O'Connor, I.; Officer, R. Microplastics in Arctic polar waters: The first reported values of particles in surface and sub-surface samples. *Sci. Rep.* **2015**, *5*, 14947. [CrossRef]
22. Woodall, L.C.; Sanchez-Vidal, A.; Canals, M.; Paterson, G.L.J.; Coppock, R.; Sleight, V.; Calafat, A.; Rogers, A.D.; Narayanaswamy, B.E.; Thompson, R.C. The deep sea is a major sink for microplastic debris. *R. Soc. Open Sci.* **2014**, *1*, 140317. [CrossRef] [PubMed]
23. Obbard, R.W.; Sadri, S.; Wong, Y.Q.; Khitun, A.A.; Baker, I.; Richard, C. Who Where Why—Wordpress blog—Community mapping examples. *Earth's Future* **2014**, *2*, 315–320. [CrossRef]
24. Neves, D.; Sobral, P.; Ferreira, J.L.; Pereira, T. Ingestion of microplastics by commercial fish off the Portuguese coast. *Mar. Pollut. Bull.* **2015**, *101*, 119–126. [CrossRef] [PubMed]
25. Remy, F.; Collard, F.; Gilbert, B.; Compère, P.; Eppe, G.; Lepoint, G. When Microplastic Is Not Plastic: The Ingestion of Artificial Cellulose Fibers by Macrofauna Living in Seagrass Macrophytodetritus. *Environ. Sci. Technol.* **2015**, *49*, 11158–11166. [CrossRef] [PubMed]
26. Bellas, J.; Martínez-Armental, J.; Martínez-Cámara, A.; Besada, V.; Martínez-Gómez, C. Ingestion of microplastics by demersal fish from the Spanish Atlantic and Mediterranean coasts. *Mar. Pollut. Bull.* **2016**, *109*, 55–60. [CrossRef] [PubMed]
27. Güven, O.; Gökdağ, K.; Jovanović, B.; Kıdeyş, A.E. Microplastic litter composition of the Turkish territorial waters of the Mediterranean Sea, and its occurrence in the gastrointestinal tract of fish. *Environ. Pollut.* **2017**, *223*, 286–294. [CrossRef] [PubMed]
28. Barrows, A.P.W.; Cathey, S.E.; Petersen, C.W. Marine environment microfiber contamination: Global patterns and the diversity of microparticle origins. *Environ. Pollut.* **2018**, *237*, 275–284. [CrossRef]
29. Bessa, F.; Barría, P.; Neto, J.M.; Frias, J.P.G.L.; Otero, V.; Sobral, P.; Marques, J.C. Occurrence of microplastics in commercial fish from a natural estuarine environment. *Mar. Pollut. Bull.* **2018**, *128*, 575–584. [CrossRef]
30. Rios-Fuster, B.; Alomar, C.; Compa, M.; Guijarro, B.; Deudero, S. Anthropogenic particles ingestion in fish species from two areas of the western Mediterranean Sea. *Mar. Pollut. Bull.* **2019**, *144*, 325–333. [CrossRef]
31. Hossain, M.S.; Rahman, M.S.; Uddin, M.N.; Sharifuzzaman, S.M.; Chowdhury, S.R.; Sarker, S.; Nawaz Chowdhury, M.S. Microplastic contamination in Penaeid shrimp from the Northern Bay of Bengal. *Chemosphere* **2020**, *238*, 124688. [CrossRef]

32. Parton, K.J.; Godley, B.J.; Santillo, D.; Tausif, M.; Omeyer, L.C.M.; Galloway, T.S. Investigating the presence of microplastics in demersal sharks of the North-East Atlantic. *Sci. Rep.* **2020**, *10*, 12204. [CrossRef] [PubMed]
33. Iliff, S.M.; Wilczek, E.R.; Harris, R.J.; Bouldin, R.; Stoner, E.W. Evidence of microplastics from benthic jellyfish (Cassiopea xamachana) in Florida estuaries. *Mar. Pollut. Bull.* **2020**, *159*, 111521. [CrossRef] [PubMed]
34. Liu, J.; Yang, Y.; Ding, J.; Zhu, B.; Gao, W. Microfibers: A preliminary discussion on their definition and sources. *Environ. Sci. Pollut. Res.* **2019**, *26*, 29497–29501. [CrossRef] [PubMed]
35. Bal, B.; Ghosh, S.; Das, A.P. Microbial recovery and recycling of manganese waste and their future application: A review. *Geomicrobiol. J.* **2018**, *36*, 85–96. [CrossRef]
36. Mohanty, S.; Ghosh, S.; Bal, B.; Prasad, A. A review of biotechnology processes applied for manganese recovery from wastes. *Rev. Environ. Sci. Bio/Technol.* **2018**, *17*, 791–811. [CrossRef]
37. Suaria, G.; Achtypi, A.; Perold, V.; Lee, J.R.; Pierucci, A.; Bornman, T.G.; Aliani, S.; Ryan, P.G. Microfibers in oceanic surface waters: A global characterization. *Sci. Adv.* **2020**, *6*, 1–9. [CrossRef]
38. Avio, C.G.; Pittura, L.; d'Errico, G.; Abel, S.; Amorello, S.; Marino, G.; Gorbi, S.; Regoli, F. Distribution and characterization of microplastic particles and textile microfibers in Adriatic food webs: General insights for biomonitoring strategies. *Environ. Pollut.* **2020**, *258*, 113766. [CrossRef]
39. Pedrotti, M.L.; Petit, S.; Eyheraguibel, B.; Kerros, M.E.; Elineau, A.; Ghiglione, J.F.; Loret, J.F.; Rostan, A.; Gorsky, G. Pollution by anthropogenic microfibers in North-West Mediterranean Sea and efficiency of microfiber removal by a wastewater treatment plant. *Sci. Total Environ.* **2021**, *758*, 144195. [CrossRef]
40. Karimah, A.; Ridho, M.R.; Munawar, S.S.; Adi, D.S.; Ismadi; Damayanti, R.; Subiyanto, B.; Fatriasari, W.; Fudholi, A. A review on natural fibers for development of eco-friendly bio-composite: Characteristics, and utilizations. *J. Mater. Res. Technol.* **2021**, *13*, 2442–2458. [CrossRef]
41. Cai, H.; Du, F.; Li, L.; Li, B.; Li, J.; Shi, H. A practical approach based on FT-IR spectroscopy for identi fi cation of semi-synthetic and natural celluloses in microplastic investigation. *Sci. Total Environ.* **2019**, *669*, 692–701. [CrossRef]
42. Conley, K.; Clum, A.; Deepe, J.; Lane, H.; Beckingham, B. Wastewater treatment plants as a source of microplastics to an urban estuary: Removal efficiencies and loading per capita over one year. *Water Res. X* **2019**, *3*, 100030. [CrossRef] [PubMed]
43. Helcoski, R.; Yonkos, L.T.; Sanchez, A.; Baldwin, A.H. Wetland soil microplastics are negatively related to vegetation cover and stem density. *Environ. Pollut.* **2019**, *256*, 113391. [CrossRef] [PubMed]
44. Athey, S.N.; Erdle, L.M. Are We Underestimating Anthropogenic Micro fi ber Pollution? A Critical Review of Occurrence, Methods, and Reporting. *Environ. Toxicol. Chem.* **2021**, *41*, 822–837. [CrossRef] [PubMed]
45. O'neill, C.; Hawkes, F.R.; Hawkes, D.L.; Lourenço, N.D.; Pinheiro, H.M.; Delée, W. Colour in textile effluents-sources, measurement, discharge consents and simulation: A review. *J. Chem. Technol. Biotechnol.* **1999**, *74*, 1009–1018. [CrossRef]
46. Kwak, J.I.; Liu, H.; Wang, D.; Lee, Y.H.; Lee, J.S.; An, Y.J. Critical review of environmental impacts of microfibers in different environmental matrices. *Comp. Biochem. Physiol. Part C Toxicol. Pharmacol.* **2022**, *251*, 109196. [CrossRef]
47. Verma, Y. Acute toxicity assessment of textile dyes and textile and dye industrial effluents using Daphnia magna bioassay. *Toxic Ind. Health* **2008**, *24*, 491–500. [CrossRef]
48. Ferraz, E.R.; Li, Z.; Boubriak, O.; de Oliveira, D.P. De Current Issues Hepatotoxicity Assessment of the Azo Dyes Disperse Orange 1 (DO1), Disperse Red 1 (DR1,) and Disperse Red 13 (DR13) in HEPG2 Cells. *J. Toxicol. Environ. Health Part A* **2012**, *75*, 991–999. [CrossRef]
49. OEKO-TEX OEKO-TEX. Available online: https://www.oeko-tex.com/en/ (accessed on 14 March 2022).
50. No, C.A.S. Agents Classified by the IARC Monographs. *Lancet Oncol.* **2016**, *1–123*, 1–37.
51. Shen, B.; Liu, H.; Ou, W.; Eilers, G.; Zhou, S. Toxicity induced by Basic Violet 14, Direct Red 28 and Acid Red 26 in zebrafish larvae. *J. Appl. Toxicol.* **2015**, *35*, 1473–1480. [CrossRef]
52. IARC. Chemical agents and related occupations. In *IARC Monographs on the Evaluation of Carcinogenic Risks to Humans*; IARC: Lyon, France, 2012; Volume 100, pp. 9–562.
53. McCormick, A.R.; Hoellein, T.J.; London, M.G.; Hittie, J.; Scott, J.W.; Kelly, J.J. Microplastic in surface waters of urban rivers: Concentration, sources, and associated bacterial assemblages. *Ecosphere* **2016**, *7*, e01556. [CrossRef]
54. Espinosa, C.; Esteban, M.Á.; Cuesta, A.; Microplastics in Aquatic Environments and and Their Toxicological Implications for Fish. Licens. *InTech 2016*, 113–145. Available online: https://www.intechopen.com/chapters/52031 (accessed on 14 March 2022).
55. Zhao, Y.; Wang, C.; Xia, S.; Jiang, J.; Hu, R.; Yuan, G.; Hu, J. Biosensor medaka for monitoring intersex caused by estrogenic chemicals. *Environ. Sci. Technol.* **2014**, *48*, 2413–2420. [CrossRef] [PubMed]
56. Rochman, C.M.; Lewison, R.L.; Eriksen, M.; Allen, H.; Cook, A.M.; Teh, S.J. Polybrominated diphenyl ethers (PBDEs) in fish tissue may be an indicator of plastic contamination in marine habitats. *Sci. Total Environ.* **2014**, *476–477*, 622–633. [CrossRef] [PubMed]
57. Meeker, J.D.; Sathyanarayana, S.; Swan, S.H. Phthalates and other additives in plastics: Human exposure and associated health outcomes. *Philos. Trans. R. Soc. B Biol. Sci.* **2009**, *364*, 2097–2113. [CrossRef]
58. Das, A.; Mishra, S. Biodegradation of the metallic carcinogen hexavalent chromium Cr(VI) by an indigenously isolated bacterial strain. *J. Carcinog.* **2010**, *9*, 6. [CrossRef]
59. Taylor, M.L.; Gwinnett, C.; Robinson, L.F.; Woodall, L.C. Plastic microfibre ingestion by deep-sea organisms. *Sci. Rep.* **2016**, *6*, 33997. [CrossRef] [PubMed]
60. Bergmann, M.; Gutow, L.; Klages, M. *Marine Anthropogenic Litter*; Springer Nature: Berlin, Germany, 2015; ISBN 9783319165103.

61. Suaria, G.; Avio, C.G.; Lattin, G.; Regoli, F.; Aliani, S.; Marche, A.I. Neustonic microplastics in the Southern Adriatic Sea. *Prelim. Results Micro* **2015**, 28. [CrossRef]
62. Avio, C.G.; Gorbi, S.; Regoli, F. Experimental development of a new protocol for extraction and characterization of microplastics in fish tissues: First observations in commercial species from Adriatic Sea. *Mar. Environ. Res.* **2015**, *111*, 18–26. [CrossRef]
63. Collard, F.; Gilbert, B.; Eppe, G.; Parmentier, E.; Das, K. Detection of Anthropogenic Particles in Fish Stomachs: An Isolation Method Adapted to Identification by Raman Spectroscopy. *Arch. Environ. Contam. Toxicol.* **2015**, *69*, 331–339. [CrossRef]
64. Alomar, C.; Sureda, A.; Capó, X.; Guijarro, B.; Tejada, S.; Deudero, S. Microplastic ingestion by Mullus surmuletus Linnaeus, 1758 fish and its potential for causing oxidative stress. *Environ. Res.* **2017**, *159*, 135–142. [CrossRef]
65. Alomar, C.; Deudero, S. Evidence of microplastic ingestion in the shark Galeus melastomus Rafinesque, 1810 in the continental shelf off the western Mediterranean Sea. *Environ. Pollut.* **2017**, *223*, 223–229. [CrossRef]
66. Avio, C.G.; Cardelli, L.R.; Gorbi, S.; Pellegrini, D.; Regoli, F. Microplastics pollution after the removal of the Costa Concordia wreck: First evidences from a biomonitoring case study. *Environ. Pollut.* **2017**, *227*, 207–214. [CrossRef] [PubMed]
67. Compa, M.; Ventero, A.; Iglesias, M.; Deudero, S. Ingestion of microplastics and natural fibres in Sardina pilchardus (Walbaum, 1792) and Engraulis encrasicolus (Linnaeus, 1758) along the Spanish Mediterranean coast. *Mar. Pollut. Bull.* **2018**, *128*, 89–96. [CrossRef]
68. Digka, N.; Tsangaris, C.; Torre, M.; Anastasopoulou, A.; Zeri, C. Microplastics in mussels and fish from the Northern Ionian Sea. *Mar. Pollut. Bull.* **2018**, *135*, 30–40. [CrossRef]
69. Duncan, E.M.; Broderick, A.C.; Fuller, W.J.; Galloway, T.S.; Godfrey, M.H.; Hamann, M.; Limpus, C.J.; Lindeque, P.K.; Mayes, A.G.; Omeyer, L.C.M.; et al. Microplastic ingestion ubiquitous in marine turtles. *Glob. Chang. Biol.* **2019**, *25*, 744–752. [CrossRef] [PubMed]
70. Anastasopoulou, A.; Kovač Viršek, M.; Bojanić Varezić, D.; Digka, N.; Fortibuoni, T.; Koren, Š.; Mandić, M.; Mytilineou, C.; Pešić, A.; Ronchi, F.; et al. Assessment on marine litter ingested by fish in the Adriatic and NE Ionian Sea macro-region (Mediterranean). *Mar. Pollut. Bull.* **2018**, *133*, 841–851. [CrossRef] [PubMed]
71. Garcia-Garin, O.; Vighi, M.; Aguilar, A.; Tsangaris, C.; Digka, N.; Kaberi, H.; Borrell, A. Boops boops as a bioindicator of microplastic pollution along the Spanish Catalan coast. *Mar. Pollut. Bull.* **2019**, *149*, 110648. [CrossRef]
72. Savoca, S.; Capillo, G.; Mancuso, M.; Faggio, C.; Panarello, G.; Crupi, R.; Bonsignore, M.; D'Urso, L.; Compagnini, G.; Neri, F.; et al. Detection of artificial cellulose microfibers in Boops boops from the northern coasts of Sicily (Central Mediterranean). *Sci. Total Environ.* **2019**, *691*, 455–465. [CrossRef]
73. Giani, D.; Baini, M.; Galli, M.; Casini, S.; Fossi, M.C. Microplastics occurrence in edible fish species (Mullus barbatus and Merluccius merluccius) collected in three different geographical sub-areas of the Mediterranean Sea. *Mar. Pollut. Bull.* **2019**, *140*, 129–137. [CrossRef]
74. Valente, T.; Sbrana, A.; Scacco, U.; Jacomini, C.; Bianchi, J.; Palazzo, L.; de Lucia, G.A.; Silvestri, C.; Matiddi, M. Exploring microplastic ingestion by three deep-water elasmobranch species: A case study from the Tyrrhenian Sea. *Environ. Pollut.* **2019**, *253*, 342–350. [CrossRef]
75. Capillo, G.; Savoca, S.; Panarello, G.; Mancuso, M.; Branca, C.; Romano, V.; D'Angelo, G.; Bottari, T.; Spanò, N. Quali-quantitative analysis of plastics and synthetic microfibers found in demersal species from Southern Tyrrhenian Sea (Central Mediterranean). *Mar. Pollut. Bull.* **2020**, *150*, 110596. [CrossRef]
76. Capone, A.; Petrillo, M.; Misic, C. Ingestion and elimination of anthropogenic fibres and microplastic fragments by the European anchovy (Engraulis encrasicolus) of the NW Mediterranean Sea. *Mar. Biol.* **2020**, *167*, 166. [CrossRef]
77. Renzi, M.; Blašković, A. Chemical fingerprint of plastic litter in sediments and holothurians from Croatia: Assessment & relation to different environmental factors. *Mar. Pollut. Bull.* **2020**, *153*, 110994. [CrossRef] [PubMed]
78. Rodríguez-Romeu, O.; Constenla, M.; Carrassón, M.; Campoy-Quiles, M.; Soler-Membrives, A. Are anthropogenic fibres a real problem for red mullets (Mullus barbatus) from the NW Mediterranean? *Sci. Total Environ.* **2020**, *733*, 139336. [CrossRef] [PubMed]
79. Savoca, S.; Bottari, T.; Fazio, E.; Bonsignore, M.; Mancuso, M.; Luna, G.M.; Romeo, T.; D'Urso, L.; Capillo, G.; Panarello, G.; et al. Plastics occurrence in juveniles of Engraulis encrasicolus and Sardina pilchardus in the Southern Tyrrhenian Sea. *Sci. Total Environ.* **2020**, *718*, 137457. [CrossRef] [PubMed]
80. Carreras-Colom, E.; Constenla, M.; Soler-Membrives, A.; Cartes, J.E.; Baeza, M.; Carrassón, M. A closer look at anthropogenic fiber ingestion in Aristeus antennatus in the NW Mediterranean Sea: Differences among years and locations and impact on health condition. *Environ. Pollut.* **2020**, *263*, 114567. [CrossRef] [PubMed]
81. Bayo, J.; Rojo, D.; Martínez-Baños, P.; López-Castellanos, J.; Olmos, S. Commercial Gilthead Seabream (Sparus aurata L.) from the Mar Menor Coastal Lagoon as Hotspots of Microplastic Accumulation in the Digestive System. *Public Health* **2021**, *18*, 6844. [CrossRef] [PubMed]
82. Yozukmaz, A. Investigation of microplastics in edible wild mussels from İzmir Bay (Aegean Sea, Western Turkey): A risk assessment for the consumers. *Mar. Pollut. Bull.* **2021**, *171*, 112733. [CrossRef]
83. Sayed, A.E.D.H.; Hamed, M.; Badrey, A.E.A.; Ismail, R.F.; Osman, Y.A.A.; Osman, A.G.M.; Soliman, H.A.M. Microplastic distribution, abundance, and composition in the sediments, water, and fishes of the Red and Mediterranean seas, Egypt. *Mar. Pollut. Bull.* **2021**, *173*, 112966. [CrossRef]

84. Faure, F.; Saini, C.; Potter, G.; Galgani, F.; de Alencastro, L.F.; Hagmann, P. An evaluation of surface micro- and mesoplastic pollution in pelagic ecosystems of the Western Mediterranean Sea. *Environ. Sci. Pollut. Res.* **2015**, *22*, 12190–12197. [CrossRef]

85. Suaria, G.; Avio, C.G.; Mineo, A.; Lattin, G.L.; Magaldi, M.G.; Belmonte, G.; Moore, C.J.; Regoli, F.; Aliani, S. The Mediterranean Plastic Soup: Synthetic polymers in Mediterranean surface waters. *Sci. Rep.* **2016**, *6*, 37551. [CrossRef]

86. Fastelli, P.; Blašković, A.; Bernardi, G.; Romeo, T.; Čižmek, H.; Andaloro, F.; Russo, G.F.; Guerranti, C.; Renzi, M. Plastic litter in sediments from a marine area likely to become protected (Aeolian Archipelago's islands, Tyrrhenian sea). *Mar. Pollut. Bull.* **2016**, *113*, 526–529. [CrossRef] [PubMed]

87. Ruiz-Orejón, L.F.; Sardá, R.; Ramis-Pujol, J. Floating plastic debris in the Central and Western Mediterranean Sea. *Mar. Environ. Res.* **2016**, *120*, 136–144. [CrossRef] [PubMed]

88. Van der Hal, N.; Ariel, A.; Angel, D.L. Exceptionally high abundances of microplastics in the oligotrophic Israeli Mediterranean coastal waters. *Mar. Pollut. Bull.* **2017**, *116*, 151–155. [CrossRef] [PubMed]

89. Cannas, S.; Fastelli, P.; Guerranti, C.; Renzi, M. Plastic litter in sediments from the coasts of south Tuscany (Tyrrhenian Sea). *Mar. Pollut. Bull.* **2017**, *119*, 372–375. [CrossRef]

90. Mistri, M.; Infantini, V.; Scoponi, M.; Granata, T.; Moruzzi, L.; Massara, F.; De Donati, M.; Munari, C. Small plastic debris in sediments from the Central Adriatic Sea: Types, occurrence and distribution. *Mar. Pollut. Bull.* **2017**, *124*, 435–440. [CrossRef]

91. Abidli, S.; Antunes, J.C.; Ferreira, J.L.; Lahbib, Y.; Sobral, P.; Trigui El Menif, N. Microplastics in sediments from the littoral zone of the north Tunisian coast (Mediterranean Sea). *Estuar. Coast. Shelf Sci.* **2018**, *205*, 1–9. [CrossRef]

92. Sanchez-Vidal, A.; Thompson, R.C.; Canals, M.; De Haan, W.P. The imprint of microfibres in Southern European deep seas. *PLoS ONE* **2018**, *13*, e0207033. [CrossRef]

93. Simon-Sánchez, L.; Grelaud, M.; Garcia-Orellana, J.; Ziveri, P. River Deltas as hotspots of microplastic accumulation: The case study of the Ebro River (NW Mediterranean). *Sci. Total Environ.* **2019**, *687*, 1186–1196. [CrossRef]

94. Suaria, G.; Musso, M.; Achtypi, A.; Bassotto, D.; Aliani, S. *Textile Fibres in Mediterranean Surface Waters: Abundance and Composition;* Springer International Publishing: Cham, Switzerland, 2020; ISBN 9783030459093.

95. Pojar, I.; Kochleus, C.; Dierkes, G.; Ehlers, S.M.; Reifferscheid, G.; Stock, F. Quantitative and qualitative evaluation of plastic particles in surface waters of the Western Black Sea. *Environ. Pollut.* **2021**, *268*, 115724. [CrossRef]

96. Bošković, N.; Joksimović, D.; Peković, M.; Perošević-Bajčeta, A.; Bajt, O. Marine Science and Engineering Microplastics in Surface Sediments along the Montenegrin Coast, Adriatic Sea: Types, Occurrence, and Distribution. *J. Mar. Sci. Eng.* **2021**, *9*, 841. [CrossRef]

97. Angiolillo, M.; Gérigny, O.; Valente, T.; Fabri, M.C.; Tambute, E.; Rouanet, E.; Claro, F.; Tunesi, L.; Vissio, A.; Daniel, B.; et al. Distribution of seafloor litter and its interaction with benthic organisms in deep waters of the Ligurian Sea (Northwestern Mediterranean). *Sci. Total Environ.* **2021**, *788*, 147745. [CrossRef] [PubMed]

98. Miller, R.Z.; Watts, A.J.R.; Winslow, B.O.; Galloway, T.S.; Barrows, A.P.W. Mountains to the sea: River study of plastic and non-plastic microfiber pollution in the northeast USA. *Mar. Pollut. Bull.* **2017**, *124*, 245–251. [CrossRef] [PubMed]

99. Rochman, C.M.; Tahir, A.; Williams, S.L.; Baxa, D.V.; Lam, R.; Miller, J.T.; Teh, F.C.; Werorilangi, S.; Teh, S.J. Anthropogenic debris in seafood: Plastic debris and fibers from textiles in fish and bivalves sold for human consumption. *Sci. Rep.* **2015**, *5*, 14340. [CrossRef] [PubMed]

100. Nadal, M.A.; Alomar, C.; Deudero, S. High levels of microplastic ingestion by the semipelagic fish bogue Boops boops (L.) around the Balearic Islands. *Environ. Pollut.* **2016**, *214*, 517–523. [CrossRef] [PubMed]

101. Naidoo, T.; Smit, A.J.; Glassom, D. Plastic ingestion by estuarine mullet Mugil cephalus (Mugilidae) in an urban harbour, KwaZulu-Natal, South Africa. *Afr. J. Mar. Sci.* **2016**, *38*, 145–149. [CrossRef]

102. Herrera, A.; Štindlová, A.; Martínez, I.; Rapp, J.; Romero-Kutzner, V.; Samper, M.D.; Montoto, T.; Aguiar-González, B.; Packard, T.; Gómez, M. Microplastic ingestion by Atlantic chub mackerel (Scomber colias) in the Canary Islands coast. *Mar. Pollut. Bull.* **2019**, *139*, 127–135. [CrossRef]

103. Le Guen, C.; Suaria, G.; Sherley, R.B.; Ryan, P.G.; Aliani, S.; Boehme, L.; Brierley, A.S. Microplastic study reveals the presence of natural and synthetic fibres in the diet of King Penguins (Aptenodytes patagonicus) foraging from South Georgia. *Environ. Int.* **2020**, *134*, 105303. [CrossRef]

104. Wesch, C.; Barthel, A.K.; Braun, U.; Klein, R.; Paulus, M. No microplastics in benthic eelpout (Zoarces viviparus): An urgent need for spectroscopic analyses in microplastic detection. *Environ. Res.* **2016**, *148*, 36–38. [CrossRef]

105. Comnea-Stancu, I.R.; Wieland, K.; Ramer, G.; Schwaighofer, A.; Lendl, B. On the Identification of Rayon/Viscose as a Major Fraction of Microplastics in the Marine Environment: Discrimination between Natural and Manmade Cellulosic Fibers Using Fourier Transform Infrared Spectroscopy. *Appl. Spectrosc.* **2017**, *71*, 939–950. [CrossRef]

106. Faruk, O.; Bledzki, A.K.; Fink, H.P.; Sain, M. Progress report on natural fiber reinforced composites. *Macromol. Mater. Eng.* **2014**, *299*, 9–26. [CrossRef]

107. Röder, T.; Moosbauer, J.; Wöss, K.; Schlader, S.; Kraft, G. Man-Made Cellulose Fibres-a Comparison Based on Morphology and Mechanical Properties. *Lenzinger Berichte* **2013**, *91*, 7–12.

108. Bredereck, K.; Hermanutz, F. Man-made cellulosics. *Color. Technol.* **2008**, *35*, 59–75. [CrossRef]

109. Morseletto, P. A new framework for policy evaluation: Targets, marine litter, Italy and the Marine Strategy Framework Directive. *Mar. Policy* **2020**, *117*, 103956. [CrossRef]

110. Henry, B.; Laitala, K.; Klepp, I.G. Microfibres from apparel and home textiles: Prospects for including microplastics in environmental sustainability assessment. *Sci. Total Environ.* **2019**, *652*, 483–494. [CrossRef] [PubMed]
111. Stagioni, M.; Montanini, S.; Vallisneri, M. Feeding habits of European hake, Merluccius merluccius (Actinopterygii: Gadiformes: Merlucciidae), from the northeastern Mediterranean sea. *Acta Ichthyol. Piscat.* **2011**, *41*, 277–284. [CrossRef]
112. Matić-Skoko, S.; Šegvić-Bubić, T.; Mandić, I.; Izquierdo-Gomez, D.; Arneri, E.; Carbonara, P.; Grati, F.; Ikica, Z.; Kolitari, J.; Milone, N.; et al. Evidence of subtle genetic structure in the sympatric species Mullus barbatus and Mullus surmuletus (Linnaeus, 1758) in the Mediterranean Sea. *Sci. Rep.* **2018**, *8*, 676. [CrossRef]
113. Anastasopoulou, A.; Mytilineou, C.; Smith, C.J.; Papadopoulou, K.N. Plastic debris ingested by deep-water fish of the Ionian Sea (Eastern Mediterranean). *Deep. Res. Part I Oceanogr. Res. Pap.* **2013**, *74*, 11–13. [CrossRef]
114. Matiddi, M.; Hochsheid, S.; Camedda, A.; Baini, M.; Cocumelli, C.; Serena, F.; Tomassetti, P.; Travaglini, A.; Marra, S.; Campani, T.; et al. Loggerhead sea turtles (Caretta caretta): A target species for monitoring litter ingested by marine organisms in the Mediterranean Sea. *Environ. Pollut.* **2017**, *230*, 199–209. [CrossRef]
115. Boerger, C.M.; Lattin, G.L.; Moore, S.L.; Moore, C.J. Plastic ingestion by planktivorous fishes in the North Pacific Central Gyre. *Mar. Pollut. Bull.* **2010**, *60*, 2275–2278. [CrossRef]
116. Cole, M.; Webb, H.; Lindeque, P.K.; Fileman, E.S.; Halsband, C.; Galloway, T.S. Isolation of microplastics in biota-rich seawater samples and marine organisms. *Sci. Rep.* **2014**, *4*, 4528. [CrossRef]
117. Gwinnett, C.; Miller, R.Z. Are we contaminating our samples? A preliminary study to investigate procedural contamination during field sampling and processing for microplastic and anthropogenic microparticles. *Mar. Pollut. Bull.* **2021**, *173*, 113095. [CrossRef] [PubMed]
118. Ng, K.L.; Obbard, J.P. Prevalence of microplastics in Singapore's coastal marine environment. *Mar. Pollut. Bull.* **2006**, *52*, 761–767. [CrossRef] [PubMed]
119. Prata, J.C.; Reis, V.; da Costa, J.P.; Mouneyrac, C.; Duarte, A.C.; Rocha-Santos, T. Contamination issues as a challenge in quality control and quality assurance in microplastics analytics. *J. Hazard. Mater.* **2021**, *403*, 123660. [CrossRef]
120. Cowger, W.; Booth, A.M.; Hamilton, B.M.; Thaysen, C.; Primpke, S.; Munno, K.; Lusher, A.L.; Dehaut, A.; Vaz, V.P.; Liboiron, M.; et al. Special Issue: Microplastics Reporting Guidelines to Increase the Reproducibility and Comparability of Research on Microplastics. *Appl. Spectrosc.* **2020**, *74*, 1066–1077. [CrossRef]
121. Miller, E.; Sedlak, M.; Lin, D.; Box, C.; Holleman, C.; Rochman, C.M.; Sutton, R. Recommended best practices for collecting, analyzing, and reporting microplastics in environmental media: Lessons learned from comprehensive monitoring of San Francisco Bay. *J. Hazard. Mater.* **2021**, *409*, 124770. [CrossRef] [PubMed]
122. Gaylarde, C.; Baptista-Neto, J.A.; da Fonseca, E.M. Plastic microfibre pollution: How important is clothes' laundering? *Heliyon* **2021**, *7*, e07105. [CrossRef]
123. Zayen, A.; Sayadi, S.; Chevalier, C.; Boukthir, M.; Ben Ismail, S.; Tedetti, M. Microplastics in surface waters of the Gulf of Gabes, southern Mediterranean Sea: Distribution, composition and influence of hydrodynamics. *Estuar. Coast. Shelf Sci.* **2020**, *242*, 106832. [CrossRef]
124. Schönlau, C.; Karlsson, T.M.; Rotander, A.; Nilsson, H.; Engwall, M.; van Bavel, B.; Kärrman, A. Microplastics in sea-surface waters surrounding Sweden sampled by manta trawl and in-situ pump. *Mar. Pollut. Bull.* **2020**, *153*, 111019. [CrossRef]
125. Kuklinski, P.; Wicikowski, L.; Koper, M.; Grala, T.; Leniec-Koper, H.; Barasiński, M.; Talar, M.; Kamiński, I.; Kibart, R.; Małecki, W. Offshore surface waters of Antarctica are free of microplastics, as revealed by a circum-Antarctic study. *Mar. Pollut. Bull.* **2019**, *149*, 110573. [CrossRef]
126. Scopetani, C.; Esterhuizen-Londt, M.; Chelazzi, D.; Cincinelli, A.; Setälä, H.; Pflugmacher, S. Self-contamination from clothing in microplastics research. *Ecotoxicol. Environ. Saf.* **2020**, *189*, 110036. [CrossRef]
127. Grigorakis, S.; Mason, S.A.; Drouillard, K.G. Determination of the gut retention of plastic microbeads and microfibers in goldfish (Carassius auratus). *Chemosphere* **2017**, *169*, 233–238. [CrossRef] [PubMed]

 *toxics*

Review

A Meta-Analysis of the Characterisations of Plastic Ingested by Fish Globally

Kok Ping Lim [1,2], Phaik Eem Lim [2,*], Sumiani Yusoff [2], Chengjun Sun [3], Jinfeng Ding [3] and Kar Hoe Loh [2]

[1] Institute for Advanced Studies, Universiti Malaya, Kuala Lumpur 50603, Malaysia; kokping@um.edu.my
[2] Institute of Ocean and Earth Science, Universiti Malaya, Kuala Lumpur 50603, Malaysia; sumiani@um.edu.my (S.Y.); khloh@um.edu.my (K.H.L.)
[3] Key Laboratory of Marine Eco-Environmental Science and Technology, Marine Bioresource and Environment Research Centre, First Institute of Oceanography, Ministry of Natural Resources, Qingdao 266061, China; csun@fio.org.cn (C.S.); dingjf1004@gmail.com (J.D.)
* Correspondence: phaikeem@um.edu.my

Abstract: Plastic contamination in the environment is common but the characterisation of plastic ingested by fish in different environments is lacking. Hence, a meta-analysis was conducted to identify the prevalence of plastic ingested by fish globally. Based on a qualitative analysis of plastic size, it was determined that small microplastics (<1 mm) are predominantly ingested by fish globally. Furthermore, our meta-analysis revealed that plastic fibres (70.6%) and fragments (19.3%) were the most prevalent plastic components ingested by fish, while blue (24.2%) and black (18.0%) coloured plastic were the most abundant. Polyethylene (15.7%) and polyester (11.6%) were the most abundant polymers. Mixed-effect models were employed to identify the effects of the moderators (sampling environment, plastic size, digestive organs examined, and sampling continents) on the prevalence of plastic shape, colour, and polymer type. Among the moderators, only the sampling environment and continent contributed to a significant difference between subgroups in plastic shape and polymer type.

Keywords: microplastic; shape; colour; polymer type

Citation: Lim, K.P.; Lim, P.E.; Yusoff, S.; Sun, C.; Ding, J.; Loh, K.H. A Meta-Analysis of the Characterisations of Plastic Ingested by Fish Globally. *Toxics* **2022**, *10*, 186. https://doi.org/10.3390/toxics10040186

Academic Editor: Costanza Scopetani

Received: 21 March 2022
Accepted: 8 April 2022
Published: 11 April 2022

Publisher's Note: MDPI stays neutral with regard to jurisdictional claims in published maps and institutional affiliations.

1. Introduction

Global plastic production has increased drastically from around 1.5 million tonnes in 1950 to 368 million tonnes in 2019, due to the high demands of consumers [1,2]. As a consequence of the large production volume of plastics and defective waste management system, it is very common for plastics to accumulate in the environment, such as in seawaters [3,4], deep sea sediments [5], artic sea ice [6], lakes [7], soils [8], and even in the atmosphere [9]. Slow degradation of the plastics has led to their accumulation in the environment. Nonetheless, radiation, heat and friction may cause fragmentation of the plastics [4] and turn them into secondary microplastics, which are plastic particles less than 5 mm in size [10]. Additionally, primary microplastics are produced purposefully to be used in various products [11] or industries [12].

It is estimated that between 1.15 and 2.41 million tonnes of mismanaged plastic waste are discharged into the oceans through rivers annually [13]. In 2014, it was estimated that at least 5.25 trillion plastic particles, weighing 268,940 tonnes, were floating in the world's oceans [14]. Hence, there is an increased risk of marine organisms ingesting plastic particles due to their high concentration in oceans. Organisms might ingest the particles by primary ingestion because they recognise the items as potential prey, or secondary ingestion via contaminated prey [15]. Many publications have shown that plastic particles are ingested by a wide variety of animal taxa in various environments, including seabirds [16], waterbirds [17], crustaceans [18,19], sharks [20] and other fish [21] and cetaceans [22,23]. Furthermore, there is trophic transfer in the ecosystem from lower to higher trophic level based on both experimental [24,25] and field studies [26–29].

Direct fatality due to the blockage of the digestive tract by larger size plastic debris has been found in many marine organisms, such as turtles [30], sea birds [31], and manatees [32]. The death of a whale shark was suspected to be caused by plastic ingestion with subsequent inflammation of the stomach mucosa triggering wounds and infections [33]. Several severe impacts due to the ingestion of plastic particles by fish in laboratory conditions have also been documented [34,35]. The plastic particles are able to promote inflammation and accumulation of lipids in zebrafish liver [36]. The growth and body condition of reef fish decreased significantly when food pieces were substituted by microplastic particles, and these effects escalated at higher microplastic concentrations [37]. Intestinal lesions in fish were observed in an experimental study and the severity increased with the concentration of microplastics [38]. Nevertheless, the exposure settings for the laboratory experiments cannot fully represent the natural environments in which the plastic types, sizes, and concentrations may fluctuate temporally and spatially.

Plastic ingestion by fish has been fairly well reviewed. The earliest review reported the incidence of plastic ingestion in 22 fish species [39]. Subsequent and more recent reviews have recorded the number of fish as follows: 90 species [40], 34 [41], 95 [42], 200 [43], 323 [44], 165 [45]; and 386 [46]. There were also various reviews on plastic ingestion by fish, but these included other marine biotas [47–49]. A systematic review of the occurrence of microplastics based on their characterisations was conducted but limited to freshwater fish species [50]. In view of the gaps in the knowledge on plastic characterisations in different environments, a meta-analysis, which included samples from all environments, was conducted to investigate the possible factors affecting plastic ingestion by fish, and to identify the abundance of plastic ingested on a global scale based on its characterisations.

2. Materials and Methods

2.1. Literature Review

In this review paper, a literature review was conducted using web-based search engines: Google Scholar and electronic databases, such as PubMed, Web of Science, Science Direct and Wiley Online Library from 1970 to December 2021 with the following keywords: "microplastic" OR "plastic" OR "plastic ingestion" OR "marine debris" AND "fish".

2.2. Quality Assessment and Data Extraction

The publications were reviewed based on the following criteria (Figure 1). Firstly, the titles and abstracts of the articles were screened to search for related studies. Studies on fish exposure to plastics in a laboratory setting were excluded. In the second step, the materials and methods section of each article was examined to ensure that the numbers of plastic shape, colour, and polymer type were reported. If the data were not reported in numbers, they were extracted from published diagrams using WebPlotDigitizer Version 4.5 (Ankit Rohatgi, Pacifica, CA, USA). Studies that assigned plastic size class and predominant size class were included for qualitative analysis. Due to the importance of contamination control in plastic research during the extraction process, the studies were checked for quality assessment/quality control (QA/QC). Studies that did not include any QA/QC were excluded from meta-analysis of plastic characterisation.

Detailed data-location, part of digestive organs examined, plastic extraction method, percentage of plastic ingested, plastic size, shape, and colour, and the polymer type were recorded. The environments where the samples were collected were retrieved from the publications based on the GPS coordinates given or sampling procedures stated in the method in each publication. The source of the samples was classified into marine, estuary, freshwater, aquaculture, and market. Samples obtained from markets were grouped into marine, estuary, or freshwater if the study specified the source of the samples [51]. Studies that purchased samples directly from the market without the source information of the samples were classified into the "market" category [52]. The plastic extraction methods were categorized into three groups, as proposed by a previous review [44]. Method 1 is a visual analysis of the GIT content with the naked eye; Method 2 is a visual analysis of the

GIT content using a microscope; and Method 3 is the chemical digestion of the GIT content, followed by filtration and microscope analysis. There are many definitions of plastic size across different guidelines and articles. For consistency, the relative size of plastic ingested by fish in this study was sorted as microplastic (<5 mm), mesoplastic (5–25 mm), and macroplastic (25–1000 mm) [53–55]. For the shape of plastics, it was standardised into five categories: fibre, film, fragment, foam, and pellet (Table 1), which is in line with several studies [7,54,56–58]. The colours of the plastics were classified into red, orange, yellow, green, blue, purple, pink, brown, grey, black, white, transparent, and others. In studies that revealed plastics from the environment or other biota, only plastics ingested by fish were considered. If samples were collected from different environments, data from the same data were documented separately.

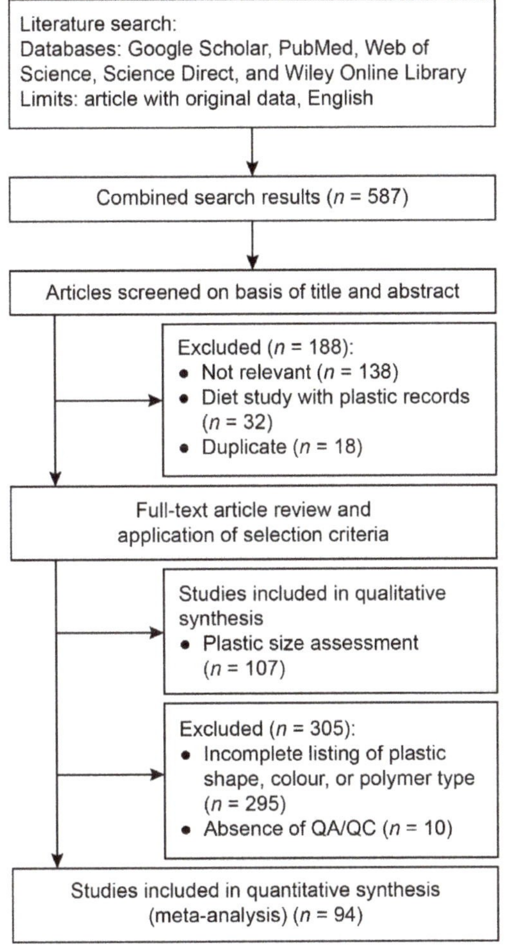

Figure 1. Flow diagram of study selection.

Table 1. Standardised shape description of plastic.

Standardised Shape	Description	Alternative Shape
Fibre	Thin or fibrous plastic that has a length longer than its width	Line, Monofilament, Thread, Polyfilament, Twine, Fibrous, Microfibre
Film	Flat and thin plane of smooth or angular edges plastic	Sheet, Plastic Packaging, Wrapper, Plastic Bag, Packet Wrap, Food Package, Strip
Fragment	Irregular, hard, and jagged plastic particle	Flake, Particle, Piece, Tag, Chip
Foam	Lightweight, sponge-like plastic	Polystyrene, Polystyrene Spherule, Styrofoam, Styrofoam Fragment, Sponge, Expanded Polystyrene Foam (EPS)
Pellet	Hard, rounded plastic particle	Bead, Granule, Microbead, Particle, Spherule

Note: Particle shape of each study was assigned to the closest standardised shape based on the appearance shown in the publications.

2.3. Statistical Analysis

Data of the number of plastic shape, colour, or polymer type (k) and the total number of plastics ingested (n) were extracted from the selected studies. Proportion of the plastic characterisation in a single study was calculated with the formula: $p = k/n$. Meta-analysis of proportions was employed to obtain a more precise estimation of the overall proportion for all plastic characterisations. Since proportions of <0.2 were common in the studies, the pooled prevalence of plastic characterisation was calculated by applying arcsine square root transformation on the proportion data. Publication bias was examined through funnel plots by trim-and-fill method and Egger's regression test with a confidence interval (CI) of 95%. Between-study heterogeneity was evaluated with I^2 statistic and tested using the Paule-Mandel estimator method. Fixed effects model was used in the case of low heterogeneity whereas random effects model was used for high heterogeneity. Mixed effects meta-regression model was employed in which the random-effects model was used to combine study effects within each subgroup and the fixed-effect model was used to test if the effects across the subgroups differed significantly from each other. In this model, assumption of different between-study variance across subgroups was applied to identify if different moderators (i.e., sampling environment, plastic size, digestive organs examined, or sampling continent) affect the prevalence of the plastics. Subgroups forest plot was created based on different moderators. Meta-regression models were used to analyse characterisations that were the most abundant: shape (fibre, fragment, film, and pellet), colour (blue, black, transparent, and white), polymer types (polyethylene (PE), polyester (PES), polypropylene (PP), and polyamide (PA). The rare characterisations were not subtracted from the total plastic numbers even though they were not included in the meta-regression models. Hence, relative abundance of each characterisations were estimated based on total plastic numbers from all of its characterisations. All statistical analyses and plotting were performed in R software (R Core Team, version 4.1.2, Vienna, Austria).

3. Results

3.1. Overview

The number of studies that reported the assessments of plastic size, shape, colour, and type were 127, 281, 195, and 153, respectively. Studies without QA/QC ($n = 107$) were excluded for the analysis of plastic size, while 94 studies with QA/QC and revealed the assessments of all three characterisations (shape, colour, and polymer type) in the same study were selected for meta-analysis. In total, data of five shapes, 13 colours, and 25 polymer types were recorded. It should be noted that the total count of plastics in polymer types was different from shape and colour, because not all of the plastics were tested with the polymer characterisation test.

3.2. Prevalence of Plastic Ingested

Only 34 out of the 107 studies (31.8%) included plastic sizes larger than 5 mm (mesoplastic and macroplastic) in their findings. Larger size particles were not included in many studies, especially recent studies, because they preferred to focus on microplastic ingestion. The most prevalent size of plastic ingested was microplastic for all the studies. Microplastics were often divided into two groups called small microplastic (<1 mm) and large microplastic (1–5 mm) [59,60]. Among the studies that reported the size class of plastic ingested, more than two-thirds of the studies (74.0%) recorded small microplastic as the predominant size class (Figure 2) [27,52,56,61–163]. Based on the pooled prevalence data, fibre plastic was the most abundant plastic ingested by the fish, with a relative abundance of 71.6% (CI 64.0–78.7%). The second most abundant plastic shape was fragment (19.4%; CI 13.8–25.7%), followed by film (0.5%; CI 0–1.5%) and pellet (0.0%; CI 0.0–0.2%) (Figure 3). Egger's regression test indicated that there was no significant publication bias for plastic shapes (Figure S1, fragment: $Z = 1.377$, $p = 0.169$, pellet: $Z = 1.491$, $p = 0.136$) except fibre ($Z = -2.256$, $p = 0.024$) and film ($Z = 2.457$, $p = 0.014$). A high heterogeneity ($I^2 = 93.6$–98.8%) was observed between studies for plastic shapes. Furthermore, blue colour plastic was predominantly ingested by fish, with a relative abundance of 24.5% (CI 20.3–28.9%). The second most abundant plastic colour was black (18.1%; CI 13.7–22.9%), followed by transparent (6.8%; CI 4.1–9.9%), and white (5.8%; CI 3.4–8.5%) (Figure 4). Egger's regression test revealed that there was no significant publication bias for plastic colours: blue ($Z = 0.300$, $p = 0.764$), black ($Z = -0.050$, $p = 0.960$), transparent ($Z = 0.418$, $p = 0.676$), and white ($Z = -0.156$, $p = 0.876$) (Figure S2). Similar to plastic shape, a high heterogeneity was found ($I^2 = 98.0$–98.6%) between studies on colour. The most abundant polymer type ingested by fish was PE, with a relative abundance of 15.7% (CI 11.3–20.6%), followed by PES (11.6%; CI 7.8–16.0%), PP (6.8%; CI 4.2–9.9%), and PA (5.6%; CI 2.9–8.8%) (Figure 5). Egger's regression test indicated that there was no significant difference for polymer types: PE ($Z = 0.738$, $p = 0.460$), PES ($Z = -0.560$, $p = 0.576$), and PA ($Z = -0.813$, $p = 0.416$), except PP ($Z = 2.128$, $p = 0.033$) (Figure S3). The between-study heterogeneity for polymer types was slightly lower than plastic shape and colour ($I^2 = 90.7$–95.1%).

A similar proportion for the dominant class size was observed in different environments, except in estuary. Seawater environments had the largest percentage, with small microplastics as the predominant size class of plastic ingested (80.6%), followed by aquaculture (75.0%), market and freshwater (71.4%), and estuary (57.1%) (Figure 4). The subgroups of continents shared similar proportion, except in Oceania (50.0%). Asia had the largest proportion of small microplastics (77.6%), followed by North America and Africa (75.0%), and Europe (72.4%). A mixed-effects model was applied to identify potential sources of heterogeneity with four categorical moderators (sampling environment, plastic size, digestive organs examined, and sampling continent). A significant difference between groups was found for two out of the four moderators, specifically, environment and continent for plastic shape and polymer type. In the case of environment, a significant subgroup difference was observed in plastic shapes: fibre ($Q_m = 16.311$, $p = 0.003$), fragment ($Q_m = 15.743$, $p = 0.003$), and pellet ($Q_m = 16.453$, $p = 0.003$), except in film ($Q_m = 0.824$, $p = 0.935$). Fibre was relatively more abundant in the market (89.7%), estuary and aquaculture (87.0%) environments than in freshwater (75.0%) and seawater (67.0%) environments. In contrast, fragments were more abundant in seawater (23.9%) than in freshwater (13.7%), aquaculture (10.7%), estuary (7.0%), and market (6.8%). The continent groups appeared to be significantly different in plastic shapes: fibre ($Q_m = 18.734$, $p = 0.002$), fragment ($Q_m = 24.886$, $p < 0.001$), film ($Q_m = 28.279$, $p < 0.001$), and pellet ($Q_m = 33.926$, $p < 0.001$). The abundance of fibre was significantly higher in North America (95.0%, $p = 0.001$) than the rest of the continent: Asia (74.8%), Europe (66.9%), Oceania (66.0%), Africa (60.6%), and South America (53.7%). The prevalence of fragment was higher in Africa (38.5%), South America (38.4%), Oceania (32.5%), Europe (23.0%), and significantly lower in Asia (14.7%, $p = 0.033$), and North America (1.5%, $p < 0.001$).

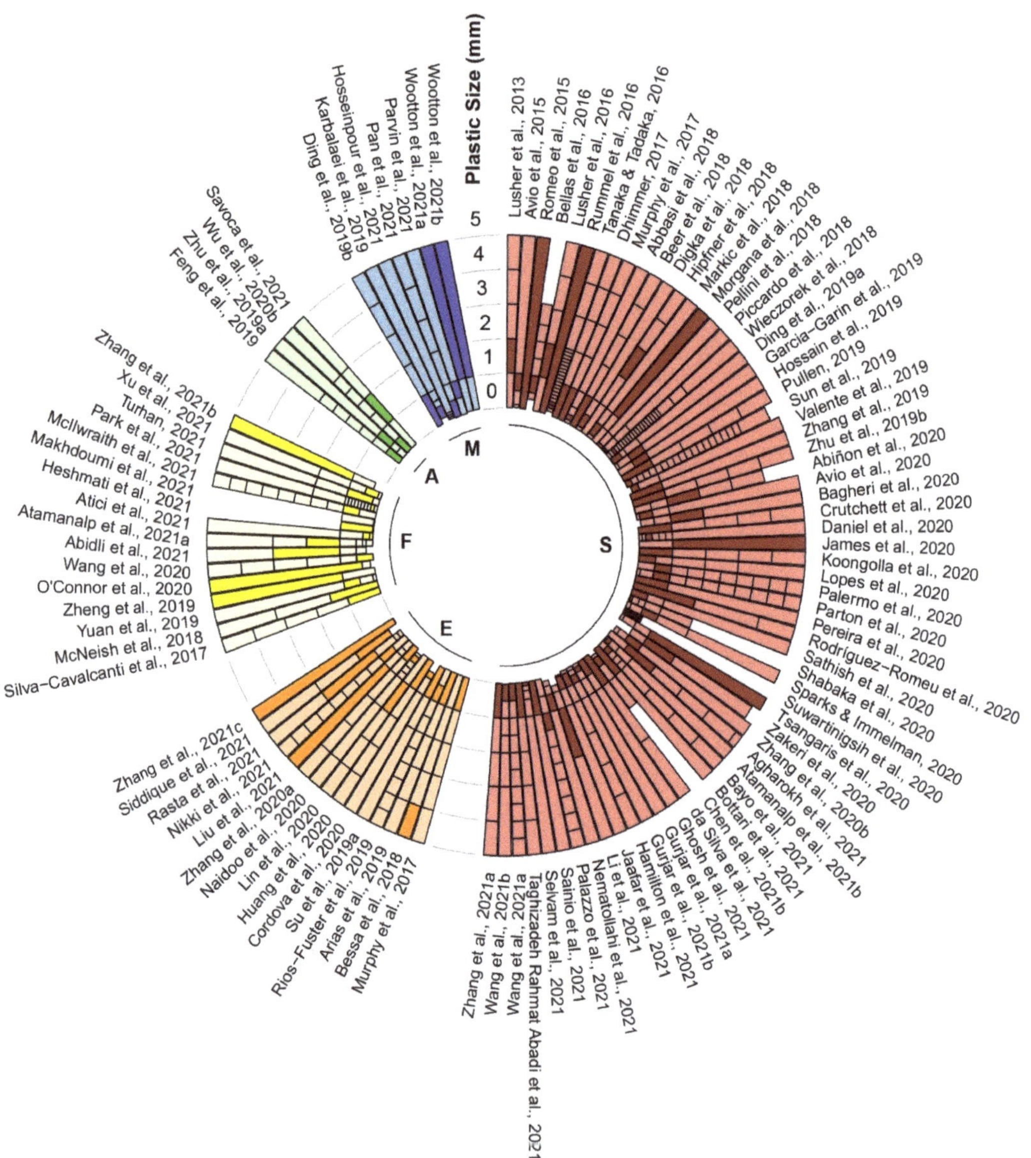

Figure 2. Overview of the assigned plastic size class and predominant size class of each study in different environments. Only size classes less than 5 mm are shown in this diagram. Each bar represents the plastic size class assigned in each study. Darker colour bars represent predominant size ingested. (S: Seawater; E: Estuarine; F: Freshwater; A: Aquaculture; M: Market). References: [27] Markic et al., 2018; [52] Ding et al., 2019a; [56] McNeish et al., 2018; [61] Abbasi et al., 2018; [62] Abidli et al., 2021; [63] Abiñon et al., 2020; [64] Agharokh et al., 2021; [65] Arias et al., 2019; [66] Atamanalp et al., 2021a; [67] Atamanalp et al., 2021b; [68] Atici et al., 2021; [69] Avio et al., 2015; [70] Avio et al., 2020; [71] Bagheri et al., 2020; [72] Bayo et al., 2021; [73] Beer et al., 2018; [74] Bellas et al., 2016; [75] Bessa et al., 2018; [76] Bottari et al., 2021; [77] Chen et al., 2021; [78] Cordova et al., 2020; [79] Crutchett et al., 2020; [80] da Silva et al., 2021; [81] Daniel et al., 2020; [82] Dhimmer, 2017; [83] Digka et al., 2018; [84] Ding et al., 2019b; [85] Feng et al., 2019; [86] Garcia-Garin et al., 2019; [87] Ghosh et al., 2021;

[88] Gurjar et al., 2021a; [89] Gurjar et al., 2021b; [90] Hamilton et al., 2021; [91] Heshmati et al., 2021; [92] Hipfner et al., 2018; [93] Hossain et al., 2019; [94] Hosseinpour et al., 2021; [95] Huang et al., 2020; [96] Jaafar et al., 2021; [97] James et al., 2020; [98] Karbalaei et al., 2019; [99] Koongolla et al., 2020; [100] Li et al., 2021; [101] Lin et al., 2020; [102] Liu et al., 2021; [103] Lopes et al., 2020; [104] Lusher et al., 2013; [105] Lusher et al., 2016; [106] Makhdoumi et al., 2021; [107] McIlwraith et al., 2021; [108] Morgana et al., 2018; [109] Murphy et al., 2017; [109] Murphy et al., 2017; [110] Naidoo et al., 2020; [111] Nematollahi et al., 2021; [112] Nikki et al., 2021; [113] O'Connor et al., 2020; [114] Palazzo et al., 2021; [115] Palermo et al., 2020; [116] Pan et al., 2021; [117] Park et al., 2021; [118] Parton et al., 2020; [119] Parvin et al., 2021; [120] Pellini et al., 2018; [121] Pereira et al., 2020; [122] Piccardo et al., 2018; [123] Pullen, 2019; [124] Rasta et al., 2021; [125] Rios-Fuster et al., 2019; [126] Rodríguez-Romeu et al., 2020; [127] Romeo et al., 2015; [128] Rummel et al., 2016; [129] Sainio et al., 2021; [130] Sathish et al., 2020; [131] Savoca et al., 2021; [132] Selvam et al., 2021; [133] Shabaka et al., 2020; [134] Siddique et al., 2021; [135] Silva-Cavalcanti et al., 2017; [136] Sparks & Immelman, 2020; [137] Su et al., 2019; [138] Sun et al., 2019; [139] Suwartinigsih et al., 2020; [140] Taghizadeh Rahmat Abadi et al., 2021; [141] Tanaka & Tadaka, 2016; [142] Tsangaris et al., 2020; [143] Turhan, 2021; [144] Valente et al., 2019; [145] Wang et al., 2021a; [146] Wang et al., 2021b; [147] Wang et al., 2020; [148] Wieczorek et al., 2018; [149] Wootton et al., 2021a; [150] Wootton et al., 2021b; [151] Wu et al., 2020; [152] Xu et al., 2021; [153] Yuan et al., 2019; [154] Zakeri et al., 2020; [155] Zhang et al., 2020a; [156] Zhang et al., 2020b; [157] Zhang et al., 2019; [158] Zhang et al., 2021a; [159] Zhang et al., 2021b; [160] Zhang et al., 2021c; [161] Zheng et al., 2019; [162] Zhu et al., 2019a; [163] Zhu et al., 2019b.

For plastic colour, no significant subgroup difference was found in the moderator of environment, except white ($Q_m = 11.020$, $p = 0.026$). The prevalence of blue plastic was highest in aquaculture (33.9%), followed by estuary (32.9%), market (25.8%), freshwater (25.6%), and seawater (22.9%) environments. In addition, the abundance of black plastic was higher in market (28.4%) and aquaculture (27.9%) than in freshwater (21.2%), seawater (17.7%), and estuary (10.3%) environments. Likewise, subgroup analysis with the moderator of continent revealed that there was no significant difference between plastic colours: blue ($Q_m = 5.156$, $p = 0.397$), black ($Q_m = 5.936$, $p = 0.313$), transparent ($Q_m = 5.259$, $p = 0.385$), and white ($Q_m = 7.747$, $p = 0.188$). In the moderator of environment, a significant difference was found in two polymer types, namely PP ($Q_m = 29.693$, $p < 0.001$) and PA ($Q_m = 21.143$, $p < 0.001$). PP had a higher abundance in freshwater (8.5%) and seawater (7.9%) than in aquaculture (5.4%), estuary (3.1%), and market (0%) environments. In contrast, PA was relatively more abundant in aquaculture (15.4%) than in seawater (7.4%), estuary (4.0%), freshwater (1.1%), and market (0.1%) environments. Subgroup analysis with the moderator of continent showed that a significant difference was found in PA ($Q_m = 50.287$, $p < 0.001$) and PES ($Q_m = 12.174$, $p = 0.033$). PE has the highest prevalence in Asia (21.6%), followed by Europe (17.2%), South America (15.1%), and Africa (14.3%), and significantly lower in North America (5.2%), and Oceania (0%). PES has a different distribution across continents, with a higher abundance in South America (22.0%), followed by Asia (14.2%), Oceania (13.6%), North America (12.2%), Europe (8.3%), and Africa (3.1%).

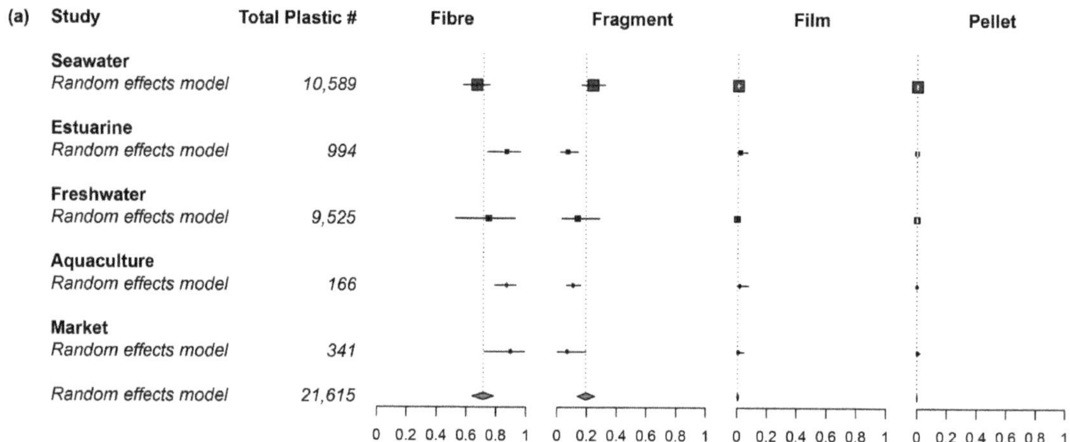

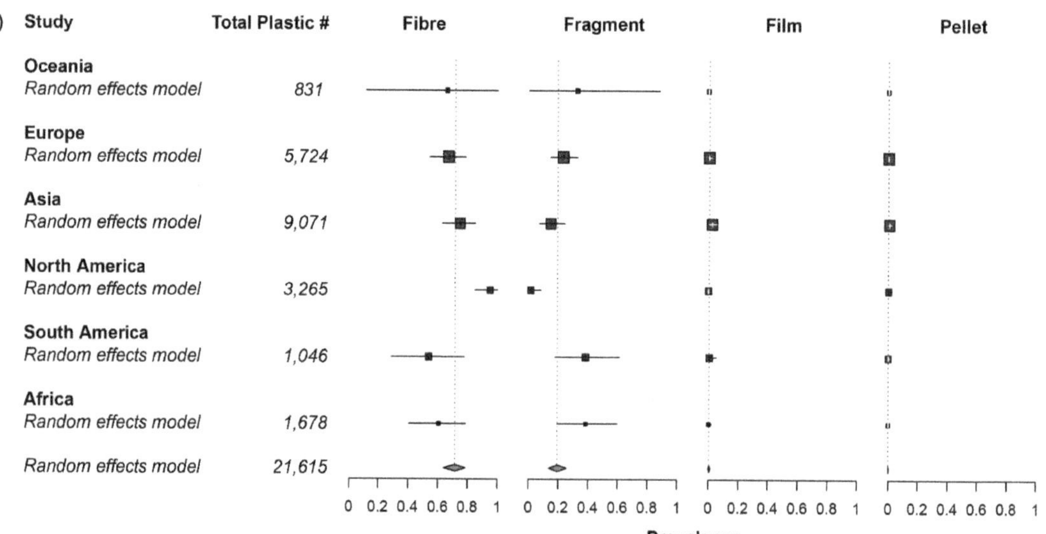

Figure 3. Prevalence forest plot for plastic shape. Blue squares represent subgroup means, while red diamonds and the dotted line represent the overall mean. (**a**) Subgroup of sampling environment. (**b**) Subgroup of sampling continent. For statistical details, see individual forest plots in supplementary information (Figures S4–S7).

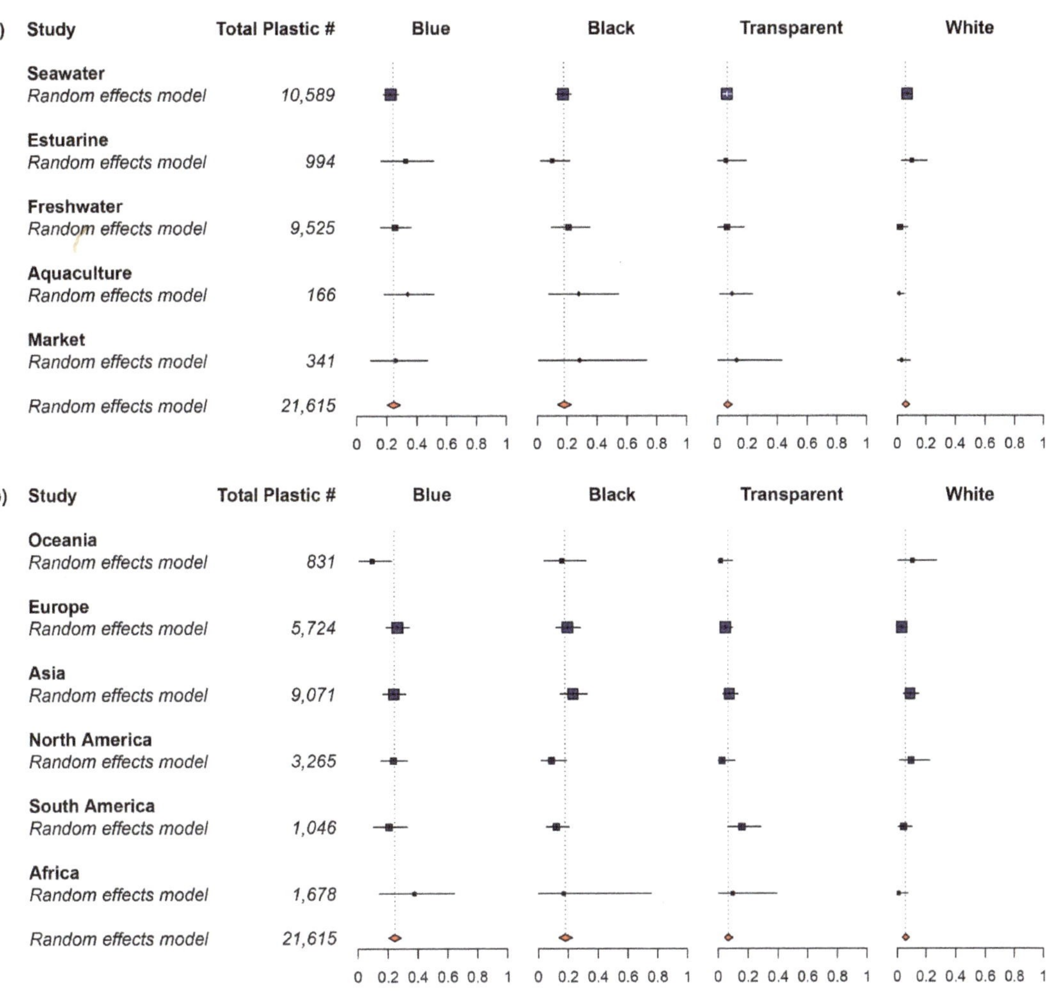

Figure 4. Prevalence forest plot for plastic colour. Blue squares represent subgroup means, while red diamonds and the dotted line represent the overall mean. (**a**) Subgroup of sampling environment. (**b**) Subgroup of sampling continent. For statistical details, see individual forest plots in supplementary information (Figures S8–S11).

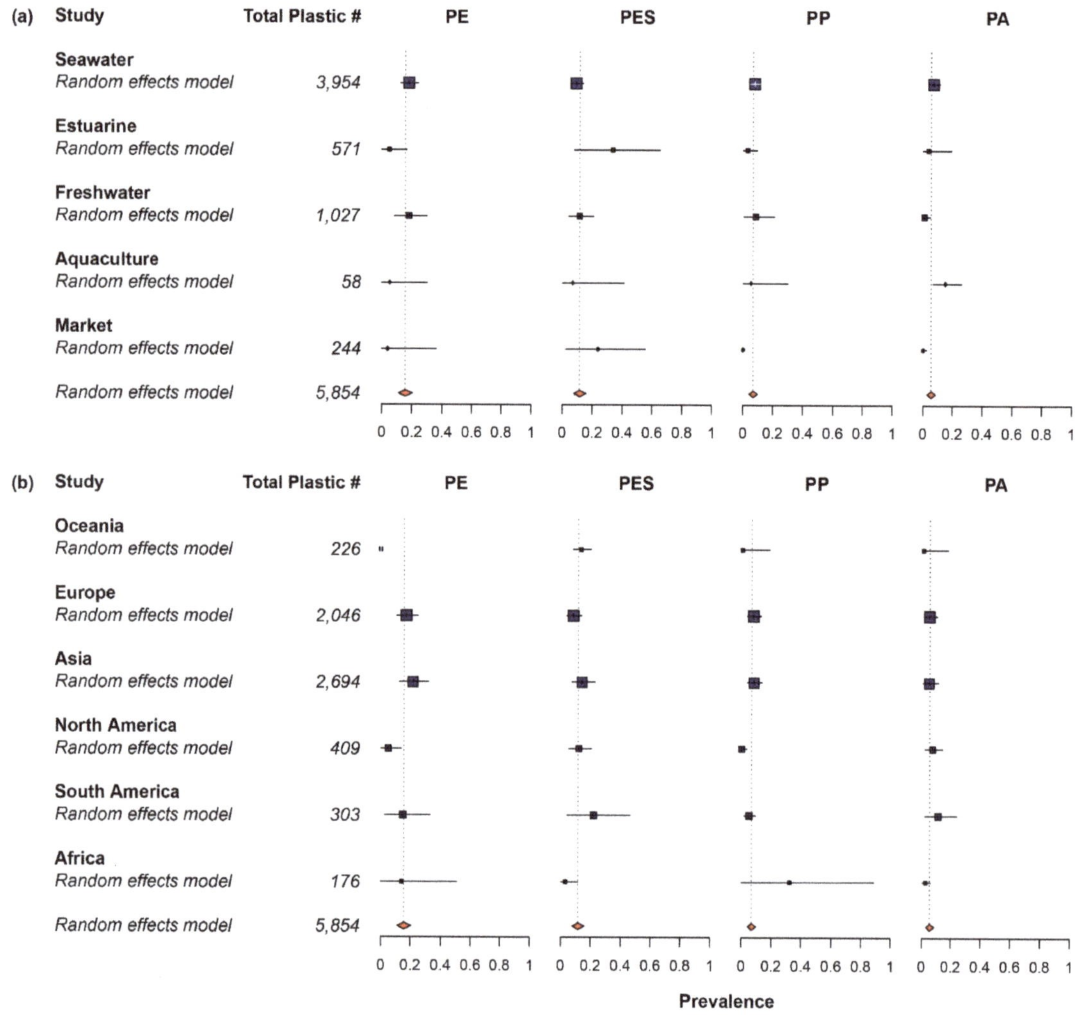

Figure 5. Prevalence forest plot for plastic polymer type. Blue squares represent subgroup means, while red diamonds and the dotted line represent the overall mean. (**a**) Subgroup of sampling environment. (**b**) Subgroup of sampling continent. PE: Polyethylene; PP: Polypropylene; PES: Polyester; PA: Polyamide. For statistical details, see individual forest plots in supplementary information (Figures S12–S15).

4. Discussion

Microplastics are widely defined as plastics with a size of <5 mm, whereas small microplastics and large microplastics are defined as plastics with a size of <1 mm and 1 to 5 mm, respectively. Small microplastics were the predominant plastic size ingested by fish in most of the reviewed studies. It was estimated that the most abundant plastic in the marine environment was microplastic (92.5%) [14]. The proportions of large and small microplastics in the marine environment were 62.3% and 37.7%, respectively. However, the concentration might be underestimated since the lower size limit of sampling and modelling used was 0.33 mm, whereby a 2.5-fold increase in microplastic contamination was observed when the lower size limit was 0.1 mm [164]. Hence, the actual concentration of small microplastics could be higher than the initial prediction. A similar concentration of microplastics can be expected in other environments since most of the microplastics in the marine environment originated from land sources such as sewage and runoff. A high concentration of small microplastics in the environment tend to be ingested by fish more easily through primary ingestion because they resemble their prey, especially zooplanktons, or secondary ingestion due to the attachment of plastics on their prey [15]. The predominance of small microplastics might be due to longer retention time in GIT, as they need longer time to be evacuated from the fish compared to larger size plastics [165]. However, several studies have excluded small microplastics during microscopic inspection and analysis, which might underestimate the actual number of plastics ingested [166–169]. It was reported that a lower detection limit would result in higher frequency of occurrence of plastic ingestion [46]. Studies with fish samples of smaller body size may influence the outcome, since they are unable to ingest larger size plastics. Therefore, there is a need to reduce the threshold size of plastic detection in order to identify all plastics, since small microplastics dominate the plastic ingested.

This meta-analysis showed that the largest percentage of plastics ingested by fish was in the form of fibre and fragment. Several studies have documented fibre plastics to be the most prevalent type of plastic in seawater, freshwater, and aquaculture environments [170–174]. Fibre plastics in the environment originate mainly from the effluent of wastewater treatment plants. An experiment illustrated that a single garment is able to produce >1900 fibres per wash and all garments can release >100 fibres per litre of effluent [12]. Similarly, it was estimated that over 700,000 fibres could be discharged from an average wash load of 6 kg fabrics [175]. Another source of fibre plastic in the environment could be from the fishery activities. The abrasion of abandoned, lost, or discarded fishing gears has contributed about 18% of the marine plastic debris in the marine environment [4]. Some fish species do not actively take up fibre plastic; instead, the fibre plastics are passively sucked in while breathing [176]. Therefore, most of the fish species may unintentionally ingest plastics that are ubiquitous in the environment. After exposure to microplastic in a laboratory study, fibre plastic accumulated the most in the gut of zebrafish, followed by fragment and pellet plastics [177]. Another study demonstrated that fibre and pellet plastics shared a similar retention time in the GIT when goldfish were fed with plastic of different shapes [178]. Shape-dependent accumulation of plastic could be another factor contributing to the prevalence of fibre plastic in fish, but more research is required. The accumulation period of plastic in GIT of fish may affect the outcome of the studies, as the plastics that have been extracted from the fish do not exactly represent the amount of plastic ingested throughout its lifetime. Instead, those samples that were found to have a relatively smaller quantity of non-fibre plastic might have egested those plastics out of their bodies when they were sampled. Hence, a larger sample size of the same species from the same sampling area should be examined to tackle this limitation.

Among the studies reviewed, blue is the most common plastic colour ingested by fish, followed by black, white, and transparent. Based on the global analysis of floating plastics in sea water, white and transparent/translucent (47%) are the most abundant plastic colours, followed by yellow and brown (26%), and blue (9%) [179]. This does not imply that the plastics in the ocean are mostly white and transparent/translucent, as the authors have excluded fibre plastic from the analysis due to the possibility of airborne contamination and fragments made up 83.6% of all the plastics collected. For studies that included fibre plastic, the predominant colours of the fibre were blue, black, transparent, and white [170]; black, grey, blue, and red [180]; transparent, blue, black, and red [181]; and transparent, white, blue, and red [182], respectively. The inconsistent results among the studies could be attributed to the differences in methodology and sampling region. Similar dominant colours such as blue, black, white, and transparent were observed in different studies. Hence, fish might accidentally consume the plastics by feeding or breathing, since the results were similar to the colour of plastics present in the environment. A study conducted in the China Sea revealed that the proportion of the plastic colour ingested by fish was similar to the proportion in water and sediment of the same sampling site [156]. Another possible explanation for the results could be related to selective feeding for the species sampled. Large pieces of plastic debris with blue and yellow colours were reported to be preferred by the fish [183]. Blue plastics were found to be predominantly ingested by Amberstripe scad, Atlantic chub mackerel, and fish larvae due to the resemblance to one of their preys: blue pigmented copepod species that were abundant in the sampling areas [184–186]. The blue pigmentation featured on zooplankton in the ocean [187] might account for them being confused with blue plastic particles. We hypothesise that only specific fish species ingest blue plastic deliberately due to the resemblance to its prey and most species consume blue plastic incidentally as a result of its abundance during feeding and breathing.

Our results confirmed that PE, PES, PP, and PA were the most prevalent polymer types ingested by fish globally. The results were not surprising, as these polymer types were widely found in marine and freshwater environments [173,188,189]. The abundance of these polymer types in the environments could be due to improper disposal of plastic waste, as they accounted for 80% of the global plastic waste generated in 2015 [190]. PE and PP might be derived from the abrasion of fishing tools, since they are widely used in fishery activities around the world, as well as the packaging used for foods and manufactured products. PE and PP are less dense polymers that will usually float on the surface of the water and are likely to be ingested by pelagic species, while demersal species tend to ingest dense plastics such as PES and PA because they usually suspend in the water column or deposition in the seabed. PA and PES are widely used in fishery activities and the clothing industry. The abundance of PA and PES in the environment is mostly originated from the effluent of washing clothes and the usage of fishery tools. For some studies, only part of the plastics extracted from the samples was tested with the polymer characterisation test, which could lead to a potential bias of these results.

5. Gaps and Recommendations

Fish are an essential component of a healthy human diet. Fish consumption increased significantly from 9.0 kg per capita in 1961 to 20.5 kg per capita in 2018 worldwide, which increased at an average annual rate of 1.5% [191]. As of 2017, fish consumption contributed 17% of animal protein intake, and 7% of all protein intake globally [191]. Although the viscera of fish are removed prior to consumption, humans still have a strong likelihood to be exposed to microplastics and even nanoplastics (<1 μm) due to the translocation of plastics to muscle tissues [192]. Meanwhile, many commercial fish species have been found to have microplastics embedded in their muscles, which are likely to be consumed by humans [61,193,194]. It was reported that seafood was one of the top three contributors of microplastics consumption by humans among the commonly consumed items [195]. Fish and bivalves were the seafood included in the study and they estimated that the total microplastics consumption of a person ranged from 39,000 to 52,000 particles per year. Lately, microplastics were detected within a small sample size of human stools, suggesting that humans had ingested these particles [196,197]. Although there was no direct evidence showing the sources of microplastics ingested by humans, it is still highly possible that part of the microplastics ingested originated from seafood, since the majority of the participants in the study consumed seafood within the study period [196,197]. Nevertheless, some fish species such as Japanese anchovy are commonly consumed by humans without the elimination of GIT, and it further increases the risk of translocation of plastic from fish to humans [141]. Furthermore, 262 out of 391 species that ingested plastic are commercial species that are frequently consumed by humans [44]. This should raise awareness of the dangers of consuming microplastics, since it poses a significant threat to human health [198]. However, research concerning plastic ingestion of fish in aquaculture environments has been overlooked and there are only a few studies on the incidence of plastic ingestion within this environment [151,162,199–201]. As of 2018, the contribution of world aquaculture to global fish production reached 82.1 million tonnes annually, which contributed 46.0% of the total fish production and increased from 25.7% in 2000 [191]. Fish cultured in aquaculture are exposed to plastic debris due to aged and shattered fishery equipment [202] and to contaminated feeds [203]. In fact, aquaculture sites are prone to accumulate plastic debris that may be ingested by fish incidentally [151,162]. There are studies showing that aquaculture fish have a lower incidence plastic ingestion than wild fish [200,201]. Hence, awareness towards them should be raised to further investigate the plastic contamination level within aquaculture fish, since they constitute almost half of the fish for human consumption globally.

Furthermore, gill and muscle tissue of the same sample should be examined together for the presence of plastic, since plastic contamination in gill was often reported [61,204] and even poses health risk towards the fish [205]. Deficiency of the record of plastic ingestion by fish is evident, as only 555 out of 22,581 known species have been investigated [46,206], comprising 2.5% compared to other taxa such as sea birds (44.0%), marine mammals (56.1%), and turtles (100.0%) [207]. Although there has been a significant improvement in ingestion records compared to previous records (fish, 0.3%; sea birds, 39.1%; marine mammals, 26.1%; and turtles, 85.7%) [49], more research on plastic ingestion in other fish species is necessary to further reveal the potential hazards in the environment.

In future research, the lowest threshold of plastic size should be mentioned in the study and threshold filter pore size must be at least 1 μm to fulfil the criteria of microplastics [208] and to capture all plastics ingested, since the predominant size of the plastic is <1 mm. It is difficult to compare the dominant size class ingested by fish across different studies because most of the studies have assigned a distinct size class (Figure 2), and the inconsistent classifications have made the comparison of plastic ingested by size more difficult. Instead, the plastic size classes should be standardised for ease of comparison of the dominant size class of plastic ingested between studies. Likewise, the shape of the plastics should be standardised, as suggested by GESAMP [54], into fibre, fragment, film, pellet, and foam. Since fibre is the dominant plastic shape ingested by fish, it should not be excluded from the analysis. Possible contamination should not be used as an exclusion criterion for plastic analysis [209]. Instead, extra care should be taken to eliminate possible contamination [210]. For studies that intend to investigate only the occurrence of microplastic in fish, any plastic that is 5 mm and above should not be excluded [211]; instead, it should be archived to record their characterisations such as size, shape, and colour, since it is still an anthropogenic particle and may pose a significant risk towards the fish. Polymer identification tests should be carried out randomly among the plastics extracted from the samples [212]. For future studies, it is essential that the size, colour, and shape of plastic ingestion be recorded and analysed to further validate if the fish species has a certain preference regarding plastic ingestion.

6. Conclusions

Our meta-analysis has revealed that the most abundant plastics ingested by fish globally was <1 mm in size, fibre shape, blue colour, and PE polymer. The results obtained were similar to the prevalence of plastics in environments where most of the fish species could ingest them passively. Hence, more research needs to be carried out in order to further validate if fish have a certain preference for ingesting plastic particles. Since fish are a one of the major protein sources, the incidence of plastic ingestion by fish, especially in aquaculture sites, should be a major cause for alarm, as it poses potential threats to human health, yet there is still a lack of information on plastic ingestion in many commercial fish species. Furthermore, it is essential that a standardised classification of plastic size, shape, and colour be established for use in future studies. A better understanding of the causes of plastic ingestion by fish can be achieved by adapting a uniform classification of plastic characterisations.

Supplementary Materials: The following supporting information can be downloaded at: https://www.mdpi.com/article/10.3390/toxics10040186/s1, Figure S1: Funnel plot for the prevalence of plastic's shapes ingested by fish from all environments. Studies are represented by full circles and imputed studies are represented by empty circles, Figure S2: Funnel plot for the prevalence of plastic's colours ingested by fish from all environments. Studies are represented by full circles and imputed studies are represented by empty circles, Figure S3: Funnel plot for the prevalence of plastic's polymer type ingested by fish from all environments. Studies are represented by full circles and imputed studies are represented by empty circles. PE: Polyethylene; PP: Polypropylene; PES: Polyester; PA: Polyamide; PS: Polystyrene, Figure S4: Forest plot for fibre subgroup analysis. Red diamonds represent subgroup means. Total: total plastics found in each study. Fibre: number of fibres found in each study, Figure S5: Forest plot for fragment subgroup analysis. Red diamonds represent subgroup means. Total: total plastics found in each study. Fragment: number of fragments found in each study, Figure S6: Forest plot for film subgroup analysis. Red diamonds represent subgroup means. Total: total plastics found in each study. Film: number of films found in each study, Figure S7: Forest plot for pellet subgroup analysis. Red diamonds represent subgroup means. Total: total plastics found in each study. Pellet: number of pellets found in each study, Figure S8: Forest plot for blue subgroup analysis. Red diamonds represent subgroup means. Total: total plastics found in each study. Blue: number of blues found in each study, Figure S9: Forest plot for black subgroup analysis. Red diamonds represent subgroup means. Total: total plastics found in each study. Black: number of blacks found in each study, Figure S10: Forest plot for transparent subgroup analysis. Red diamonds represent subgroup means. Total: total plastics found in each study. Transparent: number of transparent found in each study, Figure S11: Forest plot for white subgroup analysis. Red diamonds represent subgroup means. Total: total plastics found in each study. White: number of whites found in each study, Figure S12: Forest plot for PE subgroup analysis. Red diamonds represent subgroup means. Total: total plastics found in each study. PE: number of PE found in each study. PE: Polyethylene, Figure S13: Forest plot for PES subgroup analysis. Red diamonds represent subgroup means. Total: total plastics found in each study. PES: number of PES found in each study. PES: Polyester, Figure S14: Forest plot for PP subgroup analysis. Red diamonds represent subgroup means. Total: total plastics found in each study. PP: number of PP found in each study. PP: Polypropylene, Figure S15: Forest plot for PA subgroup analysis. Red diamonds represent subgroup means. Total: total plastics found in each study. PA: number of PA found in each study. PA: Polyamide.

Author Contributions: Writing—original draft: K.P.L.; Writing—review and editing: K.P.L., P.E.L., S.Y.; Visualisation: K.P.L.; Data curation: K.P.L.; Formal analysis: K.P.L., J.D.; Funding acquisition: P.E.L., C.S.; Supervision: P.E.L., C.S.; Methodology: K.P.L., K.H.L.; Validation: J.D., K.H.L. All authors have read and agreed to the published version of the manuscript.

Funding: This work was supported by the research fund from First Institute of Oceanography-University of Malaya Joint Center of Marine Science and Technology (FIO-UM JCMST) (IF002-2020) and Asian Countries Maritime Cooperation Fund (99950410).

Institutional Review Board Statement: Not applicable.

Informed Consent Statement: Not applicable.

Data Availability Statement: Not applicable.

Acknowledgments: Kok Ping Lim was supported by the FIO-UM JCMST.

Conflicts of Interest: The authors declare no conflict of interest.

References

1. PlasticEurope. Plastics—The Facts 2020. Available online: https://www.plasticseurope.org/en/resources/market-data (accessed on 20 February 2022).
2. PlasticEurope. The Compelling Facts About Plastics. Available online: https://www.plasticseurope.org/en/resources/market-data (accessed on 20 February 2022).
3. Derraik, J.G. The pollution of the marine environment by plastic debris: A review. *Mar. Pollut Bull.* **2002**, *44*, 842–852. [CrossRef]
4. Andrady, A.L. Microplastics in the marine environment. *Mar. Pollut. Bull.* **2011**, *62*, 1596–1605. [CrossRef] [PubMed]
5. Van Cauwenberghe, L.; Vanreusel, A.; Mees, J.; Janssen, C.R. Microplastic pollution in deep-sea sediments. *Env. Pollut.* **2013**, *182*, 495–499. [CrossRef] [PubMed]

6. Obbard, R.W.; Sadri, S.; Wong, Y.Q.; Khitun, A.A.; Baker, I.; Thompson, R.C. Global warming releases microplastic legacy frozen in Arctic Sea ice. *Earths Future* **2014**, *2*, 315–320. [CrossRef]
7. Free, C.M.; Jensen, O.P.; Mason, S.A.; Eriksen, M.; Williamson, N.J.; Boldgiv, B. High-levels of microplastic pollution in a large, remote, mountain lake. *Mar. Pollut. Bull.* **2014**, *85*, 156–163. [CrossRef]
8. Zhu, F.; Zhu, C.; Wang, C.; Gu, C. Occurrence and Ecological Impacts of Microplastics in Soil Systems: A Review. *Bull. Env. Contam. Toxicol.* **2019**, *102*, 741–749. [CrossRef]
9. Cai, L.; Wang, J.; Peng, J.; Tan, Z.; Zhan, Z.; Tan, X.; Chen, Q. Characteristic of microplastics in the atmospheric fallout from Dongguan city, China: Preliminary research and first evidence. *Env. Sci. Pollut. Res. Int.* **2017**, *24*, 24928–24935. [CrossRef]
10. Arthur, C.; Baker, J.; Bamford, H. Proceedings of the International Research Workshop on the Occurrence, Effects and Fate of Microplastic Marine Debris. NOAA Technical Memorandum NOS-OR&R-30, Tacoma, WA, USA, 9–11 September 2008.
11. Fendall, L.S.; Sewell, M.A. Contributing to marine pollution by washing your face: Microplastics in facial cleansers. *Mar. Pollut. Bull.* **2009**, *58*, 1225–1228. [CrossRef]
12. Browne, M.A.; Crump, P.; Niven, S.J.; Teuten, E.; Tonkin, A.; Galloway, T.; Thompson, R. Accumulation of microplastic on shorelines worldwide: Sources and sinks. *Env. Sci. Technol.* **2011**, *45*, 9175–9179. [CrossRef]
13. Lebreton, L.C.M.; van der Zwet, J.; Damsteeg, J.W.; Slat, B.; Andrady, A.; Reisser, J. River plastic emissions to the world's oceans. *Nat. Commun.* **2017**, *8*, 15611. [CrossRef]
14. Eriksen, M.; Lebreton, L.C.; Carson, H.S.; Thiel, M.; Moore, C.J.; Borerro, J.C.; Galgani, F.; Ryan, P.G.; Reisser, J. Plastic pollution in the world's oceans: More than 5 trillion plastic pieces weighing over 250,000 tons afloat at sea. *PLoS ONE* **2014**, *9*, e111913. [CrossRef] [PubMed]
15. Ryan, P.G. Ingestion of Plastics by Marine Organisms. In *Hazardous Chemicals Associated with Plastics in the Marine Environment*; Takada, H., Karapanagioti, H.K., Eds.; Springer International Publishing: Cham, Switzerland, 2016; pp. 235–266.
16. Kuhn, S.; van Franeker, J.A. Plastic ingestion by the northern fulmar (*Fulmarus glacialis*) in Iceland. *Mar. Pollut. Bull.* **2012**, *64*, 1252–1254. [CrossRef] [PubMed]
17. Reynolds, C.; Ryan, P.G. Micro-plastic ingestion by waterbirds from contaminated wetlands in South Africa. *Mar. Pollut Bull.* **2018**, *126*, 330–333. [CrossRef] [PubMed]
18. Murray, F.; Cowie, P.R. Plastic contamination in the decapod crustacean *Nephrops norvegicus* (Linnaeus, 1758). *Mar. Pollut. Bull.* **2011**, *62*, 1207–1217. [CrossRef] [PubMed]
19. Iannilli, V.; Di Gennaro, A.; Lecce, F.; Sighicelli, M.; Falconieri, M.; Pietrelli, L.; Poeta, G.; Battisti, C. Microplastics in *Talitrus saltator* (Crustacea, Amphipoda): New evidence of ingestion from natural contexts. *Environ. Sci. Pollut. Res. Int.* **2018**, *25*, 28725–28729. [CrossRef]
20. Fernandez, C.; Anastasopoulou, A. Plastic ingestion by blue shark *Prionace glauca* in the South Pacific Ocean (south of the Peruvian Sea). *Mar. Pollut. Bull.* **2019**, *149*, 110501. [CrossRef]
21. Foekema, E.M.; De Gruijter, C.; Mergia, M.T.; van Franeker, J.A.; Murk, A.J.; Koelmans, A.A. Plastic in north sea fish. *Env. Sci. Technol.* **2013**, *47*, 8818–8824. [CrossRef]
22. Lusher, A.L.; Hernandez-Milian, G.; O'Brien, J.; Berrow, S.; O'Connor, I.; Officer, R. Microplastic and macroplastic ingestion by a deep diving, oceanic cetacean: The True's beaked whale *Mesoplodon mirus*. *Environ. Pollut.* **2015**, *199*, 185–191. [CrossRef]
23. Stamper, M.A.; Whitaker, B.R.; Schofield, T.D. Case Study: Morbidity in a Pygmy Sperm Whale *Kogia breviceps* due to ocean-bourne plastic. *Mar. Mammal. Sci.* **2006**, *22*, 719–722. [CrossRef]
24. Setälä, O.; Fleming-Lehtinen, V.; Lehtiniemi, M. Ingestion and transfer of microplastics in the planktonic food web. *Env. Pollut.* **2014**, *185*, 77–83. [CrossRef]
25. Nelms, S.E.; Galloway, T.S.; Godley, B.J.; Jarvis, D.S.; Lindeque, P.K. Investigating microplastic trophic transfer in marine top predators. *Env. Pollut.* **2018**, *238*, 999–1007. [CrossRef] [PubMed]
26. Chagnon, C.; Thiel, M.; Antunes, J.; Ferreira, J.L.; Sobral, P.; Ory, N.C. Plastic ingestion and trophic transfer between Easter Island flying fish (*Cheilopogon rapanouiensis*) and yellowfin tuna (*Thunnus albacares*) from Rapa Nui (Easter Island). *Env. Pollut.* **2018**, *243*, 127–133. [CrossRef] [PubMed]
27. Markic, A.; Niemand, C.; Bridson, J.H.; Mazouni-Gaertner, N.; Gaertner, J.C.; Eriksen, M.; Bowen, M. Double trouble in the South Pacific subtropical gyre: Increased plastic ingestion by fish in the oceanic accumulation zone. *Mar. Pollut. Bull.* **2018**, *136*, 547–564. [CrossRef] [PubMed]
28. Welden, N.A.; Abylkhani, B.; Howarth, L.M. The effects of trophic transfer and environmental factors on microplastic uptake by plaice, *Pleuronectes plastessa*, and spider crab, *Maja squinado*. *Environ. Pollut.* **2018**, *239*, 351–358. [CrossRef]
29. Ferreira, G.V.B.; Barletta, M.; Lima, A.R.A.; Morley, S.A.; Costa, M.F. Dynamics of Marine Debris Ingestion by Profitable Fishes Along the Estuarine Ecocline. *Sci. Rep.* **2019**, *9*, 13514. [CrossRef]
30. Santos, R.G.; Andrades, R.; Boldrini, M.A.; Martins, A.S. Debris ingestion by juvenile marine turtles: An underestimated problem. *Mar. Pollut. Bull.* **2015**, *93*, 37–43. [CrossRef]
31. Pierce, K.E.; Harris, R.J.; Larned, L.S.; Pokras, M. Obstruction and starvation associated with plastic ingestion in a Northern Gannet Morus bassanus and a Greater Shearwater Puffinus gravis. *Mar. Ornithol.* **2004**, *32*, 187–189.
32. Beck, C.A.; Barros, N.B. The impact of debris on the Florida manatee. *Mar. Pollut. Bull.* **1991**, *22*, 508–510. [CrossRef]

33. Haetrakul, T.; Munanansup, S.; Assawawongkasem, N.; Chansue, N. A Case Report: Stomach Foreign Object in Whaleshark (Rhincodon Types) Stranded in Thailand. In Proceedings of the 4th International Symposium on SEASTAR2000 and Asian Bio-logging Science (The 8th SEASTAR2000 Workshop), Phuket, Thailand, 15–17 December 2007.
34. Jabeen, K.; Li, B.; Chen, Q.; Su, L.; Wu, C.; Hollert, H.; Shi, H. Effects of virgin microplastics on goldfish (*Carassius auratus*). *Chemosphere* **2018**, *213*, 323–332. [CrossRef]
35. Naidoo, T.; Glassom, D. Decreased growth and survival in small juvenile fish, after chronic exposure to environmentally relevant concentrations of microplastic. *Mar. Pollut. Bull.* **2019**, *145*, 254–259. [CrossRef]
36. Lu, Y.; Zhang, Y.; Deng, Y.; Jiang, W.; Zhao, Y.; Geng, J.; Ding, L.; Ren, H. Uptake and Accumulation of Polystyrene Microplastics in Zebrafish (*Danio rerio*) and Toxic Effects in Liver. *Environ. Sci. Technol.* **2016**, *50*, 4054–4060. [CrossRef] [PubMed]
37. Critchell, K.; Hoogenboom, M.O. Effects of microplastic exposure on the body condition and behaviour of planktivorous reef fish (*Acanthochromis polyacanthus*). *PLoS ONE* **2018**, *13*, e0193308. [CrossRef] [PubMed]
38. Ahrendt, C.; Perez-Venegas, D.J.; Urbina, M.; Gonzalez, C.; Echeveste, P.; Aldana, M.; Pulgar, J.; Galban-Malagon, C. Microplastic ingestion cause intestinal lesions in the intertidal fish *Girella laevifrons*. *Mar. Pollut. Bull.* **2020**, *151*, 110795. [CrossRef] [PubMed]
39. Hoss, D.E.; Settle, L.R. Ingestion of Plastics by Teleost Fishes. In Proceedings of the Second International Conference on Marine Debris. NOAA Technical Memorandum. NOAA-TM-NMFS-SWFSC-154, Honolulu, HI, USA, 2–7 April 1989; pp. 693–709.
40. Cannon, S.M.E.; Lavers, J.L.; Figueiredo, B. Plastic ingestion by fish in the Southern Hemisphere: A baseline study and review of methods. *Mar. Pollut. Bull.* **2016**, *107*, 286–291. [CrossRef] [PubMed]
41. Pinheiro, C.; Oliveira, U.; Vieira, M. Occurrence and impacts of microplastics in freshwater fish. *J. Aquac. Mar. Biol.* **2017**, *5*, 00138. [CrossRef]
42. Liboiron, F.; Ammendolia, J.; Saturno, J.; Melvin, J.; Zahara, A.; Richard, N.; Liboiron, M. A zero percent plastic ingestion rate by silver hake (*Merluccius bilinearis*) from the south coast of Newfoundland, Canada. *Mar. Pollut. Bull.* **2018**, *131*, 267–275. [CrossRef]
43. Kroon, F.J.; Motti, C.E.; Jensen, L.H.; Berry, K.L.E. Classification of marine microdebris: A review and case study on fish from the Great Barrier Reef, Australia. *Sci. Rep.* **2018**, *8*, 16422. [CrossRef]
44. Markic, A.; Gaertner, J.-C.; Gaertner-Mazouni, N.; Koelmans, A.A. Plastic ingestion by marine fish in the wild. *Crit. Rev. Env. Sci. Tec.* **2020**, *50*, 657–697. [CrossRef]
45. Wootton, N.; Reis-Santos, P.; Gillanders, B.M. Microplastic in fish—A global synthesis. *Rev. Fish Biol. Fish.* **2021**, *31*, 753–771. [CrossRef]
46. Savoca, M.S.; McInturf, A.G.; Hazen, E.L. Plastic ingestion by marine fish is widespread and increasing. *Glob. Chang. Biol.* **2021**, *27*, 2188–2199. [CrossRef]
47. Garrido Gamarro, E.; Ryder, J.; Elevoll, E.O.; Olsen, R.L. Microplastics in fish and shellfish—A threat to seafood safety? *J. Aquat. Food Prod. Technol.* **2020**, *29*, 417–425. [CrossRef]
48. Provencher, J.F.; Bond, A.L.; Avery-Gomm, S.; Borrelle, S.B.; Rebolledo, E.L.B.; Hammer, S.; Kühn, S.; Lavers, J.L.; Mallory, M.L.; Trevail, A. Quantifying ingested debris in marine megafauna: A review and recommendations for standardization. *Anal. Methods* **2017**, *9*, 1454–1469. [CrossRef]
49. Gall, S.C.; Thompson, R.C. The impact of debris on marine life. *Mar. Pollut. Bull.* **2015**, *92*, 170–179. [CrossRef] [PubMed]
50. Azizi, N.; Khoshnamvand, N.; Nasseri, S. The quantity and quality assessment of microplastics in the freshwater fishes: A systematic review and meta-analysis. *Reg. Stud. Mar. Sci.* **2021**, *47*, 101955. [CrossRef]
51. Rochman, C.M.; Tahir, A.; Williams, S.L.; Baxa, D.V.; Lam, R.; Miller, J.T.; Teh, F.C.; Werorilangi, S.; Teh, S.J. Anthropogenic debris in seafood: Plastic debris and fibers from textiles in fish and bivalves sold for human consumption. *Sci. Rep.* **2015**, *5*, 14340. [CrossRef] [PubMed]
52. Ding, J.; Li, J.; Sun, C.; Jiang, F.; Ju, P.; Qu, L.; Zheng, Y.; He, C. Detection of microplastics in local marine organisms using a multi-technology system. *Anal. Methods* **2019**, *11*, 78–87. [CrossRef]
53. Lee, J.; Lee, J.S.; Jang, Y.C.; Hong, S.Y.; Shim, W.J.; Song, Y.K.; Hong, S.H.; Jang, M.; Han, G.M.; Kang, D.; et al. Distribution and Size Relationships of Plastic Marine Debris on Beaches in South Korea. *Arch. Env. Contam. Toxicol.* **2015**, *69*, 288–298. [CrossRef]
54. GESAMP. *Guidelines for the Monitoring and Assessment of Plastic Litter and Microplastics in the Ocean*; United Nations Environment Programme: Nairobi, Kenya, 2019; p. 130.
55. Fossi, M.C.; Romeo, T.; Baini, M.; Panti, C.; Marsili, L.; Campani, T.; Canese, S.; Galgani, F.; Druon, J.-N.; Airoldi, S. Plastic debris occurrence, convergence areas and fin whales feeding ground in the Mediterranean marine protected area Pelagos sanctuary: A modeling approach. *Front. Mar. Sci.* **2017**, *4*, 167. [CrossRef]
56. McNeish, R.E.; Kim, L.H.; Barrett, H.A.; Mason, S.A.; Kelly, J.J.; Hoellein, T.J. Microplastic in riverine fish is connected to species traits. *Sci. Rep.* **2018**, *8*, 11639. [CrossRef]
57. Eriksen, M.; Mason, S.; Wilson, S.; Box, C.; Zellers, A.; Edwards, W.; Farley, H.; Amato, S. Microplastic pollution in the surface waters of the Laurentian Great Lakes. *Mar. Pollut. Bull.* **2013**, *77*, 177–182. [CrossRef]
58. Acharya, S.; Rumi, S.S.; Hu, Y.; Abidi, N. Microfibers from synthetic textiles as a major source of microplastics in the environment: A review. *Text. Res. J.* **2021**, *91*, 2136–2156. [CrossRef]
59. Imhof, H.K.; Schmid, J.; Niessner, R.; Ivleva, N.P.; Laforsch, C. A novel, highly efficient method for the separation and quantification of plastic particles in sediments of aquatic environments. *Limnol. Oceanogr. Meth.* **2012**, *10*, 524–537. [CrossRef]
60. Naji, A.; Nuri, M.; Amiri, P.; Niyogi, S. Small microplastic particles (S-MPPs) in sediments of mangrove ecosystem on the northern coast of the Persian Gulf. *Mar. Pollut. Bull.* **2019**, *146*, 305–311. [CrossRef] [PubMed]

61. Abbasi, S.; Soltani, N.; Keshavarzi, B.; Moore, F.; Turner, A.; Hassanaghaei, M. Microplastics in different tissues of fish and prawn from the Musa Estuary, Persian Gulf. *Chemosphere* **2018**, *205*, 80–87. [CrossRef] [PubMed]
62. Abidli, S.; Akkari, N.; Lahbib, Y.; El Menif, N.T. First evaluation of microplastics in two commercial fish species from the lagoons of Bizerte and Ghar El Melh (Northern Tunisia). *Reg. Stud. Mar. Sci.* **2021**, *41*, 101581. [CrossRef]
63. Abiñon, B.S.F.; Camporedondo, B.S.; Mercadal, E.M.B.; Olegario, K.M.R.; Palapar, E.M.H.; Ypil, C.W.R.; Tambuli, A.E.; Lomboy, C.A.L.M.; Garces, J.J.C. Abundance and characteristics of microplastics in commercially sold fishes from Cebu Island, Philippines. *Int. J. Aquat. Biol.* **2020**, *8*, 424–433. [CrossRef]
64. Agharokh, A.; S Taleshi, M.; Bibak, M.; Rasta, M.; Torabi Jafroudi, H.; Rubio Armesto, B. Assessing the relationship between the abundance of microplastics in sediments, surface waters, and fish in the Iran southern shores. *Environ. Sci. Pollut. Res. Int.* **2021**, *29*, 18546–18558. [CrossRef]
65. Arias, A.H.; Ronda, A.C.; Oliva, A.L.; Marcovecchio, J.E. Evidence of Microplastic Ingestion by Fish from the Bahia Blanca Estuary in Argentina, South America. *Bull. Environ. Contam. Toxicol.* **2019**, *102*, 750–756. [CrossRef]
66. Atamanalp, M.; Köktürk, M.; Parlak, V.; Ucar, A.; Arslan, G.; Alak, G. A new record for the presence of microplastics in dominant fish species of the Karasu River Erzurum, Turkey. *Environ. Sci. Pollut. Res. Int.* **2021**, *29*, 7866–7876. [CrossRef]
67. Atamanalp, M.; Köktürk, M.; Uçar, A.; Duyar, H.A.; Özdemir, S.; Parlak, V.; Esenbuǧa, N.; Alak, G. Microplastics in Tissues (Brain, Gill, Muscle and Gastrointestinal) of *Mullus barbatus* and *Alosa immaculata*. *Arch. Environ. Contam. Toxicol.* **2021**, *81*, 460–469. [CrossRef]
68. Atici, A.A.; Sepil, A.; Sen, F. High levels of microplastic ingestion by commercial, planktivorous *Alburnus tarichi* in Lake Van, Turkey. *Food Addit. Contam. Part A Chem. Anal. Control. Expo. Risk Assess* **2021**, *38*, 1767–1777. [CrossRef] [PubMed]
69. Avio, C.G.; Gorbi, S.; Regoli, F. Experimental development of a new protocol for extraction and characterization of microplastics in fish tissues: First observations in commercial species from Adriatic Sea. *Mar. Environ. Res.* **2015**, *111*, 18–26. [CrossRef] [PubMed]
70. Avio, C.G.; Pittura, L.; d'Errico, G.; Abel, S.; Amorello, S.; Marino, G.; Gorbi, S.; Regoli, F. Distribution and characterization of microplastic particles and textile microfibers in Adriatic food webs: General insights for biomonitoring strategies. *Environ. Pollut.* **2020**, *258*, 113766. [CrossRef]
71. Bagheri, T.; Gholizadeh, M.; Abarghouei, S.; Zakeri, M.; Hedayati, A.; Rabaniha, M.; Aghaeimoghadam, A.; Hafezieh, M. Microplastics distribution, abundance and composition in sediment, fishes and benthic organisms of the Gorgan Bay, Caspian sea. *Chemosphere* **2020**, *257*, 127201. [CrossRef] [PubMed]
72. Bayo, J.; Rojo, D.; Martinez-Banos, P.; Lopez-Castellanos, J.; Olmos, S. Commercial Gilthead Seabream (*Sparus aurata* L.) from the Mar Menor Coastal Lagoon as Hotspots of Microplastic Accumulation in the Digestive System. *Int. J. Environ. Res. Public Health* **2021**, *18*, 6844. [CrossRef]
73. Beer, S.; Garm, A.; Huwer, B.; Dierking, J.; Nielsen, T.G. No increase in marine microplastic concentration over the last three decades—A case study from the Baltic Sea. *Sci. Total Environ.* **2018**, *621*, 1272–1279. [CrossRef] [PubMed]
74. Bellas, J.; Martinez-Armental, J.; Martinez-Camara, A.; Besada, V.; Martinez-Gomez, C. Ingestion of microplastics by demersal fish from the Spanish Atlantic and Mediterranean coasts. *Mar. Pollut. Bull.* **2016**, *109*, 55–60. [CrossRef]
75. Bessa, F.; Barria, P.; Neto, J.M.; Frias, J.; Otero, V.; Sobral, P.; Marques, J.C. Occurrence of microplastics in commercial fish from a natural estuarine environment. *Mar. Pollut. Bull.* **2018**, *128*, 575–584. [CrossRef]
76. Bottari, T.; Savoca, S.; Mancuso, M.; Capillo, G.; GiuseppePanarello, G.; MartinaBonsignore, M.; Crupi, R.; Sanfilippo, M.; D'Urso, L.; Compagnini, G.; et al. Plastics occurrence in the gastrointestinal tract of *Zeus faber* and *Lepidopus caudatus* from the Tyrrhenian Sea. *Mar. Pollut. Bull.* **2019**, *146*, 408–416. [CrossRef]
77. Chen, J.C.; Fang, C.; Zheng, R.H.; Hong, F.K.; Jiang, Y.L.; Zhang, M.; Li, Y.; Hamid, F.S.; Bo, J.; Lin, L.S. Microplastic pollution in wild commercial nekton from the South China Sea and Indian Ocean, and its implication to human health. *Mar. Environ. Res.* **2021**, *167*, 105295. [CrossRef]
78. Cordova, M.R.; Riani, E.; Shiomoto, A. Microplastics ingestion by blue panchax fish (*Aplocheilus sp.*) from Ciliwung Estuary, Jakarta, Indonesia. *Mar. Pollut. Bull.* **2020**, *161*, 111763. [CrossRef] [PubMed]
79. Crutchett, T.; Paterson, H.; Ford, B.M.; Speldewinde, P. Plastic Ingestion in Sardines (*Sardinops sagax*) From Frenchman Bay, Western Australia, Highlights a Problem in a Ubiquitous Fish. *Front. Mar. Sci.* **2020**, *7*, 526. [CrossRef]
80. Da Silva, J.M.; Alves, L.M.F.; Laranjeiro, M.I.; Bessa, F.; Silva, A.V.; Norte, A.C.; Lemos, M.F.L.; Ramos, J.A.; Novais, S.C.; Ceia, F.R. Accumulation of chemical elements and occurrence of microplastics in small pelagic fish from a neritic environment. *Environ. Pollut.* **2022**, *292*, 118451. [CrossRef] [PubMed]
81. Daniel, D.B.; Ashraf, P.M.; Thomas, S.N. Microplastics in the edible and inedible tissues of pelagic fishes sold for human consumption in Kerala, India. *Environ. Pollut.* **2020**, *266*, 115365. [CrossRef]
82. Dhimmer, V.R. Microplastics in Gastrointestinal Tracts of *Trachurus trachurus* and *Scomber colias* from the Portuguese Coastal Waters. Ph.D. Thesis, Universidade Nova de Lisboa, Lisbon, Portugal, 2017.
83. Digka, N.; Tsangaris, C.; Torre, M.; Anastasopoulou, A.; Zeri, C. Microplastics in mussels and fish from the Northern Ionian Sea. *Mar. Pollut. Bull.* **2018**, *135*, 30–40. [CrossRef]
84. Ding, J.; Jiang, F.; Li, J.; Wang, Z.; Sun, C.; Wang, Z.; Fu, L.; Ding, N.X.; He, C. Microplastics in the Coral Reef Systems from Xisha Islands of South China Sea. *Environ. Sci. Technol.* **2019**, *53*, 8036–8046. [CrossRef]
85. Feng, Z.; Zhang, T.; Li, Y.; He, X.; Wang, R.; Xu, J.; Gao, G. The accumulation of microplastics in fish from an important fish farm and mariculture area, Haizhou Bay, China. *Sci. Total Environ.* **2019**, *696*, 133948. [CrossRef]

86. Garcia-Garin, O.; Vighi, M.; Aguilar, A.; Tsangaris, C.; Digka, N.; Kaberi, H.; Borrell, A. *Boops boops* as a bioindicator of microplastic pollution along the Spanish Catalan coast. *Mar. Pollut. Bull.* **2019**, *149*, 110648. [CrossRef]

87. Ghosh, G.C.; Akter, S.M.; Islam, R.M.; Habib, A.; Chakraborty, T.K.; Zaman, S.; Kabir, A.E.; Shipin, O.V.; Wahid, M.A. Microplastics contamination in commercial marine fish from the Bay of Bengal. *Reg. Stud. Mar. Sci.* **2021**, *44*, 101728. [CrossRef]

88. Gurjar, U.R.; Xavier, K.A.M.; Shukla, S.P.; Deshmukhe, G.; Jaiswar, A.K.; Nayak, B.B. Incidence of microplastics in gastrointestinal tract of golden anchovy (*Coilia dussumieri*) from north east coast of Arabian Sea: The ecological perspective. *Mar. Pollut. Bull.* **2021**, *169*, 112518. [CrossRef]

89. Gurjar, U.R.; Xavier, K.A.M.; Shukla, S.P.; Jaiswar, A.K.; Deshmukhe, G.; Nayak, B.B. Microplastic pollution in coastal ecosystem off Mumbai coast, India. *Chemosphere* **2021**, *288*, 132484. [CrossRef] [PubMed]

90. Hamilton, B.M.; Rochman, C.M.; Hoellein, T.J.; Robison, B.H.; Van Houtan, K.S.; Choy, C.A. Prevalence of microplastics and anthropogenic debris within a deep-sea food web. *Mar. Ecol. Prog. Ser.* **2021**, *675*, 23–33. [CrossRef]

91. Heshmati, S.; Makhdoumi, P.; Pirsaheb, M.; Hossini, H.; Ahmadi, S.; Fattahi, H. Occurrence and characterization of microplastic content in the digestive system of riverine fishes. *J. Environ. Manag.* **2021**, *299*, 113620. [CrossRef] [PubMed]

92. Hipfner, J.M.; Galbraith, M.; Tucker, S.; Studholme, K.R.; Domalik, A.D.; Pearson, S.F.; Good, T.P.; Ross, P.S.; Hodum, P. Two forage fishes as potential conduits for the vertical transfer of microfibres in Northeastern Pacific Ocean food webs. *Env. Pollut.* **2018**, *239*, 215–222. [CrossRef]

93. Hossain, M.S.; Sobhan, F.; Uddin, M.N.; Sharifuzzaman, S.M.; Chowdhury, S.R.; Sarker, S.; Chowdhury, M.S.N. Microplastics in fishes from the Northern Bay of Bengal. *Sci. Total Environ.* **2019**, *690*, 821–830. [CrossRef]

94. Hosseinpour, A.; Chamani, A.; Mirzaei, R.; Mohebbi-Nozar, S.L. Occurrence, abundance and characteristics of microplastics in some commercial fish of northern coasts of the Persian Gulf. *Mar. Pollut. Bull.* **2021**, *171*, 112693. [CrossRef]

95. Huang, J.S.; Koongolla, J.B.; Li, H.X.; Lin, L.; Pan, Y.F.; Liu, S.; He, W.H.; Maharana, D.; Xu, X.R. Microplastic accumulation in fish from Zhanjiang mangrove wetland, South China. *Sci. Total Sci. Total Environ.* **2020**, *708*, 134839. [CrossRef]

96. Jaafar, N.; Azfaralariff, A.; Musa, S.M.; Mohamed, M.; Yusoff, A.H.; Lazim, A.M. Occurrence, distribution and characteristics of microplastics in gastrointestinal tract and gills of commercial marine fish from Malaysia. *Sci. Total Environ.* **2021**, *799*, 149457. [CrossRef]

97. James, K.; Vasant, K.; Padua, S.; Gopinath, V.; Abilash, K.S.; Jeyabaskaran, R.; Babu, A.; John, S. An assessment of microplastics in the ecosystem and selected commercially important fishes off Kochi, south eastern Arabian Sea, India. *Mar. Pollut. Bull.* **2020**, *154*, 111027. [CrossRef]

98. Karbalaei, S.; Golieskardi, A.; Hamzah, H.B.; Abdulwahid, S.; Hanachi, P.; Walker, T.R.; Karami, A. Abundance and characteristics of microplastics in commercial marine fish from Malaysia. *Mar. Pollut. Bull.* **2019**, *148*, 5–15. [CrossRef]

99. Koongolla, J.B.; Lin, L.; Pan, Y.F.; Yang, C.P.; Sun, D.R.; Liu, S.; Xu, X.R.; Maharana, D.; Huang, J.S.; Li, H.X. Occurrence of microplastics in gastrointestinal tracts and gills of fish from Beibu Gulf, South China Sea. *Environ. Pollut.* **2020**, *258*, 113734. [CrossRef] [PubMed]

100. Li, W.; Pan, Z.; Xu, J.; Liu, Q.; Zou, Q.; Lin, H.; Wu, L.; Huang, H. Microplastics in a pelagic dolphinfish (*Coryphaena hippurus*) from the Eastern Pacific Ocean and the implications for fish health. *Sci. Total Environ.* **2021**, *809*, 151126. [CrossRef] [PubMed]

101. Lin, L.; Ma, L.S.; Li, H.X.; Pan, Y.F.; Liu, S.; Zhang, L.; Peng, J.P.; Fok, L.; Xu, X.R.; He, W.H. Low level of microplastic contamination in wild fish from an urban estuary. *Mar. Pollut. Bull.* **2020**, *160*, 111650. [CrossRef] [PubMed]

102. Liu, S.; Chen, H.; Wang, J.; Su, L.; Wang, X.; Zhu, J.; Lan, W. The distribution of microplastics in water, sediment, and fish of the Dafeng River, a remote river in China. *Ecotoxicol. Environ. Saf.* **2021**, *228*, 113009. [CrossRef] [PubMed]

103. Lopes, C.; Raimundo, J.; Caetano, M.; Garrido, S. Microplastic ingestion and diet composition of planktivorous fish. *Limnol. Oceanogr. Lett.* **2020**, *5*, 103–112. [CrossRef]

104. Lusher, A.L.; McHugh, M.; Thompson, R.C. Occurrence of microplastics in the gastrointestinal tract of pelagic and demersal fish from the English Channel. *Mar. Pollut. Bull.* **2013**, *67*, 94–99. [CrossRef]

105. Lusher, A.L.; O'Donnell, C.; Officer, R.; O'Connor, I. Microplastic interactions with North Atlantic mesopelagic fish. *ICES J. Mar. Sci.* **2016**, *73*, 1214–1225. [CrossRef]

106. Makhdoumi, P.; Hossini, H.; Nazmara, Z.; Mansouri, K.; Pirsaheb, M. Occurrence and exposure analysis of microplastic in the gut and muscle tissue of riverine fish in Kermanshah province of Iran. *Mar. Pollut. Bull.* **2021**, *173*, 112915. [CrossRef]

107. McIlwraith, H.K.; Kim, J.; Helm, P.; Bhavsar, S.P.; Metzger, J.S.; Rochman, C.M. Evidence of Microplastic Translocation in Wild-Caught Fish and Implications for Microplastic Accumulation Dynamics in Food Webs. *Environ. Sci. Technol.* **2021**, *55*, 12372–12382. [CrossRef]

108. Morgana, S.; Ghigliotti, L.; Estevez-Calvar, N.; Stifanese, R.; Wieckzorek, A.; Doyle, T.; Christiansen, J.S.; Faimali, M.; Garaventa, F. Microplastics in the Arctic: A case study with sub-surface water and fish samples off Northeast Greenland. *Environ. Pollut.* **2018**, *242*, 1078–1086. [CrossRef]

109. Murphy, F.; Russell, M.; Ewins, C.; Quinn, B. The uptake of macroplastic & microplastic by demersal & pelagic fish in the Northeast Atlantic around Scotland. *Mar. Pollut. Bull.* **2017**, *122*, 353–359. [CrossRef] [PubMed]

110. Naidoo, T.; Sershen; Thompson, R.C.; Rajkaran, A. Quantification and characterisation of microplastics ingested by selected juvenile fish species associated with mangroves in KwaZulu-Natal, South Africa. *Environ. Pollut.* **2020**, *257*, 113635. [CrossRef] [PubMed]

111. Nematollahi, M.J.; Keshavarzi, B.; Moore, F.; Esmaeili, H.R.; Nasrollahzadeh Saravi, H.; Sorooshian, A. Microplastic fibers in the gut of highly consumed fish species from the southern Caspian Sea. *Mar. Pollut. Bull.* **2021**, *168*, 112461. [CrossRef] [PubMed]
112. Nikki, R.; Abdul Jaleel, K.U.; Ragesh, S.; Shini, S.; Saha, M.; Dinesh Kumar, P.K. Abundance and characteristics of microplastics in commercially important bottom dwelling finfishes and shellfish of the Vembanad Lake, India. *Mar. Pollut. Bull.* **2021**, *172*, 112803. [CrossRef] [PubMed]
113. O'Connor, J.D.; Murphy, S.; Lally, H.T.; O'Connor, I.; Nash, R.; O'Sullivan, J.; Bruen, M.; Heerey, L.; Koelmans, A.A.; Cullagh, A.; et al. Microplastics in brown trout (*Salmo trutta* Linnaeus, 1758) from an Irish riverine system. *Environ. Pollut.* **2020**, *267*, 115572. [CrossRef] [PubMed]
114. Palazzo, L.; Coppa, S.; Camedda, A.; Cocca, M.; De Falco, F.; Vianello, A.; Massaro, G.; de Lucia, G.A. A novel approach based on multiple fish species and water column compartments in assessing vertical microlitter distribution and composition. *Env. Pollut.* **2021**, *272*, 116419. [CrossRef]
115. Palermo, J.; Labrador, K.; Follante, J.; Agmata, A.; Pante, M.; Rollon, R.; David, L. Susceptibility of *Sardinella lemuru* to emerging marine microplastic pollution. *Glob. J. Environ. Sci. Manag.* **2020**, *6*, 373–384. [CrossRef]
116. Pan, Z.; Zhang, C.; Wang, S.; Sun, D.; Zhou, A.; Xie, S.; Xu, G.; Zou, J. Occurrence of Microplastics in the Gastrointestinal Tract and Gills of Fish from Guangdong, South China. *J. Mar. Sci. Eng.* **2021**, *9*, 981. [CrossRef]
117. Park, T.J.; Kim, M.K.; Lee, S.H.; Lee, Y.S.; Kim, M.J.; Song, H.Y.; Park, J.H.; Zoh, K.D. Occurrence and characteristics of microplastics in fish of the Han River, South Korea: Factors affecting microplastic abundance in fish. *Environ. Res.* **2021**, *206*, 112647. [CrossRef]
118. Parton, K.J.; Godley, B.J.; Santillo, D.; Tausif, M.; Omeyer, L.C.M.; Galloway, T.S. Investigating the presence of microplastics in demersal sharks of the North-East Atlantic. *Sci. Rep.* **2020**, *10*, 12204. [CrossRef]
119. Parvin, F.; Jannat, S.; Tareq, S.M. Abundance, characteristics and variation of microplastics in different freshwater fish species from Bangladesh. *Sci. Total Environ.* **2021**, *784*, 147137. [CrossRef] [PubMed]
120. Pellini, G.; Gomiero, A.; Fortibuoni, T.; Ferra, C.; Grati, F.; Tassetti, A.N.; Polidori, P.; Fabi, G.; Scarcella, G. Characterization of microplastic litter in the gastrointestinal tract of *Solea solea* from the Adriatic Sea. *Environ. Pollut.* **2018**, *234*, 943–952. [CrossRef] [PubMed]
121. Pereira, J.M.; Rodriguez, Y.; Blasco-Monleon, S.; Porter, A.; Lewis, C.; Pham, C.K. Microplastic in the stomachs of open-ocean and deep-sea fishes of the North-East Atlantic. *Environ. Pollut.* **2020**, *265*, 115060. [CrossRef] [PubMed]
122. Piccardo, M.; Felline, S.; Terlizzi, A. *Preliminary Assessment of Microplastic Accumulation in Wild Mediterranean Species. Proceedings of the International Conference on Microplastic Pollution in the Mediterranean Sea*; Springer Water: Cham. Switzerland, 2018; pp. 115–120.
123. Pullen, E.V. Microplastics in the Digestive System of the Atlantic Sharpnose Shark (*Rhizoprionodon terraenovae*) in Winyah Bay, SC. Master's Thesis, Coastal Carolina University, Conway, SC, USA, 2019.
124. Rasta, M.; Sattari, M.; Taleshi, M.S.; Namin, J.I. Microplastics in different tissues of some commercially important fish species from Anzali Wetland in the Southwest Caspian Sea, Northern Iran. *Mar. Pollut. Bull.* **2021**, *169*, 112479. [CrossRef]
125. Rios-Fuster, B.; Alomar, C.; Compa, M.; Guijarro, B.; Deudero, S. Anthropogenic particles ingestion in fish species from two areas of the western Mediterranean Sea. *Mar. Pollut. Bull.* **2019**, *144*, 325–333. [CrossRef]
126. Rodriguez-Romeu, O.; Constenla, M.; Carrasson, M.; Campoy-Quiles, M.; Soler-Membrives, A. Are anthropogenic fibres a real problem for red mullets (*Mullus barbatus*) from the NW Mediterranean? *Sci. Total Environ.* **2020**, *733*, 139336. [CrossRef]
127. Romeo, T.; Pietro, B.; Peda, C.; Consoli, P.; Andaloro, F.; Fossi, M.C. First evidence of presence of plastic debris in stomach of large pelagic fish in the Mediterranean Sea. *Mar. Pollut. Bull.* **2015**, *95*, 358–361. [CrossRef]
128. Rummel, C.D.; Loder, M.G.; Fricke, N.F.; Lang, T.; Griebeler, E.M.; Janke, M.; Gerdts, G. Plastic ingestion by pelagic and demersal fish from the North Sea and Baltic Sea. *Mar. Pollut. Bull.* **2016**, *102*, 134–141. [CrossRef]
129. Sainio, E.; Lehtiniemi, M.; Setälä, O. Microplastic ingestion by small coastal fish in the northern Baltic Sea, Finland. *Mar. Pollut. Bull.* **2021**, *172*, 112814. [CrossRef]
130. Sathish, M.N.; Jeyasanta, I.; Patterson, J. Occurrence of microplastics in epipelagic and mesopelagic fishes from Tuticorin, Southeast coast of India. *Sci. Total Environ.* **2020**, *720*, 137614. [CrossRef]
131. Savoca, S.; Matanović, K.; D'Angelo, G.; Vetri, V.; Anselmo, S.; Bottari, T.; Mancuso, M.; Kužir, S.; Spanò, N.; Capillo, G. Ingestion of plastic and non-plastic microfibers by farmed gilthead sea bream (*Sparus aurata*) and common carp (*Cyprinus carpio*) at different life stages. *Sci. Total Environ.* **2021**, *782*, 146851. [CrossRef]
132. Selvam, S.; Manisha, A.; Roy, P.D.; Venkatramanan, S.; Chung, S.; Muthukumar, P.; Jesuraja, K.; Elgorban, A.M.; Ahmed, B.; Elzain, H.E. Microplastics and trace metals in fish species of the Gulf of Mannar (Indian Ocean) and evaluation of human health. *Environ. Pollut.* **2021**, *291*, 118089. [CrossRef] [PubMed]
133. Shabaka, S.H.; Marey, R.S.; Ghobashy, M.; Abushady, A.M.; Ismail, G.A.; Khairy, H.M. Thermal analysis and enhanced visual technique for assessment of microplastics in fish from an Urban Harbor, Mediterranean Coast of Egypt. *Mar. Pollut. Bull.* **2020**, *159*, 111465. [CrossRef] [PubMed]
134. Siddique, M.A.M.; Uddin, A.; Rahman, S.M.A.; Rahman, M.; Islam, M.S.; Kibria, G. Microplastics in an anadromous national fish, Hilsa shad *Tenualosa ilisha* from the Bay of Bengal, Bangladesh. *Mar. Pollut. Bull.* **2021**, *174*, 113236. [CrossRef]
135. Silva-Cavalcanti, J.S.; Silva, J.D.B.; Franca, E.J.; Araujo, M.C.B.; Gusmao, F. Microplastics ingestion by a common tropical freshwater fishing resource. *Environ. Pollut.* **2017**, *221*, 218–226. [CrossRef]

136. Sparks, C.; Immelman, S. Microplastics in offshore fish from the Agulhas Bank, South Africa. *Mar. Pollut. Bull.* **2020**, *156*, 111216. [CrossRef]
137. Su, L.; Deng, H.; Li, B.; Chen, Q.; Pettigrove, V.; Wu, C.; Shi, H. The occurrence of microplastic in specific organs in commercially caught fishes from coast and estuary area of east China. *J. Hazard. Mater.* **2019**, *365*, 716–724. [CrossRef]
138. Sun, X.; Li, Q.; Shi, Y.; Zhao, Y.; Zheng, S.; Liang, J.; Liu, T.; Tian, Z. Characteristics and retention of microplastics in the digestive tracts of fish from the Yellow Sea. *Environ. Pollut.* **2019**, *249*, 878–885. [CrossRef]
139. Suwartiningsih, N.; Setyowati, I.; Astuti, R. Microplastics in pelagic and demersal fishes of Pantai Baron, Yogyakarta, Indonesia. *J. Biodjati.* **2020**, *5*, 33–49. [CrossRef]
140. Taghizadeh Rahmat Abadi, Z.; Abtahi, B.; Grossart, H.P.; Khodabandeh, S. Microplastic content of Kutum fish, *Rutilus frisii kutum* in the southern Caspian Sea. *Sci. Total Environ.* **2021**, *752*, 141542. [CrossRef]
141. Tanaka, K.; Takada, H. Microplastic fragments and microbeads in digestive tracts of planktivorous fish from urban coastal waters. *Sci. Rep.* **2016**, *6*, 34351. [CrossRef] [PubMed]
142. Tsangaris, C.; Digka, N.; Valente, T.; Aguilar, A.; Borrell, A.; de Lucia, G.A.; Gambaiani, D.; Garcia-Garin, O.; Kaberi, H.; Martin, J.; et al. Using *Boops boops* (osteichthyes) to assess microplastic ingestion in the Mediterranean Sea. *Mar. Pollut. Bull.* **2020**, *158*, 111397. [CrossRef] [PubMed]
143. Turhan, D.Ö. Evaluation of Microplastics in the Surface Water, Sediment and Fish of Sürgü Dam Reservoir (Malatya) in Turkey. *Turk. J. Fish Aquat. Sci.* **2021**, *22*, TRJFAS20157. [CrossRef]
144. Valente, T.; Sbrana, A.; Scacco, U.; Jacomini, C.; Bianchi, J.; Palazzo, L.; de Lucia, G.A.; Silvestri, C.; Matiddi, M. Exploring microplastic ingestion by three deep-water elasmobranch species: A case study from the Tyrrhenian Sea. *Environ. Pollut.* **2019**, *253*, 342–350. [CrossRef]
145. Wang, F.; Wu, H.; Wu, W.; Wang, L.; Liu, J.; An, L.; Xu, Q. Microplastic characteristics in organisms of different trophic levels from Liaohe Estuary, China. *Sci. Total Env.* **2021**, *789*, 148027. [CrossRef]
146. Wang, Q.; Zhu, X.; Hou, C.; Wu, Y.; Teng, J.; Zhang, C.; Tan, H.; Shan, E.; Zhang, W.; Zhao, J. Microplastic uptake in commercial fishes from the Bohai Sea, China. *Chemosphere* **2021**, *263*, 127962. [CrossRef]
147. Wang, S.; Zhang, C.; Pan, Z.; Sun, D.; Zhou, A.; Xie, S.; Wang, J.; Zou, J. Microplastics in wild freshwater fish of different feeding habits from Beijiang and Pearl River Delta regions, south China. *Chemosphere* **2020**, *258*, 127345. [CrossRef]
148. Wieczorek, A.M.; Morrison, L.; Croot, P.L.; Allcock, A.L.; MacLoughlin, E.; Savard, O.; Brownlow, H.; Doyle, T.K. Frequency of microplastics in mesopelagic fishes from the Northwest Atlantic. *Front. Mar. Sci.* **2018**, *5*, 39. [CrossRef]
149. Wootton, N.; Ferreira, M.; Reis-Santos, P.; Gillanders, B.M. A Comparison of Microplastic in Fish from Australia and Fiji. *Front. Mar. Sci.* **2021**, *8*, 677. [CrossRef]
150. Wootton, N.; Reis-Santos, P.; Dowsett, N.; Turnbull, A.; Gillanders, B.M. Low abundance of microplastics in commercially caught fish across southern Australia. *Environ. Pollut.* **2021**, *290*, 118030. [CrossRef]
151. Wu, F.; Wang, Y.; Leung, J.Y.S.; Huang, W.; Zeng, J.; Tang, Y.; Chen, J.; Shi, A.; Yu, X.; Xu, X.; et al. Accumulation of microplastics in typical commercial aquatic species: A case study at a productive aquaculture site in China. *Sci. Total Env.* **2020**, *708*, 135432. [CrossRef] [PubMed]
152. Xu, X.; Zhang, L.; Xue, Y.; Gao, Y.; Wang, L.; Peng, M.; Jiang, S.; Zhang, Q. Microplastic pollution characteristic in surface water and freshwater fish of Gehu Lake, China. *Environ. Sci Pollut. Res. Int.* **2021**, *28*, 67203–67213. [CrossRef] [PubMed]
153. Yuan, W.; Liu, X.; Wang, W.; Di, M.; Wang, J. Microplastic abundance, distribution and composition in water, sediments, and wild fish from Poyang Lake, China. *Ecotoxicol. Environ. Saf.* **2019**, *170*, 180–187. [CrossRef] [PubMed]
154. Zakeri, M.; Naji, A.; Akbarzadeh, A.; Uddin, S. Microplastic ingestion in important commercial fish in the southern Caspian Sea. *Mar. Pollut. Bull.* **2020**, *160*, 111598. [CrossRef] [PubMed]
155. Zhang, C.; Wang, S.; Pan, Z.; Sun, D.; Xie, S.; Zhou, A.; Wang, J.; Zou, J. Occurrence and distribution of microplastics in commercial fishes from estuarine areas of Guangdong, South China. *Chemosphere* **2020**, *260*, 127656. [CrossRef] [PubMed]
156. Zhang, D.; Cui, Y.; Zhou, H.; Jin, C.; Yu, X.; Xu, Y.; Li, Y.; Zhang, C. Microplastic pollution in water, sediment, and fish from artificial reefs around the Ma'an Archipelago, Shengsi, China. *Sci. Total Environ.* **2020**, *703*, 134768. [CrossRef] [PubMed]
157. Zhang, F.; Wang, X.; Xu, J.; Zhu, L.; Peng, G.; Xu, P.; Li, D. Food-web transfer of microplastics between wild caught fish and crustaceans in East China Sea. *Mar. Pollut. Bull.* **2019**, *146*, 173–182. [CrossRef]
158. Zhang, F.; Xu, J.; Zhu, L.; Peng, G.; Jabeen, K.; Wang, X.; Li, D. Seasonal distributions of microplastics and estimation of the microplastic load ingested by wild caught fish in the East China Sea. *J. Hazard. Mater.* **2021**, *419*, 126456. [CrossRef]
159. Zhang, L.; Xie, Y.; Zhong, S.; Liu, J.; Qin, Y.; Gao, P. Microplastics in freshwater and wild fishes from Lijiang River in Guangxi, Southwest China. *Sci. Total Environ.* **2021**, *755*, 142428. [CrossRef]
160. Zhang, S.; Sun, Y.; Liu, B.; Li, R. Full size microplastics in crab and fish collected from the mangrove wetland of Beibu Gulf: Evidences from Raman Tweezers (1–20 mum) and spectroscopy (20–5000 mum). *Sci. Total Environ.* **2021**, *759*, 143504. [CrossRef]
161. Zheng, K.; Fan, Y.; Zhu, Z.; Chen, G.; Tang, C.; Peng, X. Occurrence and Species-Specific Distribution of Plastic Debris in Wild Freshwater Fish from the Pearl River Catchment, China. *Environ. Toxicol. Chem.* **2019**, *38*, 1504–1513. [CrossRef] [PubMed]
162. Zhu, J.; Zhang, Q.; Li, Y.; Tan, S.; Kang, Z.; Yu, X.; Lan, W.; Cai, L.; Wang, J.; Shi, H. Microplastic pollution in the Maowei Sea, a typical mariculture bay of China. *Sci. Total Environ.* **2019**, *658*, 62–68. [CrossRef] [PubMed]
163. Zhu, L.; Wang, H.; Chen, B.; Sun, X.; Qu, K.; Xia, B. Microplastic ingestion in deep-sea fish from the South China Sea. *Sci. Total Environ.* **2019**, *677*, 493–501. [CrossRef] [PubMed]

164. Lindeque, P.K.; Cole, M.; Coppock, R.L.; Lewis, C.N.; Miller, R.Z.; Watts, A.J.R.; Wilson-McNeal, A.; Wright, S.L.; Galloway, T.S. Are we underestimating microplastic abundance in the marine environment? A comparison of microplastic capture with nets of different mesh-size. *Environ. Pollut.* **2020**, *265*, 114721. [CrossRef] [PubMed]

165. Roch, S.; Ros, A.F.; Friedrich, C.; Brinker, A. Microplastic evacuation in fish is particle size-dependent. *Freshw. Biol.* **2021**, *66*, 926–935. [CrossRef]

166. Jantz, L.A.; Morishige, C.L.; Bruland, G.L.; Lepczyk, C.A. Ingestion of plastic marine debris by longnose lancetfish (*Alepisaurus ferox*) in the North Pacific Ocean. *Mar. Pollut. Bull.* **2013**, *69*, 97–104. [CrossRef]

167. Liboiron, M.; Melvin, J.; Richard, N.; Saturno, J.; Ammendolia, J.; Liboiron, F.; Charron, L.; Mather, C. Low incidence of plastic ingestion among three fish species significant for human consumption on the island of Newfoundland, Canada. *Mar. Pollut. Bull.* **2019**, *141*, 244–248. [CrossRef]

168. Santos, T.d.; Bastian, R.; Felden, J.; Rauber, A.M.; Reynalte-Tataje, D.A.; Mello, F.T.d. First record of microplastics in two freshwater fish species (*Iheringhthys labrosus* and *Astyanax lacustris*) from the middle section of the Uruguay River, Brazil. *Acta Limnol. Bras.* **2020**, *32*, e26. [CrossRef]

169. Saturno, J.; Liboiron, M.; Ammendolia, J.; Healey, N.; Earles, E.; Duman, N.; Schoot, I.; Morris, T.; Favaro, B. Occurrence of plastics ingested by Atlantic cod (*Gadus morhua*) destined for human consumption (Fogo Island, Newfoundland and Labrador). *Mar. Pollut. Bull.* **2020**, *153*, 110993. [CrossRef]

170. Gago, J.; Carretero, O.; Filgueiras, A.V.; Vinas, L. Synthetic microfibers in the marine environment: A review on their occurrence in seawater and sediments. *Mar. Pollut. Bull.* **2018**, *127*, 365–376. [CrossRef]

171. Xu, Y.; Chan, F.K.S.; Stanton, T.; Johnson, M.F.; Kay, P.; He, J.; Wang, J.; Kong, C.; Wang, Z.; Liu, D.; et al. Synthesis of dominant plastic microfibre prevalence and pollution control feasibility in Chinese freshwater environments. *Sci. Total Environ.* **2021**, *783*, 146863. [CrossRef] [PubMed]

172. Wang, W.; Ndungu, A.W.; Li, Z.; Wang, J. Microplastics pollution in inland freshwaters of China: A case study in urban surface waters of Wuhan, China. *Sci. Total Environ.* **2017**, *575*, 1369–1374. [CrossRef] [PubMed]

173. Burns, E.E.; Boxall, A.B.A. Microplastics in the aquatic environment: Evidence for or against adverse impacts and major knowledge gaps. *Environ. Toxicol. Chem.* **2018**, *37*, 2776–2796. [CrossRef] [PubMed]

174. Chen, G.; Li, Y.; Wang, J. Occurrence and ecological impact of microplastics in aquaculture ecosystems. *Chemosphere* **2021**, *274*, 129989. [CrossRef]

175. Napper, I.E.; Thompson, R.C. Release of synthetic microplastic plastic fibres from domestic washing machines: Effects of fabric type and washing conditions. *Mar. Pollut. Bull.* **2016**, *112*, 39–45. [CrossRef]

176. Li, B.; Liang, W.; Liu, Q.X.; Fu, S.; Ma, C.; Chen, Q.; Su, L.; Craig, N.J.; Shi, H. Fish Ingest Microplastics Unintentionally. *Env. Sci Technol.* **2021**, *55*, 10471–10479. [CrossRef]

177. Qiao, R.; Deng, Y.; Zhang, S.; Wolosker, M.B.; Zhu, Q.; Ren, H.; Zhang, Y. Accumulation of different shapes of microplastics initiates intestinal injury and gut microbiota dysbiosis in the gut of zebrafish. *Chemosphere* **2019**, *236*, 124334. [CrossRef]

178. Grigorakis, S.; Mason, S.A.; Drouillard, K.G. Determination of the gut retention of plastic microbeads and microfibers in goldfish (*Carassius auratus*). *Chemosphere* **2017**, *169*, 233–238. [CrossRef]

179. Marti, E.; Martin, C.; Galli, M.; Echevarria, F.; Duarte, C.M.; Cozar, A. The Colors of the Ocean Plastics. *Env. Sci. Technol.* **2020**, *54*, 6594–6601. [CrossRef]

180. Suaria, G.; Achtypi, A.; Perold, V.; Lee, J.R.; Pierucci, A.; Bornman, T.G.; Aliani, S.; Ryan, P.G. Microfibers in oceanic surface waters: A global characterization. *Sci. Adv.* **2020**, *6*, eaay8493. [CrossRef]

181. Barrows, A.P.W.; Cathey, S.E.; Petersen, C.W. Marine environment microfiber contamination: Global patterns and the diversity of microparticle origins. *Environ. Pollut.* **2018**, *237*, 275–284. [CrossRef] [PubMed]

182. Lu, H.-C.; Ziajahromi, S.; Neale, P.A.; Leusch, F.D. A systematic review of freshwater microplastics in water and sediments: Recommendations for harmonisation to enhance future study comparisons. *Sci. Total Environ.* **2021**, *781*, 146693. [CrossRef]

183. Carson, H.S. The incidence of plastic ingestion by fishes: From the prey's perspective. *Mar. Pollut. Bull.* **2013**, *74*, 170–174. [CrossRef] [PubMed]

184. Ory, N.C.; Sobral, P.; Ferreira, J.L.; Thiel, M. Amberstripe scad *Decapterus muroadsi* (Carangidae) fish ingest blue microplastics resembling their copepod prey along the coast of Rapa Nui (Easter Island) in the South Pacific subtropical gyre. *Sci. Total Environ.* **2017**, *586*, 430–437. [CrossRef] [PubMed]

185. Gove, J.M.; Whitney, J.L.; McManus, M.A.; Lecky, J.; Carvalho, F.C.; Lynch, J.M.; Li, J.; Neubauer, P.; Smith, K.A.; Phipps, J.E.; et al. Prey-size plastics are invading larval fish nurseries. *Proc. Natl. Acad. Sci. USA* **2019**, *116*, 24143–24149. [CrossRef] [PubMed]

186. Herrera, A.; Stindlova, A.; Martinez, I.; Rapp, J.; Romero-Kutzner, V.; Samper, M.D.; Montoto, T.; Aguiar-Gonzalez, B.; Packard, T.; Gomez, M. Microplastic ingestion by Atlantic chub mackerel (*Scomber colias*) in the Canary Islands coast. *Mar. Pollut. Bull.* **2019**, *139*, 127–135. [CrossRef]

187. Herring, P. Blue pigment of a surface-living oceanic copepod. *Nature* **1965**, *205*, 103–104. [CrossRef]

188. Erni-Cassola, G.; Zadjelovic, V.; Gibson, M.I.; Christie-Oleza, J.A. Distribution of plastic polymer types in the marine environment; A meta-analysis. *J. Hazard. Mater.* **2019**, *369*, 691–698. [CrossRef]

189. Yang, L.; Zhang, Y.; Kang, S.; Wang, Z.; Wu, C. Microplastics in freshwater sediment: A review on methods, occurrence, and sources. *Sci. Total Environ.* **2021**, *754*, 141948. [CrossRef]

190. Geyer, R.; Jambeck, J.R.; Law, K.L. Production, use, and fate of all plastics ever made. *Sci. Adv.* **2017**, *3*, e1700782. [CrossRef]

191. Food and Agriculture Organization (FAO). *The State of World Fisheries and Aquaculture 2020*; Food and Agriculture Organization (FAO): Rome, Italy, 2020. [CrossRef]
192. Zeytin, S.; Wagner, G.; Mackay-Roberts, N.; Gerdts, G.; Schuirmann, E.; Klockmann, S.; Slater, M. Quantifying microplastic translocation from feed to the fillet in European sea bass *Dicentrarchus labrax*. *Mar. Pollut. Bull.* **2020**, *156*, 111210. [CrossRef] [PubMed]
193. Akhbarizadeh, R.; Moore, F.; Keshavarzi, B. Investigating microplastics bioaccumulation and biomagnification in seafood from the Persian Gulf: A threat to human health? *Food Addit. Contam. Part A* **2019**, *36*, 1696–1708. [CrossRef] [PubMed]
194. Barboza, L.G.A.; Lopes, C.; Oliveira, P.; Bessa, F.; Otero, V.; Henriques, B.; Raimundo, J.; Caetano, M.; Vale, C.; Guilhermino, L. Microplastics in wild fish from North East Atlantic Ocean and its potential for causing neurotoxic effects, lipid oxidative damage, and human health risks associated with ingestion exposure. *Sci. Total Environ.* **2020**, *717*, 134625. [CrossRef] [PubMed]
195. Cox, K.D.; Covernton, G.A.; Davies, H.L.; Dower, J.F.; Juanes, F.; Dudas, S.E. Human Consumption of Microplastics. *Environ. Sci Technol.* **2019**, *53*, 7068–7074. [CrossRef] [PubMed]
196. Schwabl, P.; Koppel, S.; Konigshofer, P.; Bucsics, T.; Trauner, M.; Reiberger, T.; Liebmann, B. Detection of Various Microplastics in Human Stool: A Prospective Case Series. *Ann. Intern. Med.* **2019**, *171*, 453–457. [CrossRef]
197. Ibrahim, Y.S.; Tuan Anuar, S.; Azmi, A.A.; Wan Mohd Khalik, W.M.A.; Lehata, S.; Hamzah, S.R.; Ismail, D.; Ma, Z.F.; Dzulkarnaen, A.; Zakaria, Z.; et al. Detection of microplastics in human colectomy specimens. *JGH Open* **2021**, *5*, 116–121. [CrossRef]
198. Smith, M.; Love, D.C.; Rochman, C.M.; Neff, R.A. Microplastics in Seafood and the Implications for Human Health. *Curr. Env. Health Rep.* **2018**, *5*, 375–386. [CrossRef]
199. Lv, W.; Zhou, W.; Lu, S.; Huang, W.; Yuan, Q.; Tian, M.; Lv, W.; He, D. Microplastic pollution in rice-fish co-culture system: A report of three farmland stations in Shanghai, China. *Sci. Total Environ.* **2019**, *652*, 1209–1218. [CrossRef]
200. Cheung, L.T.O.; Lui, C.Y.; Fok, L. Microplastic Contamination of Wild and Captive Flathead Grey Mullet (*Mugil cephalus*). *Int. J. Environ. Res. Public Health* **2018**, *15*, 597. [CrossRef]
201. Ibrahim, Y.S.; Rathnam, R.; Anuar, S.T.; Khalik, W.M.A.W.M. Isolation and Charaterisation of Microplastic Abundance in *Lates calcarifer* from Setiu Wetlands, Malaysia. *Malays. J. Anal. Sci.* **2017**, *21*, 1054–1064. [CrossRef]
202. Chen, B.; Fan, Y.; Huang, W.; Rayhan, A.; Chen, K.; Cai, M. Observation of microplastics in mariculture water of Longjiao Bay, southeast China: Influence by human activities. *Mar. Pollut. Bull.* **2020**, *160*, 111655. [CrossRef] [PubMed]
203. Hanachi, P.; Karbalaei, S.; Walker, T.R.; Cole, M.; Hosseini, S.V. Abundance and properties of microplastics found in commercial fish meal and cultured common carp (*Cyprinus carpio*). *Environ. Sci. Pollut. Res. Int.* **2019**, *26*, 23777–23787. [CrossRef] [PubMed]
204. Hurt, R.; O'Reilly, C.M.; Perry, W.L. Microplastic prevalence in two fish species in two US reservoirs. *Limnol. Oceanohr. Lett.* **2020**, *5*, 147–153. [CrossRef]
205. Yin, L.; Chen, B.; Xia, B.; Shi, X.; Qu, K. Polystyrene microplastics alter the behavior, energy reserve and nutritional composition of marine jacopever (*Sebastes schlegelii*). *J. Hazard. Mater.* **2018**, *360*, 97–105. [CrossRef] [PubMed]
206. IUCN. The IUCN Red List of Threatened Species. Available online: https://www.iucnredlist.org (accessed on 20 February 2022).
207. Kuhn, S.; van Franeker, J.A. Quantitative overview of marine debris ingested by marine megafauna. *Mar. Pollut. Bull.* **2020**, *151*, 110858. [CrossRef]
208. Frias, J.; Nash, R. Microplastics: Finding a consensus on the definition. *Mar. Pollut. Bull.* **2019**, *138*, 145–147. [CrossRef]
209. Fernandez-Ojeda, C.; Muniz, M.C.; Cardoso, R.P.; Dos Anjos, R.M.; Huaringa, E.; Nakazaki, C.; Henostroza, A.; Garces-Ordonez, O. Plastic debris and natural food in two commercially important fish species from the coast of Peru. *Mar. Pollut. Bull.* **2021**, *173*, 113039. [CrossRef]
210. Song, Z.; Liu, K.; Wang, X.; Wei, N.; Zong, C.; Li, C.; Jiang, C.; He, Y.; Li, D. To what extent are we really free from airborne microplastics? *Sci. Total Environ.* **2021**, *754*, 142118. [CrossRef]
211. Güven, O.; Gokdag, K.; Jovanovic, B.; Kideys, A.E. Microplastic litter composition of the Turkish territorial waters of the Mediterranean Sea, and its occurrence in the gastrointestinal tract of fish. *Environ. Pollut.* **2017**, *223*, 286–294. [CrossRef]
212. Hermsen, E.; Mintenig, S.M.; Besseling, E.; Koelmans, A.A. Quality Criteria for the Analysis of Microplastic in Biota Samples: A Critical Review. *Environ. Sci. Technol.* **2018**, *52*, 10230–10240. [CrossRef]

Review

Microplastics in the Marine Environment: Sources, Fates, Impacts and Microbial Degradation

Huirong Yang [1,2], Guanglong Chen [1,2] and Jun Wang [1,2,*]

[1]   Joint Laboratory of Guangdong Province and Hong Kong Region on Marine Bioresource Conservation and Exploitation, College of Marine Sciences, South China Agricultural University, Guangzhou 510642, China; hry@scau.edu.cn (H.Y.); glchen@scau.edu.cn (G.C.)

[2]   Guangdong Laboratory for Lingnan Modern Agriculture, South China Agricultural University, Guangzhou 510642, China

*   Correspondence: wangjun2016@scau.edu.cn; Tel./Fax: +86-20-8757-1321

Abstract: The serious global microplastic pollution has attracted public concern in recent years. Microplastics are widely distributed in various environments and their pollution is already ubiquitous in the ocean system, which contributes to exponential concern in the past decade and different research areas. Due to their tiny size coupled with the various microbial communities in aquatic habitats capable of accumulating organic pollutants, abundant literature is available for assessing the negative impact of MPs on the physiology of marine organisms and eventually on the human health. This study summarizes the current literature on MPs in the marine environment to obtain a better knowledge about MP contamination. This review contains three sections: (1) sources and fates of MPs in the marine environment, (2) impacts of MPs on marine organisms, and (3) bacteria for the degradation of marine MPs. Some measures and efforts must be taken to solve the environmental problems caused by microplastics. The knowledge in this review will provide background information for marine microplastics studies and management strategies in future.

Keywords: source; fate; bacterial degradation; marine environment; microplastics

Citation: Yang, H.; Chen, G.; Wang, J. Microplastics in the Marine Environment: Sources, Fates, Impacts and Microbial Degradation. *Toxics* **2021**, *9*, 41. https://doi.org/10.3390/toxics9020041

Academic Editors: Costanza Scopetani, Tania Martellini, Diana Campos and Víctor Manuel León

Received: 27 January 2021
Accepted: 19 February 2021
Published: 22 February 2021

Publisher's Note: MDPI stays neutral with regard to jurisdictional claims in published maps and institutional affiliations.

1. Introduction

Plastics have brought a lot of benefits to modern life, driving the tremendous growth in plastic demand, because of their low cost, light weight, and durable character [1,2]. It was reported that 3 billion tons of plastic were manufactured in 2016, and every year, some 8 million tons of plastics will eventually enter the marine environment [3,4]. One of the consequences of this accumulation in the marine environment is the low percentage of recycled plastics [5,6] as just 9.4 million tonnes of plastic postconsumer waste were collected in Europe to be recycled in 2018 (both inside and outside the Europe) [7]. Plastic pollution is already ubiquitous in the ocean environment. Most worrying of all, it was estimated that the weight of plastics in the ocean will be more than that of the fish by 2050 [8].

Microplastics (MPs) are plastic fragments or particles with a diameter of less than 5 mm formed by fragmentation of larger plastics [9–14]. Plastics can fragment into smaller particles in the marine environment [15,16]. Microplastics appear in various shapes, such as foils, foams, fibers, pellets, fragments and microbeads [17,18]. Generally, plastics are chemically diverse. The density of polyamide (PA), polyvinylchloride (PVC), and polyethylene terephthalate (PET) are higher than that of seawater, increasing the settlement rates in sediments, while polystyrene (PS), high-density polyethylene (HDPE), low-density polyethylene (LDPE), polypropylene (PP) and polyurethane (PUR) with lower densities might float mainly on seawater [19–22] (Figure 1).

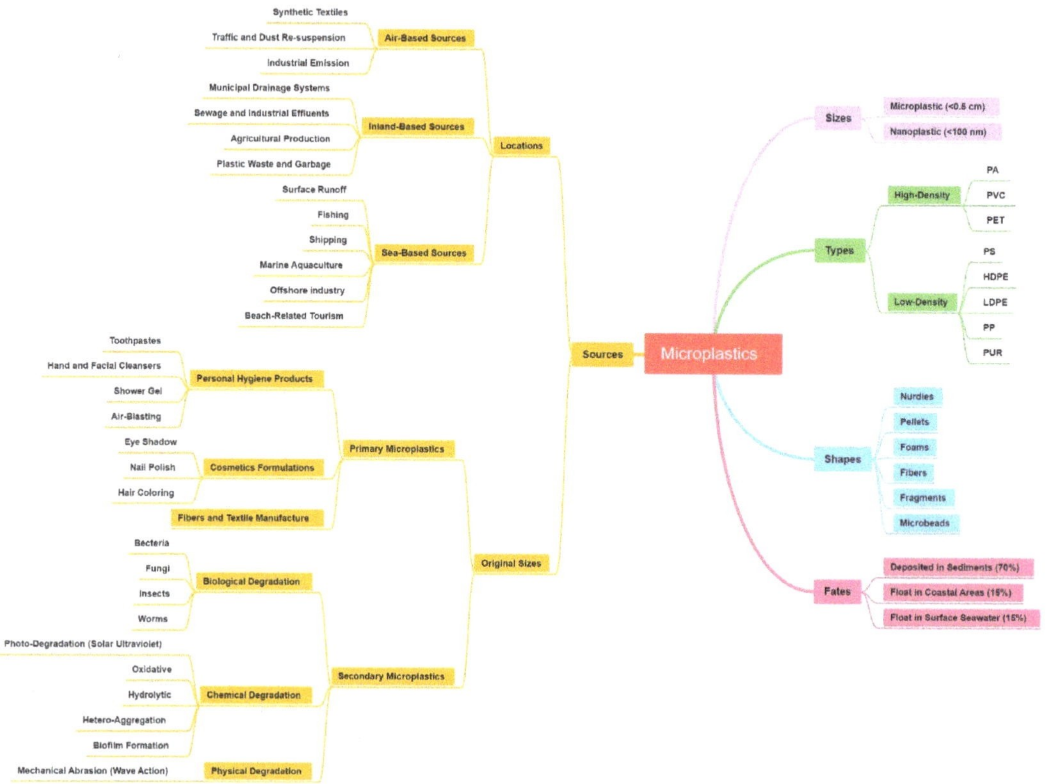

Figure 1. The basic characteristics of microplastic about size, type, shape, source and fate.

Microplastics are prevalent in the environment, especially the marine environment, due to hydrodynamic processes, transportation by wind and ocean currents, ranging from the large ocean gyres such as the Pacific Ocean [9,23], the Atlantic Ocean [24], Indian Ocean [25], polar regions [26–28], and the equator [29], and from coasts [30,31] to open seas [32,33]. It was estimated that more than 15 trillion microplastics were present in the global ocean in 2014, weighing more than 93 thousand metric tons [34]. MPs are abundant in the Great Pacific Garbage Patch, with about 1.69 trillion (94%) floating pieces [10] that are microplastics. Generally, microplastics pollution is already a ubiquitous presence in the ocean environment, which contributes to exponential public and scientific concern in last decade and different research areas (Figure 2).

Due to their tiny size, MPs can be ingested accidentally by marine species [35,36], such as fish [37], mussels [38–40], zooplankton [41], seabirds [42], sand hoppers [43] and worms [44].

The ecological threat of MPs to the oceanic environment and their health risk to organisms have not been fully clarified, but given the sharply increasing amount of evidence about the presence and effects of MPs in the marine environment, MP pollution has become a great environmental concern [45–55]. Some measures and efforts must be taken to solve the problems caused by microplastics and improve plastic waste management.

The present review will summarize existing research on MPs in the marine environment to provide a better understanding about MPs contamination in marine environment. This review contains three sections: (1) sources and fates of MPs in marine environment, (2) impacts of MPs on marine organisms, and (3) bacteria for the degradation of marine MPs.

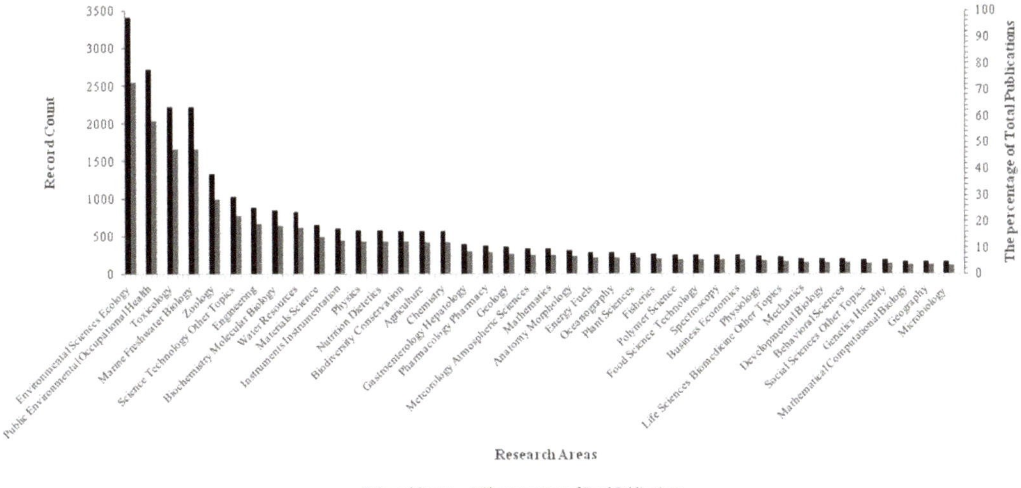

Figure 2. The record count and the percentage of total publications in the top 40 research areas related to the assessment of the microplastic effects on organisms and bacterial degradation over time. Source: Web of Science; Period: 1944–2020; Total Publications: 4685; h-index: 162; Average citations per item: 29.31; Sum of Times Cited: 137,315 (without self-citations: 53,749); Citing articles: 32,830 (without self-citations: 29,560). TS = (microplastic * OR micro-plastic * OR plastic particle * OR plastic particulate OR plastic debris OR plastisphere * OR microplastic pollution *) AND (source * OR fate * OR occurrence * OR distribute * OR influence * OR impact * OR affect OR risk * OR effect * OR exposure * OR exposed OR colonize OR colonization OR bacteria * OR germ * OR microbiological OR microorganisms OR microbial OR microbiota OR macrobiotic OR biotechnological OR degrade * OR degradation * OR biodegradation * OR biodegrade * OR organisms * OR creature * OR biota * OR habitat *) AND (marine * OR ocean * OR sea * OR seawater * OR beach * OR shore * OR coast * OR seacoast * OR seaboard *).

2. Sources and Fates of MPs in Marine Environment

2.1. Sources of marine MPs

Marine microplastic pollution originates from a variety of sources and can generally be divided into inland-based, sea-based and air-based sources [19,56–58] (Figure 1). Rivers are considered to be the most important pathways for microplastics to be transported from inland areas to the ocean [59]. About 80% of the plastic pieces in the ocean originated from the terrestrial environment [12,56,60]. Plastic debris in municipal drainage systems and sewage effluents, or improper management of inland areas is blown into the sea through rivers, and plastic waste from beach-related tourism is discarded directly into the environment [18,56,57,61]. Sea-based sources originate from fishing, shipping and offshore industries [62,63]. The emissions and leaks of large shipping are considered as an important source of microplastics [64]. Loss and damage of fishing and aquaculture equipment can easily introduce plastic particles into the ocean [9,65,66]. Followed by marine aquaculture, the main offshore source is the world's fishing fleet [67], garbage illegally discarded from ships or offshore platforms [68], and a large proportion of items comes from lost containers [56,69]. In addition, airborne MPs are also important sources [70].

According to their original sizes, microplastics can be divided into two groups. Originally designed plastic microbeads, industrially produced particles and powders (<5 mm in diameter) could enter the ocean directly through sewage effluent, which is called primary microplastics [57,71]. When subjected to the combined effects of physical, biological and chemical processes, large plastic fragments are broken down and degraded into tiny fragments, which are secondary microplastics and can be transported to the marine environment [72–74]. Primary microplastics are widely used in personal hygiene products containing abrasives and scrubs (like toothpastes, hand and facial cleansers; shower gels and air-blasting aids, etc.) [28,75–78], cosmetics formulations (such as eye shadow, nail polish, hair coloring, etc.) [79,80], and also fiber and textile manufacture [81].

Generally, secondary microplastics imply the breakdown of large plastic debris due to biological, chemical and physical degradation, which are representative of microbial species biodegradation, photodegradation (solar ultraviolet radiation) and mechanical abrasion (wave action), respectively. Plastic debris in the ocean are subject to mechanical damage and photodegradation well as oxidative degradation, which break down fragile plastics into microplastics [82,83]. Besides, microplastics can further degrade to nano-scale plastic pieces [40]. These microplastics and nanoplastics are more easily ingested and will have long-term adverse impacts on the marine environment, making them become a public concern in the future [40,83–85] (Figure 1).

2.2. Fates of Marine MPs

Generally, debris in any water body will ultimately enter the ocean. Transported by water power and wind power, microplastics gradually migrate and diffuse through the ocean, eventually becoming as ubiquitous as they are today, ranging from the large ocean gyres (e.g., the Pacific Ocean [9,23]; the Atlantic Ocean [24]; Indian Ocean [25]) to the polar regions and equator, from densely populated areas to remote islands, and from beaches down to the abysses of the sea [26,27,29,30,33]. They come in various shapes, with fibers being the most common form, followed by fragments. Marine circulation, estuaries and other coastal areas where humans are active are the ecosystems most seriously polluted by microplastics [86–88]. Approximately 70% of marine plastic debris is deposited in sediments, 15% floats in coastal areas and the remainder float on the surface seawater (Figure 1). Microplastics will be accumulated in the global ocean circulation, since some of them are less dense than seawater and float on the sea surface, and the converging sea currents concentrate and retain debris for a long time [23,35,89,90]. According to the surveys, there are only at least 7000 tonnes of plastic debris on the surface of the high seas [89], but at least 4.8 million tonnes of plastic debris enter the marine environment each year [91], which is inconsistent with data on surface plastics, suggesting that a significant number of plastics sinks to unknown depths. Microplastics have even been found on the seafloor at 2200–10,000 m depth, containing both high [92] and low [93] density (relative to seawater) microplastics. This indicates that the migration of microplastics is a dynamic process, which may not only be carried to every part of the marine through physical effects such as crushing and coastal deposition, but also through chemical processes such as oxidation or hydrolysis [62,94], and may also be carried to every part of the ocean through biological absorption, digestion and excretion [95].

Weathering processes, biodegradation processes, oxidative and hydrolytic degradation [62,93] and hetero-aggregation and biofilm formation [96,97] could significantly affect the fate of microplastic pieces in the oceanic environment (Figure 1). Biological pollution and subsequent chemical deposition of plastics, could dominate migration in seawater environments [98–100]. Therefore, according to biofilm growth, sedimentation and marine depth distribution of various physical factors such as light, salinity, water density, temperature, and viscosity, a theoretical predicted model was established to simulate the impact of biological pollution on the migration of microplastics, and forecast the size-dependent vertical migration of sea microplastics [101].

In addition to the origin and fate of MPs, many papers have also focused on the particle size, shape, type, color and mesh size of MPS and how to sample it to fully understand the characteristics of MPS in marine ecosystems (Table 1). This information will be helpful for further evaluation of plastic production plans and for more scientific and effective control of plastic products [102–117].

Table 1. The characterization of MPs in marine ecosystem.

Location	Sample Type	Mesh Size	Concentration	Particle Size	MP Type	Poly Type	Reference
North African coasts of Mediterranean European seas	surface sediments beach litters	NA	182.66 ± 27.32 649.33 ± 184.02/kg sediment DW	NA	fibers (70%), fragments (21%), pellets (5%), films (2%) and foams (2%)	PE (48%), PP (16%), PET (14%), PS (9%), butyl branham (7%), EPM (3%), TCA (3%)	[102]
Nordic Seas	seawater	NA	1.19 ± 0.28 items/L (EGC), 2.43 ± 0.84 items/L (GSG)	0.1–0.5 mm	fiber (76.1%), transparent (76.2%), small microplastics (48.1%)	PA, PE, PET, PMMA, PP, PS,	[103]
Terra Nova Bay, Antarctica	macrobenthic species	NA	1.0 items/individual, 0.7 items/mg DW	50 and 100 μm	nylon (86%), polyethylene (5%)	PAA, PARA, PA, PP, PS, PTFE	[104]
Southern Caspian coastal northern, Iran	coastal sediment	250–500 μm	25 items/kg 330 items/kg	<2 μm	fiber, fragment, film	PS, PE	[105]
Banten Bay, Indonesia	sediment	0.45 μm	mean: 267 ± 98 particles/kg DW, min: 101 particles/kg DW, max: 431particles/kg DW	500 and 1000 μm (>50%)	foam (30.4%), fragments (26.5%), granules (24.4%), and fibers (18.7%)	Cellophane, PS	[106]
Caspian Sea	surface waters sediments	50 μm 0.3 mm	34,490 particles/km², 210 particles/kg	1–4.75 mm in surface water (68%) and coastal sediments (50%)	fragment (38%), styrofoam (31%), film (20%), lines (9%) (surface water) styrofoam (35%), fragment (31%) (sediment)	PE, PP, PS	[107]
Boknafjord, Norway	sediments	10–250 μm	11 to 140 μg/kg DW	40–100 μm	NA	PE: 32.3–139.2 μg/kg PVC: 9–120 μg/kg PET: 12–136.5 μg/kg PP: 10–78.4 μg/kg PA: 16–73.1 μg/kg	[108]
Kingston Harbour	surface waters	335 μm	mean: 674.13 particles/km², min: 5.73 particles/m³, max: 2697 particles/m³	1–2.5 mm	fragment	PE, PP	[109]
Northwestern Pacific Ocean	surface waters	330 μm	mean: 1.0 × 10⁴ items/km² (6.4 × 10²–4.2 × 10⁴ items/km²)	0.5–1.0 mm (50%), 1–2.5 mm (29.8%), 2.5–5.0 mm (17.6%)	granules (39.7%) sheets (26.7%), films (24.7%), and lines (8.9%)	PE (57.8%), PP (36.0%), PA (3.4%)	[110]
Qinzhou Bay, China	sediment	5 mm	15–12,852 items/kg	0.16–5.0 mm	fragment (94%), sphere (5.2%), fiber (0.5%)	PS, PP, PE, oxidized PE, LDPE	[111]
Irish Continental shelf, Atlantic	sediment bottom water	250 μm	Max: 0.5 cm	250 μm– 5 mm	fibers (85%) fragments (15%)	23% PA, 11%PET, 3% PP, 2% acrylic	[31]
Deep–sea of Mid–Atlantic and Indian Ocean	organisms		found inside oral or stomach area	NA	100% Fibers	Modified acrylic, PP, PET, viscose, acrylic	[32]
Polar waters, Arctic	surface and subsurface water	333 μm	NA	Surface: 0.34/m³, Subsurface: 2.86/m³	fibers (95%) fragments (4.9%)	30% Rayon 15% PET 15% PA 5% PE	[112]
NE Pacific Ocean	subsurface seawater	7.8 × 7.5 mm	8–9200 particles/m³	Mean: 606 ± 221 μm (62–5000 μm)	fibres, fragments	NA	[113]
Beach Continental shelf (Belgium)	sediment	38 μm	92.8/kg 97.2/kg	38 μm–1 mm	fibers (59%), granules (25%)	PP, PA, PVA	[30]
Beach Continental shelf (UK)	sediment		0.4/50 mL 5.6/50 mL	~20 μm	fibers	9 polymers	[64]

PE: polyethylene; PP: polypropylene; PS: polystyrene; EPM: ethylene propylene diene; TCA: tricellulose acetate; PA: polyamide; PET: polyethylene terephthalate; PMMA: poly methyl methacrylate; PAA: poly (acrylic acid); PARA: polyaryl amide; PTFE: polytetrafluoroethylene; LDPE: low-density polyethylene; PVC: polyvinyl chloride. NA: Not available.

3. Impacts on Marine Organisms of MPs

Recently, abundant literature has assessed the accumulation of microplastics in marine organisms through direct contact [36] or food chain exposure [37] to MPs. MPs are ingested by organisms and have negative effects on their development, metabolism, reproduction and cellular response, and so on [118–134].

3.1. Exposure

Basically, there are two primary modes of MP exposure for marine organisms: bathing contact and ingestion. Bathing, of course, is the most common contact method in MP bioassays of natural marine environments, making it possible to study the various adverse effects caused by microplastics on the aquatic organisms through contact [36]. For example, microplastics could attach to the surface of skin, crust and ectoderm of *Artemia franciscana* [55]. Besides, microplastics could be ingested by low-nutrient organisms (like zooplankton such as artemia [55,118,135] and larvae of various marine animals such as shellfish and sea squirts [118,135,136], which are more readily available and easily exposed to suspended microplastics, since microplastics are similar than planktonic organisms and sediments in size and density [38,55,137–139].

3.2. Translocation

Microplastics are found in the circulatory system and tissues of some marine organisms because they could pass through epithelial tissues and even cell membranes. This phenomenon was called "translocation" [36,140]. For example, after a 3 h exposure, HDPE was detected in mussels' stomachs and accumulated in the lysosomal system [39]. Since microplastics cannot be digested or absorbed, they can pass through cell membranes, transport through the inner layer of intestinal epithelium into the circulatory system and enter tissues after ingestion [38,56]. Therefore, MPs could be translocated and accumulated in cells and specialized tissues, such as gills and guts [141], liver [142], lysosomal system and hemolymph in blood cells [39].

Translocation efficiency depends mainly on the size of the MPs, but is also biologically affected by other factors, such as shape, concentration and the related organisms [143,144]. MP < 10 μm may be compatible with the use of membrane surface recognition elements through the epithelium [145]. As the size of microplastics decreases, the ability for microplastics to accumulate in marine organisms may increase, because the smaller the microplastics, the easier their transport. Currently, one of the main techniques for studying translocation is to expose organisms to fluorescently labeled plastic particles and then use a microscope (e.g., fluorescence and confocal microscopy) to observe MPs in the tissue, as well as do the quantitative analysis through flow cytometry [146,147].

3.3. Bioaccumulation and Bioavailability

The two important indexes to access the impacts of MPs to organisms are bioaccumulation and bioavailability [36]. There are interactions between MPs and organisms in the marine environment [148]. Microplastics can be ingested directly by marine organisms or transferred and accumulated in the food web from lower trophic organisms to higher trophic organisms, and the higher the trophic level, the more microplastics may be enriched in the organism [149]. In addition, toxic pollutants could be transported and accumulated in organisms along with microplastics through the ingestion, which has been demonstrated during experimental exposure tests. It has been speculated that POPs could be significantly bioaccumulated in the food web via microplastics [137,150,151].

The bioaccumulation of MPs has been identified in the digestive tract such as the oral area [33], gastrointestinal tract [37,116,142,152] and liver [153] of marine organisms, and followed by translocation to the circulatory system, other specific tissues and cells [39,141,142]. According to Bottari et al., fibrous microplastics are found in the digestive systems of *Zeus faber* and *Lepidopus caudatus* [152]. Microplastics have been reported to be found in fish populations at the bottom of the Mediterranean, with PE accounting for the largest proportion [153]. Furthermore, it has been reported that when *Dicentrarchus labrax* ingest microplastics, the particles accumulate in the liver, accompanied by oxidative stress [154]. Even some endangered species, such as bluefin tuna, have been found to have microplastics in their bodies, which raises concerns about the extent of microplastics pollution in marine species [142].

Bioavailability strongly relies on the physiochemical properties of microplastics, like their size, shape, and density [11,138]. The conclusion is that the size of microplastics is the most important factor. As the size decreases, the potential of bioaccumulation and bioavailability increase [9,138], because microplastics with smaller size are similar to planktonic organisms, and could be easily mistakenly ingested by zooplankton [36]. The irregular shape of plastic particles or fibers results in different bioavailability [155].

Additionally, biological factors could increase the microplastic bioavailability. MPs egested within fecal matter might be ingested by subsequent detritivores and suspension feeders [156], then be cast up on the benthos, attracted to the sediment, and MPs could be available for infauna, sediment-dwelling organisms capable of bioturbation [30,57,137]. Furthermore, their bioavailability in the water column is also influenced by biological fouling and aggregation, and after decontamination, they float at the sea-air interface [56] or sink below the marine surface, due to reduced buoyancy [96].

Microplastics could enhance the bioavailability of adsorbed pollutants, which has attracted more interest from scientists [135,136]. Unfortunately, due to the very high number of possible interaction factors, including physical (e.g., salinity, pH, and temperature), chemical (e.g., hydrolysis, oxidation, reduction and enrichment) and biological factors (e.g., organisms variables), it is difficult to assess how the bioavailability of pollutants enhanced by microplastics [136].

3.4. Toxic Effects

Microplastics have toxic effects on marine organisms. Different types and sizes of microplastics have different toxic effects on marine species, which are ultimately reflected in the physiological response of organisms and the damage they are subjected to [118–134] (Table 2). In addition, different microplastics also adsorb different pollutants, which combine to further damage the health of living marine organisms [150,157–162] (Table 2).

3.4.1. Physiological Impacts

Some morphological changes were detected in the marine phytoplankton when they ingested microplastics. For example, some thylakoids were deformed and cell walls were thickened [118], algae homo-aggregation and algae-microplastics hetero-aggregation [118], as well as expression of certain chloroplast genes was reduced [119].

As for the development, studies examining the impact of MPs have reported significant effects on the development of marine zooplankton and other invertebrates, such as dry weight loss in lugworms [120], intergenerational developmental responses in copepods [121], anomalous growth delays in juvenile [122] and larval [123] development in sea urchins and ascidians, development parameter alteration in shellfish [124], malformations or dead embryos [105], embryonic development abnormalities [125] in a dose- [120,124,126], time- [127], and size- [128] dependent manner in larvae and adults of different invertebrates. Particularly, the microplastics in the larvae of marine organisms will seriously affect the normal growth of the organism and sometimes microplastics might even cause death, due to their limited abilities to control their internal environment [127]. It was reported that the molting times of the larvae increased significantly in a short period of time after ingesting microparticles [55] and that microparticles had a restrictive effect on their feeding, that is, the microparticles had a sublethal effect on the larvae [55]. Studies have shown that after worms' ingestion of microplastics, their energy reserves are significantly reduced and particles accumulate in the intestines where they induce inflammation [36].

Table 2. Effects of microplastics and nanoplastics on marine organisms

Phyla	Species	Development	MP Size	Adsorption	MP Types	Negative Effects	References
Bacillariophyta	*Chaetoceros neogracile*	spore, adult	50 µm	NA	PS	Particles decrease chlorophyll content, esterase activity, cell growth and photosynthetic efficiency of diatoms.	[163]
Aschelminthes	*Brachionus koreanus*	adult	0.05, 0.5, 6 µm	NA	PS	Inhibition of multiple resistance to p-glycoproteins and multidrug resistant proteins leads to increased toxicity and oxidative stress damage to membrane lipids.	[164]
Mollusca	*Crassostrea gigas*	embryo, larva, adult	50 µm	NA	PS	Particles reduce fertilization rate and development ability of embryo and larva.	[48]
	Mytilus galloprovincialis	larva	140 ± 34.6 nm	Cbz	PS	Increased total oxidant status of digestive glands, influence neurotransmission, genotoxicity and lipid peroxidation.	[45]
		adult	0.1–1 mm	pyrene	PE, PS	Alter immune response, lysosomal compartment, peroxisome, antioxidant system, and neurotoxic effects	[46]
Arthropoda	*Artemia franciscana*	larva	40, 50 µm	NA	PS	Impairment of feeding ability, behavioral ability and physiological conditions.	[55]
	Calanus finmarchicus	adult	particles: 10–30 µm fibers: 10 × 30 µm average: 398 ± 54 µm	NA	PA	Alter predation behavior, reduce fat storage, and affect growth and development.	[47]
		embryo	minimum: 10 ± 2 µm	NA	PE	Produce cell death and affect energy metabolism.	[49]
Chordata	*Danio rerio*		50, 200, 500 µm	Au	PS	Oxidative stress and inflammation reaction.	[50]
			44 nm	PAHs	PS	Energy metabolism.	[51]
		larva	25 µm	NA	PS	Glucodermatin receptors disrupt glucose homeostasis, leading to abnormal larval activity.	[52]
		adult	44 nm	PAHs	PS	Energy metabolism.	[51]
			25 µm	Cu	PS	Inflammatory reaction.	[53]
			50 µm	BPA	PS	Neurotoxicity.	[54]
	Fish cell lines (SAF-1, DLB-1)	/	100 nm	NA	PS	Change the activity of superoxide dismutase and Glutathione S-transferase and the toxicity of drugs.	[165]

PS: polystyrene; PE: polyethylene; LDPE: low-density polyethylene; HDPE: high-density polyethylene; PA: polyamide; PP: polypropylene; PUR: polyurethane; PET: polyethylene terephthalate; PVC: polyvinyl chloride. PAHs: polycyclic Aromatic Hydrocarbons; BPA: bisphenol A; Cbz: carbamazepine. NA: Not available.

The effects of microplastics on oxidative stress, inflammatory reactions and metabolic disorders of marine animals were studied. For example, the accumulation of MPs may result in inflammation, lipid accumulation and energy metabolism in fish [128], while oxidative stress and enzyme activity reductions occur in crabs [129].

The adverse impact of microplastic on the reproduction in marine animals, such as egg production [130], fecundity [121], fertilization rates [125], oocyte number [127], population size [130,131] and population growth rate [131] were assessed with significant dose-dependent [130] and distinct size-dependent effects [98,107] being observed in marine invertebrates studies.

At the cellular level, exposure marine animals to MPs induced comprehensive cellular responses. Microplastics could significantly down-regulate histone 3 gene expression [130], and up-regulate Abcb1, cas-8 [132], sod, gpx, idp, pk [133] gene expression. Besides, the activity of phagocytes and mitochondria is significantly increased, and the proportion of oxy radical and immune cells is also up-regulated [134].

3.4.2. Joint Toxicity

Due to the high adsorption capacity of microplastics, many hydrophobic pollutants could adsorb and accumulate on microplastics and accompanied by biomagnification (e.g., PAHs, PCBs, nonylphenols, pesticides, dioxins) [150,157]. Studies have shown that millimeter-sized microplastics have no obvious adsorption toxicity, while micron-sized or even nanosized microplastics have a relatively strong ability to absorb pollutants [131]. For heavy metal pollutants, 32–40 µm plastic particles exposed to heavy metals induce oxidative stress in fish and stimulate their innate immunity [158]. As for organic pollutants, there are studies that have shown that 50 nm plastic particles exposed to PAHs are obviously toxic to aquatic zooplankton and cause significant chemical damage [159]. The biological amplification of organic pollutants becomes higher because plastics reduce the metabolism of pollutants, and the combined toxicity presents an additive effect [160].

In addition to the original monomer, many microplastic products also contain a variety of additives, such as flame retardants, plasticizers, dyes and antioxidants, which make microplastics display joint toxicity with the additives [157,161].

The accumulation and biomagnification of microplastics and their surface-adsorbed pollutants need to be further studied. The joint toxicity may pose a persistent threat to marine ecosystems, due to the durability of microplastics and toxic chemicals [17,162]. Because the toxicity mechanism of microplastics is not fully clear, understanding toxic effects caused by microplastics is important to assess their environmental impacts.

4. Bacteria for Degradation of Marine MPs

4.1. Bacteria Colonizing Microplastics

Some studies highlight the differences between the bacteria living on organic particles with seawater [166], on microplastics and in a free state [167]. The bacterial community that settles on the surfaces of marine microplastic is significantly different from that in surrounding middle and upper waters or other particle types [166]. If these bacteria have been established enzymatic mechanism for degrading plastic, they would be of particular interest for bioremediation and bioengineering.

Studies show that some bacterial groups such as the phyla *Bacteroidetes*, *Proteobacteria*, *Cyanobacteria* and *Firmicutes* appear to colonize microplastics more often than others, indicating that the specific taxonomic bacteria consider microplastics as a beneficially ecological niche and a potential metabolic adaptation to the material (e.g., attachment, additive resistance, chemotaxis, and degradation). Similar taxa belonging to Bacteroidetes and Proteobacteria seem to be shared by the core bacteria of the seafloor and subsurface plastisphere share, and some photoautotrophic bacteria dominated the sub-surface communities [168,169].

4.2. Plastisphere Served as a New Niche for Marine Environment

Recently, the first study using the modern technology of large-scale DNA sequencing gave a detailed image of the microbial communities that inhabit microplastics [128]. Debris is usually described by the term "plastisphere" in marine biology research [169], they serve as various habitats for microbial colonies in aquatic environments besides accumulating organic pollutants [168–171].

Based on morphological data and DNA sequencing technology, the factors that drive the composition of plastisphere are complex and comprehensive. In addition to the main factors, season and surrounding environment, polymer type, surface feature, and size also affected the diversity and abundance of the colonizing bacterial groups [168,172]. For example, studies highlighted significant differences in microbiota communities on microplastics from the two different oceans, and the diversity of bacteria living in water columns and bacteria attached to microplastic debris [173]. Studies show that plastic surfaces could be rapidly colonized by heterotrophic bacteria, which can survive longer than in the surrounding aquatic environments [174].

4.3. Biodegradation of Bacteria in Marine Environment

Microbial biodegradation is a process in which microbial communities (bacteria, actinomycetes and fungi) use organic matter as a carbon source to metabolize, resulting in a transformation from organic carbon to biogas and biomass [175,176]. Generally, the biodegradation process of MPs is proposed to consist of four main basic stages and continuous successive steps: biodeterioration, biofragmentation, assimilation and mineralization [168].

Interest in plastic biodegradation is also growing, and bacteria are considered to be one of the most important ways to solve marine plastic pollution, because of their potential capacity for biodegradation of plastic wastes. *Corynebacterium*, *Arthrobacter*, *Pseudomonas*, *Micrococcus*, *Streptomyces* and *Rhodococcus* are the main bacterial groups in this context, and they can use plastics as sole carbon source under lab conditions [176]. Interestingly, it was discovered that significant differences exist in the diversity, abundance and activity of bacterial and physiochemical characters of plastics between biodegradable and non-biodegradable plastics, indicating the presence of plastic-degrading microbes [177]. Nowadays, there is an increasing number of anecdotal evidence that bacteria can show the capability to degrade ocean plastic pieces [169,172,174] (Table 3).

The factors involved in plastic biodegradability depend not only on the ability of microorganisms but also on the characteristics and surface structure of the material, such as the roughness, electrostatic interactions, topography, hydrophobicity, and free energy [106]. In addition, various environmental factors, such as oxygen level, temperature, humidity, salinity, and limitation of light have an important impact on the biodegradation of plastics [186]. The additives in the polymer could increase the rate of biodegradability. These additives will affect their chemical and thermal sensitivity as well as their ability to absorb ultraviolet light and lead to the loss of stable properties that are more suitable for microbial attachment [187].

The current test standards for assessing plastic biodegradability of marine plastics tend to use to use optical, atomic force and scanning electron microscopy to confirm the results of major tests based on respirometers, since each of them has limitations, and none of these techniques are sufficient by itself [188]. To date, standard guidelines and methods for conducting these experiments have not been established.

Our understanding of metabolic mechanisms of biodegradable marine bacteria and their enzymes is very limited. Furthermore, the biodegradation mechanics of marine plastic debris and its potential impact processes need further research to make full use of its impact.

Table 3. Outstanding plastic-degrading bacteria in existing research.

Plastic Types	Year	Strains	Source	Plastic Forms	Weight Loss	Principle	References
PS	2015	*Exiguobacterium* sp. *YT2*	Intestines of *Tenebrio molitor*	sheet	(7.4% ± 0.4%)/60 days	NA	[178,179]
LDPE	2014	*Bacillus* sp. *YP1*	Intestines of *Plodia interpunctella Hübner*	film	(10.7% ± 0.2%)/60 days	NA	[180]
HDPE	2010	*GMB7*	Plastic waste landfill in Mannar, India	film	15%/30 days	NA	[181]
PA	2000	*Flavobacterium* sp. *KI72*	NA	NA	NA	Hydrolysis of polymer hydrolases	[177]
PP				None			
PUR	1995	*Comamonas acidovorans TB-35*	Soil	film	100%/7 days	Hydrolysis of esterase encoded by gene *PudA*	[182,183]
	2014	*Pseudomonas putida A12*	Soil	emulsion	92%/4 days	Hydrolysis of a 45 kDa esterase	[184]
	2017	*Bacillus* sp. *S10-2*	Spacecraft	emulsion, film	19%/60 days	Hydrolysis of esterase	[185]
PET	2011	*Bacillus subtilis*	Laboratory	film	NA	Hydrolysis of p-nitrobenzylesterase	[45]
PVC				None			

PS: polystyrene; PE: polyethylene; LDPE: low-density polyethylene; HDPE: high-density polyethylene; PA: polyamide; PP: polypropylene; PUR: polyurethane; PET: polyethylene terephthalate; PVC: polyvinyl chloride. NA: not available.

5. Conclusions

The accumulation of microplastics in the marine environment is a serious threat to the health of marine organisms, which may eventually affect the survival of human beings. Therefore, it has attracted extensive attention from society and researchers. Many studies have shown that different bacterial communities colonize microplastics in the marine environment, which has inspired us to investigate the bacterial degradation of marine microplastics. However, until now, we don't know much about how these bacteria work. The rich diversity and activity of these bacteria indicate their potential in the biogeochemical cycling of plastics, but further research is needed. Contact experiments must be carefully designed to test the ability of these bacteria to react with plastics and adapt to changing marine environments, so it is important to integrate research approaches from multiple disciplines. In order to take full advantage of the influence of bacterial communities on MPs, more controlled experiments are needed to simulate real marine ecosystems. Further studies of bacteria associated with plastic degradation will help develop situ biodegradable methods and materials. According to the current technology and methods, it is impossible to completely remove all the microplastics in the ocean, but we can still try to partially reduce marine microplastic pollution. Bacterial degradation is an appropriate choice for this. While developing methods for degrading plastics, relevant stakeholders such as governments, the public, manufacturers and scientists should pay high attention to the problem of marine microplastics pollution. We should take responsibility and working together to reduce unnecessary plastic production and reduce plastic waste by recycling plastic to tackle increasing MP issues.

Author Contributions: Conceptualization, J.W. and H.Y.; formal analysis, G.C.; investigation, G.C.; data curation, H.Y.; writing—original draft preparation, H.Y.; writing—review and editing, J.W. and G.C.; visualization, J.W.; supervision, J.W. and H.Y.; project administration, J.W. and H.Y.; funding acquisition, J.W. and H.Y. All authors have read and agreed to the published version of the manuscript.

Funding: This research was funded by the National Key Research and Development Program of China (2018YFD0900604), Guangdong Forestry Science and Technology Innovation Project, Provincial Projects with Special Funds for Promoting Economic Development of Marine and Fisheries Department of Guangdong (SDYY-2018-05), Project of Guangzhou Association for Science & Technology (K20200102008); Guangdong Province Universities and Colleges Pearl River Scholar Funded Scheme

(2018), the National Natural Science Foundation of China (42077364), and Key Research Projects of Universities in Guangdong Province (2019KZDXM003 and 2020KZDZX1040).

Acknowledgments: We appreciate the provision of SCAU Wushan Campus Teaching & Research Base.

Conflicts of Interest: The authors declare no conflict of interest. The authors themselves are responsible for the content and writing of the paper.

References

1. Andrady, A.L.; Neal, M.A. Applications and societal benefits of plastics. *Philos. Trans. R. Soc. Lond. B Biol. Sci.* **2009**, *364*, 1977–1984. [CrossRef]
2. Thompson, R.C.; Swan, S.H.; Moore, C.J.; vom Saal, F.S. Our plastic age. *Philos. Trans. R. Soc. B* **2009**, *364*, 1973–1976. [CrossRef]
3. Imran, M.; Das, K.R.; Naik, M.M. Co-selection of multi-antibiotic resistance in bacterial pathogens in metal and microplastic contaminated environments: An emerging health threat. *Chemosphere* **2019**, *215*, 846–857. [CrossRef]
4. Law, K.L. Plastics in the Marine Environment. *Ann. Rev. Mar. Sci.* **2017**, *9*, 205–229. [CrossRef]
5. Dris, R.; Imhof, H.; Sanchez, W.; Gasperi, J.; Galgani, F.; Tassin, B.; Laforsch, C. Beyond the ocean: Contamination of freshwater ecosystems with (micro-)plastic particles. *Environ. Chem.* **2015**, *12*. [CrossRef]
6. Hahladakis, J.N.; Velis, C.A.; Weber, R.; Iacovidou, E.; Purnell, P. An overview of chemical additives present in plastics: Migration, release, fate and environmental impact during their use, disposal and recycling. *J. Hazard. Mater.* **2018**, *344*, 179–199. [CrossRef] [PubMed]
7. Plastics Europe. Plastics—The Facts 2020 An Analysis of European Plastics Production, Demand and Waste Data. Plastics Europe. 2020. Available online: https://www.plasticseurope.org/en/resources/publications/4312-plastics-facts-2020 (accessed on 22 January 2021).
8. Rocha-Santos, T.A.P. Editorial overview: Micro and nano-plastics. *Curr. Opin. Environ. Sci. Health* **2018**, *1*, 52–54. [CrossRef]
9. Law, K.L.; Thompson, R.C. Microplastics in the seas. *Oceans* **2014**, *345*, 144–145. [CrossRef] [PubMed]
10. Lebreton, L.; Slat, B.; Ferrari, F.; Sainte-Rose, B.; Aitken, J.; Marthouse, R.; Hajbane, S.; Cunsolo, S.; Schwarz, A.; Levivier, A.; et al. Evidence that the Great Pacific Garbage Patch is rapidly accumulating plastic. *Sci. Rep.* **2018**, *8*. [CrossRef] [PubMed]
11. Wright, S.L.; Thompson, R.C.; Galloway, T.S. The physical impacts of microplastics on marine organisms: A review. *Environ. Pollut.* **2013**, *178*, 483–492. [CrossRef]
12. GESAMP. Sources, Fate and Effects of Microplastics in the Marine Environment: Part 2 of A Global Assessment. 2016. Available online: http://www.gesamp.org/publications/microplastics-in-the-marine-environment-part-2 (accessed on 21 February 2021).
13. Galgani, F.; Hanke, G.; Werner, S.; De Vrees, L. Marine litter within the European Marine Strategy Framework Directive. *ICES J. Mar. Sci.* **2013**, *70*, 1055–1064. [CrossRef]
14. Pellini, G.; Gomiero, A.; Fortibuoni, T.; Ferra, C.; Grati, F.; Tassetti, A.N.; Polidori, P.; Fabi, G.; Scarcella, G. Characterization of microplastic litter in the gastrointestinal tract of Solea solea from the Adriatic Sea. *Environ. Pollut.* **2018**, *234*, 943–952. [CrossRef]
15. Galloway, T.S.; Cole, M.; Lewis, C. Interactions of microplastic debris throughout the marine ecosystem. *Nat. Ecol. Evol.* **2017**, *1*, 116. [CrossRef]
16. Gigault, J.; ter Halle, A.; Baudrimont, M.; Pascal, P.Y.; Gauffre, F.; Phi, T.L.; El Hadri, H.; Grassl, B.; Reynaud, S. Current opinion: What is a nanoplastic? *Environ. Pollut.* **2018**, *235*, 1030–1034. [CrossRef]
17. Hidalgo-Ruz, V.; Gutow, L.; Thompson, R.C.; Thiel, M. Microplastics in the marine environment: A review of the methods used for identification and quantification. *Environ. Sci. Technol.* **2012**, *46*, 3060–3075. [CrossRef]
18. Klein, S.; Worch, E.; Knepper, T.P. Occurrence and Spatial Distribution of Microplastics in River Shore Sediments of the Rhine-Main Area in Germany. *Environ. Sci. Technol.* **2015**, *49*, 6070–6076. [CrossRef]
19. Auta, H.S.; Emenike, C.U.; Fauziah, S.H. Distribution and importance of microplastics in the marine environment: A review of the sources, fate, effects, and potential solutions. *Environ. Int.* **2017**, *102*, 165–176. [CrossRef]
20. Duis, K.; Coors, A. Microplastics in the aquatic and terrestrial environment: Sources (with a specific focus on personal care products), fate and effects. *Environ. Sci. Eur.* **2016**, *28*, 2. [CrossRef]
21. Revel, M.; Chatel, A.; Mouneyrac, C. Micro (nano) plastics: A threat to human health? *Curr. Opin. Environ. Sci. Health* **2018**, *1*, 17–23. [CrossRef]
22. Rauscher, H.; Sokull-Kluttgen, B.; Stamm, H. The European Commission's recommendation on the definition of nanomaterial makes an impact. *Nanotoxicology* **2013**, *7*, 1195–1197. [CrossRef] [PubMed]
23. Eriksen, M.; Mason, S.; Wilson, S.; Box, C.; Zellers, A.; Edwards, W.; Farley, H.; Amato, S. Microplastic pollution in the surface waters of the Laurentian Great Lakes. *Mar. Pollut. Bull.* **2013**, *77*, 177–182. [CrossRef] [PubMed]
24. Cozar, A.; Marti, E.; Duarte, C.M.; Garcia-de-Lomas, J.; van Sebille, E.; Ballatore, T.J.; Eguiluz, V.M.; Gonzalez-Gordillo, J.I.; Pedrotti, M.L.; Echevarria, F.; et al. The Arctic Ocean as a dead end for floating plastics in the North Atlantic branch of the Thermohaline Circulation. *Sci. Adv.* **2017**, *3*. [CrossRef]
25. Reddy, M.S.; Basha, S.; Adimurthy, S.; Ramachandraiah, G. Description of the small plastics fragments in marine sediments along the Alang-Sosiya ship-breaking yard, India. *Estuar. Coast. Shelf Sci.* **2006**, *68*, 656–660. [CrossRef]
26. Bergmann, M.; Sandhop, N.; Schewe, I.; D'Hert, D. Observations of floating anthropogenic litter in the Barents Sea and Fram Strait, Arctic. *Polar Biol.* **2016**, *39*, 553–560. [CrossRef]

27. Peeken, I.; Primpke, S.; Beyer, B.; Gutermann, J.; Katlein, C.; Krumpen, T.; Bergmann, M.; Hehemann, L.; Gerdts, G. Arctic sea ice is an important temporal sink and means of transport for microplastic. *Nat. Commun.* **2018**, *9*. [CrossRef]
28. Waller, C.L.; Griffiths, H.J.; Waluda, C.M.; Thorpe, S.E.; Loaiza, I.; Moreno, B.; Pacherres, C.O.; Hughes, K.A. Microplastics in the Antarctic marine system: An emerging area of research. *Sci. Total Environ.* **2017**, *598*, 220–227. [CrossRef]
29. do Sul, J.A.I.; Spengler, A.; Costa, M.F. Here, there and everywhere. Small plastic fragments and pellets on beaches of Fernando de Noronha (Equatorial Western Atlantic). *Mar. Pollut. Bull.* **2009**, *58*, 1236–1238. [CrossRef] [PubMed]
30. Claessens, M.; De Meester, S.; Van Landuyt, L.; De Clerck, K.; Janssen, C.R. Occurrence and distribution of microplastics in marine sediments along the Belgian coast. *Mar. Pollut. Bull.* **2011**, *62*, 2199–2204. [CrossRef]
31. Martin, J.; Lusher, A.; Thompson, R.C.; Morley, A. The Deposition and Accumulation of Microplastics in Marine Sediments and Bottom Water from the Irish Continental Shelf. *Sci. Rep.* **2017**, *7*, 10772. [CrossRef] [PubMed]
32. Taylor, M.L.; Gwinnett, C.; Robinson, L.F.; Woodall, L.C. Plastic microfibre ingestion by deep-sea organisms. *Sci. Rep.* **2016**, *6*, 33997. [CrossRef] [PubMed]
33. Van Cauwenberghe, L.; Vanreusel, A.; Mees, J.; Janssen, C.R. Microplastic pollution in deep-sea sediments. *Environ. Pollut.* **2013**, *182*, 495–499. [CrossRef] [PubMed]
34. van Sebille, E.; Wilcox, C.; Lebreton, L.; Maximenko, N.; Hardesty, B.D.; van Franeker, J.A.; Eriksen, M.; Siegel, D.; Galgani, F.; Law, K.L. A global inventory of small floating plastic debris. *Environ. Res. Lett.* **2015**, *10*. [CrossRef]
35. Goldstein, M.C.; Titmus, A.J.; Ford, M. Scales of spatial heterogeneity of plastic marine debris in the northeast pacific ocean. *PLoS ONE* **2013**, *8*, e80020. [CrossRef]
36. Wright, S.L.; Rowe, D.; Thompson, R.C.; Galloway, T.S. Microplastic ingestion decreases energy reserves in marine worms. *Curr. Biol.* **2013**, *23*, R1031–R1033. [CrossRef] [PubMed]
37. Lusher, A.L.; McHugh, M.; Thompson, R.C. Occurrence of microplastics in the gastrointestinal tract of pelagic and demersal fish from the English Channel. *Mar. Pollut. Bull.* **2013**, *67*, 94–99. [CrossRef] [PubMed]
38. Browne, M.A.; Dissanayake, A.; Galloway, T.S.; Lowe, D.M.; Thompson, R.C. Ingested microscopic plastic translocates to the circulatory system of the mussel, *Mytilus edulis* (L). *Environ. Sci. Technol.* **2008**, *42*, 5026–5031. [CrossRef] [PubMed]
39. von Moos, N.; Burkhardt-Holm, P.; Kohler, A. Uptake and effects of microplastics on cells and tissue of the blue mussel *Mytilus edulis* L. after an experimental exposure. *Environ. Sci. Technol.* **2012**, *46*, 11327–11335. [CrossRef]
40. Wegner, A.; Besseling, E.; Foekema, E.M.; Kamermans, P.; Koelmans, A.A. Effects of nanopolystyrene on the feeding behavior of the blue mussel (*Mytilus edulis* L.). *Environ. Toxicol. Chem.* **2012**, *31*, 2490–2497. [CrossRef] [PubMed]
41. Cole, M.; Lindeque, P.; Fileman, E.; Halsband, C.; Goodhead, R.; Moger, J.; Galloway, T.S. Microplastic ingestion by zooplankton. *Environ. Sci. Technol.* **2013**, *47*, 6646–6655. [CrossRef]
42. Rodriguez, A.; Rodriguez, B.; Nazaret Carrasco, M. High prevalence of parental delivery of plastic debris in Cory's shearwaters (*Calonectris diomedea*). *Mar. Pollut. Bull.* **2012**, *64*, 2219–2223. [CrossRef]
43. Ugolini, A.; Ungherese, G.; Ciofini, M.; Lapucci, A.; Camaiti, M. Microplastic debris in sandhoppers. *Estuar. Coast. Shelf Sci.* **2013**, *129*, 19–22. [CrossRef]
44. Browne, M.A.; Niven, S.J.; Galloway, T.S.; Rowland, S.J.; Thompson, R.C. Microplastic moves pollutants and additives to worms, reducing functions linked to health and biodiversity. *Curr. Biol.* **2013**, *23*, 2388–2392. [CrossRef]
45. Brandts, I.; Teles, M.; Goncalves, A.P.; Barreto, A.; Franco-Martinez, L.; Tvarijonaviciute, A.; Martins, M.A.; Soares, A.; Tort, L.; Oliveira, M. Effects of nanoplastics on Mytilus galloprovincialis after individual and combined exposure with carbamazepine. *Sci. Total Environ.* **2018**, *643*, 775–784. [CrossRef] [PubMed]
46. Avio, C.G.; Gorbi, S.; Milan, M.; Benedetti, M.; Fattorini, D.; d'Errico, G.; Pauletto, M.; Bargelloni, L.; Regoli, F. Pollutants bioavailability and toxicological risk from microplastics to marine mussels. *Environ. Pollut.* **2015**, *198*, 211–222. [CrossRef] [PubMed]
47. Cole, M.; Coppock, R.; Lindeque, P.K.; Altin, D.; Reed, S.; Pond, D.W.; Sorensen, L.; Galloway, T.S.; Booth, A.M. Effects of Nylon Microplastic on Feeding, Lipid Accumulation, and Moulting in a Coldwater Copepod. *Environ. Sci. Technol.* **2019**, *53*, 7075–7082. [CrossRef] [PubMed]
48. Tallec, K.; Huvet, A.; Di Poi, C.; Gonzalez-Fernandez, C.; Lambert, C.; Petton, B.; Le Goic, N.; Berchel, M.; Soudant, P.; Paul-Pont, I. Nanoplastics impaired oyster free living stages, gametes and embryos. *Environ. Pollut.* **2018**, *242*, 1226–1235. [CrossRef] [PubMed]
49. Enfrin, M.; Lee, J.; Gibert, Y.; Basheer, F.; Kong, L.; Dumee, L.F. Release of hazardous nanoplastic contaminants due to microplastics fragmentation under shear stress forces. *J. Hazard. Mater.* **2020**, *384*, 121393. [CrossRef]
50. Lee, W.S.; Cho, H.J.; Kim, E.; Huh, Y.H.; Kim, H.J.; Kim, B.; Kang, T.; Lee, J.S.; Jeong, J. Bioaccumulation of polystyrene nanoplastics and their effect on the toxicity of Au ions in zebrafish embryos. *Nanoscale* **2019**, *11*, 3173–3185. [CrossRef] [PubMed]
51. Trevisan, R.; Voy, C.; Chen, S.; Di Giulio, R.T. Nanoplastics Decrease the Toxicity of a Complex PAH Mixture but Impair Mitochondrial Energy Production in Developing Zebrafish. *Environ. Sci. Technol.* **2019**, *53*, 8405–8415. [CrossRef]
52. Brun, N.R.; van Hage, P.; Hunting, E.R.; Haramis, A.G.; Vink, S.C.; Vijver, M.G.; Schaaf, M.J.M.; Tudorache, C. Polystyrene nanoplastics disrupt glucose metabolism and cortisol levels with a possible link to behavioural changes in larval zebrafish. *Commun. Biol.* **2019**, *2*, 382. [CrossRef]
53. Brun, N.R.; Koch, B.E.V.; Varela, M.; Peijnenburg, W.J.G.M.; Spaink, H.P.; Vijver, M.G. Nanoparticles induce dermal and intestinal innate immune system responses in zebrafish embryos. *Environ. Sci. Nano* **2018**, *5*, 904–916. [CrossRef]

54. Chen, Q.; Yin, D.; Jia, Y.; Schiwy, S.; Legradi, J.; Yang, S.; Hollert, H. Enhanced uptake of BPA in the presence of nanoplastics can lead to neurotoxic effects in adult zebrafish. *Sci. Total Environ.* **2017**, *609*, 1312–1321. [CrossRef] [PubMed]
55. Bergami, E.; Bocci, E.; Vannuccini, M.L.; Monopoli, M.; Salvati, A.; Dawson, K.A.; Corsi, I. Nano-sized polystyrene affects feeding, behavior and physiology of brine shrimp Artemia franciscana larvae. *Ecotoxicol. Environ. Saf.* **2016**, *123*, 18–25. [CrossRef]
56. Andrady, A.L. Microplastics in the marine environment. *Mar. Pollut. Bull.* **2011**, *62*, 1596–1605. [CrossRef] [PubMed]
57. Browne, M.A.; Crump, P.; Niven, S.J.; Teuten, E.; Tonkin, A.; Galloway, T.; Thompson, R. Accumulation of microplastic on shorelines worldwide: Sources and sinks. *Environ. Sci. Technol.* **2011**, *45*, 9175–9179. [CrossRef]
58. Saal, F.S.; Parmigiani, S.; Palanza, P.L.; Everett, L.G.; Ragaini, R. The plastic world: Sources, amounts, ecological impacts and effects on development, reproduction, brain and behavior in aquatic and terrestrial animals and humans Introduction. *Environ. Res.* **2008**, *108*, 127–130. [CrossRef]
59. Lebreton, L.C.M.; van der Zwet, J.; Damsteeg, J.W.; Slat, B.; Andrady, A.; Reisser, J. River plastic emissions to the world's oceans. *Nat. Commun.* **2017**, *8*, 15611. [CrossRef]
60. Mani, T.; Hauk, A.; Walter, U.; Burkhardt-Holm, P. Microplastics profile along the Rhine River. *Sci. Rep.* **2015**, *5*, 17988. [CrossRef] [PubMed]
61. Barnes, D.K.; Galgani, F.; Thompson, R.C.; Barlaz, M. Accumulation and fragmentation of plastic debris in global environments. *Philos. Trans. R. Soc. Lond. B Biol. Sci.* **2009**, *364*, 1985–1998. [CrossRef]
62. Bell, J.D.; Watson, R.A.; Ye, Y. Global fishing capacity and fishing effort from 1950–2012 (vol 18, pg 489, 2017). *Fish Fish.* **2017**, *18*, 792–793. [CrossRef]
63. Watson, R.A.; Cheung, W.W.L.; Anticamara, J.A.; Sumaila, R.U.; Zeller, D.; Pauly, D. Global marine yield halved as fishing intensity redoubles. *Fish Fish.* **2013**, *14*, 493–503. [CrossRef]
64. Thompson, R.C.; Olsen, Y.; Mitchell, R.P.; Davis, A.; Rowland, S.J.; John, A.W.G.; McGonigle, D.; Russell, A.E. Lost at sea: Where is all the plastic? *Science* **2004**, *304*, 838. [CrossRef]
65. Al-Oufi, H.; McLean, E.; Kumar, A.S.; Claereboudt, M.; Al-Habsi, M. The effects of solar radiation upon breaking strength and elongation of fishing nets. *Fish. Res.* **2004**, *66*, 115–119. [CrossRef]
66. Thomas, S.N.; Hridayanathan, C. The effect of natural sunlight on the strength of polyamide 6 multifilament and monofilament fishing net materials. *Fish. Res.* **2006**, *81*, 326–330. [CrossRef]
67. Hinojosa, I.A.; Thiel, M. Floating marine debris in fjords, gulfs and channels of southern Chile. *Mar. Pollut. Bull.* **2009**, *58*, 341–350. [CrossRef]
68. Sheavly, S.B.; Register, K.M. Marine debris & plastics: Environmental concerns, sources, impacts and solutions. *J. Polym. Environ.* **2007**, *15*, 301–305. [CrossRef]
69. Derraik, J.G. The pollution of the marine environment by plastic debris: A review. *Mar. Pollut. Bull.* **2002**, *44*, 842–852. [CrossRef]
70. Cai, L.; Wang, J.; Peng, J.; Tan, Z.; Zhan, Z.; Tan, X.; Chen, Q. Characteristic of microplastics in the atmospheric fallout from Dongguan city, China: Preliminary research and first evidence. *Environ. Sci. Pollut. Res. Int.* **2017**, *24*, 24928–24935. [CrossRef] [PubMed]
71. Cole, M.; Lindeque, P.; Halsband, C.; Galloway, T.S. Microplastics as contaminants in the marine environment: A review. *Mar. Pollut. Bull.* **2011**, *62*, 2588–2597. [CrossRef]
72. Arias-Villamizar, C.A.; Vazquez-Morillas, A. Degradation of Conventional and Oxodegradable High Density Polyethylene in Tropical Aqueous and Outdoor Environments. *Rev. Int. Contam. Ambient.* **2018**, *34*, 137–147. [CrossRef]
73. Cooper, D.A.; Corcoran, P.L. Effects of mechanical and chemical processes on the degradation of plastic beach debris on the island of Kauai, Hawaii. *Mar. Pollut. Bull.* **2010**, *60*, 650–654. [CrossRef]
74. Veiga, J.M.; Fleet, D.; Kinsey, S.; Nilsson, P.; Vlachogianni, T.; Werner, S.; Galgani, F.; Thompson, R.C.; Dagevos, J.; Gago, J.; et al. Identifying Sources of Marine Litter. MSFD GES TG Marine Litter Thematic Report; JRC Technical Reports. 2016. Available online: https://core.ac.uk/download/pdf/81685372.pdf (accessed on 21 February 2021).
75. Chang, M. Reducing microplastics from facial exfoliating cleansers in wastewater through treatment versus consumer product decisions. *Mar. Pollut. Bull.* **2015**, *101*, 330–333. [CrossRef] [PubMed]
76. Fendall, L.S.; Sewell, M.A. Contributing to marine pollution by washing your face: Microplastics in facial cleansers. *Mar. Pollut. Bull.* **2009**, *58*, 1225–1228. [CrossRef]
77. Gregory, M.R. Environmental implications of plastic debris in marine settings–entanglement, ingestion, smothering, hangers-on, hitch-hiking and alien invasions. *Philos. Trans. R. Soc. Lond. B Biol. Sci.* **2009**, *364*, 2013–2025. [CrossRef]
78. Lei, K.; Qiao, F.; Liu, Q.; Wei, Z.; Qi, H.; Cui, S.; Yue, X.; Deng, Y.; An, L. Microplastics releasing from personal care and cosmetic products in China. *Mar. Pollut. Bull.* **2017**, *123*, 122–126. [CrossRef] [PubMed]
79. Castaneda, R.A.; Avlijas, S.; Simard, M.A.; Ricciardi, A. Microplastic pollution in St. Lawrence River sediments. *Can. J. Fish. Aquat. Sci.* **2014**, *71*, 1767–1771. [CrossRef]
80. Napper, I.E.; Bakir, A.; Rowland, S.J.; Thompson, R.C. Characterisation, quantity and sorptive properties of microplastics extracted from cosmetics. *Mar. Pollut. Bull.* **2015**, *99*, 178–185. [CrossRef]
81. Cesa, F.S.; Turra, A.; Baruque-Ramos, J. Synthetic fibers as microplastics in the marine environment: A review from textile perspective with a focus on domestic washings. *Sci. Total Environ.* **2017**, *598*, 1116–1129. [CrossRef]
82. Feldman, D. Polymer weathering: Photo-oxidation. *J. Polym. Environ.* **2002**, *10*, 163–173. [CrossRef]

83. Wagner, M.; Scherer, C.; Alvarez-Munoz, D.; Brennholt, N.; Bourrain, X.; Buchinger, S.; Fries, E.; Grosbois, C.; Klasmeier, J.; Marti, T.; et al. Microplastics in freshwater ecosystems: What we know and what we need to know. *Environ. Sci. Eur.* **2014**, *26*, 12. [CrossRef]

84. Rosenkranz, P.; Chaudhry, Q.; Stone, V.; Fernandes, T.F. A comparison of nanoparticle and fine particle uptake by Daphnia magna. *Environ. Toxicol. Chem.* **2009**, *28*, 2142–2149. [CrossRef] [PubMed]

85. Velzeboer, I.; Kwadijk, C.J.; Koelmans, A.A. Strong sorption of PCBs to nanoplastics, microplastics, carbon nanotubes, and fullerenes. *Environ. Sci. Technol.* **2014**, *48*, 4869–4876. [CrossRef]

86. Eriksen, M.; Lebreton, L.C.; Carson, H.S.; Thiel, M.; Moore, C.J.; Borerro, J.C.; Galgani, F.; Ryan, P.G.; Reisser, J. Plastic Pollution in the World's Oceans: More than 5 Trillion Plastic Pieces Weighing over 250,000 Tons Afloat at Sea. *PLoS ONE* **2014**, *9*, e111913. [CrossRef]

87. Galgani, F.; Hanke, G.; Maes, T. Global Distribution, Composition and Abundance of Marine Litter. In *Marine Anthropogenic Litter*; Springer: Cham, Switzerland, 2015; pp. 29–56. [CrossRef]

88. Peters, C.A.; Bratton, S.P. Urbanization is a major influence on microplastic ingestion by sunfish in the Brazos River Basin, Central Texas, USA. *Environ. Pollut.* **2016**, *210*, 380–387. [CrossRef] [PubMed]

89. Cozar, A.; Echevarria, F.; Gonzalez-Gordillo, J.I.; Irigoien, X.; Ubeda, B.; Hernandez-Leon, S.; Palma, A.T.; Navarro, S.; Garcia-de-Lomas, J.; Ruiz, A.; et al. Plastic debris in the open ocean. *Proc. Natl. Acad. Sci. USA* **2014**, *111*, 10239–10244. [CrossRef]

90. Reisser, J.; Slat, B.; Noble, K.; du Plessis, K.; Epp, M.; Proietti, M.; de Sonneville, J.; Becker, T.; Pattiaratchi, C. The vertical distribution of buoyant plastics at sea: An observational study in the North Atlantic Gyre. *Biogeosciences* **2015**, *12*, 1249–1256. [CrossRef]

91. Jambeck, J.R.; Geyer, R.; Wilcox, C.; Siegler, T.R.; Perryman, M.; Andrady, A.; Narayan, R.; Law, K.L. Plastic waste inputs from land into the ocean. *Science* **2015**, *347*, 768–771. [CrossRef]

92. Courtene-Jones, W.; Quinn, B.; Gary, S.F.; Mogg, A.O.M.; Narayanaswamy, B.E. Microplastic pollution identified in deep-sea water and ingested by benthic invertebrates in the rockall trough, north atlantic ocean. *Environ. Pollut.* 2017. [CrossRef] [PubMed]

93. Peng, X.; Chen, M.; Chen, S.; Dasgupta, S.; Xu, H.; Ta, K.; Du, M.; Li, J.; Guo, Z.; Bai, S. Microplastics contaminate the deepest part of the world's ocean. *Geochem. Perspect. Lett.* **2018**, 1–5. [CrossRef]

94. Lambert, S.; Wagner, M. Environmental performance of bio-based and biodegradable plastics: The road ahead. *Chem. Soc. Rev.* **2017**, *46*, 6855–6871. [CrossRef]

95. Law, K.L.; Moret-Ferguson, S.; Maximenko, N.A.; Proskurowski, G.; Peacock, E.E.; Hafner, J.; Reddy, C.M. Plastic accumulation in the North Atlantic subtropical gyre. *Science* **2010**, *329*, 1185–1188. [CrossRef] [PubMed]

96. Rummel, C.D.; Jahnke, A.; Gorokhova, E.; Kuhnel, D.; Schmitt-Jansen, M. Impacts of Biofilm Formation on the Fate and Potential Effects of Microplastic in the Aquatic Environment. *Environ. Sci. Technol.* **2017**, *4*, 258–267. [CrossRef]

97. Woodall, L.C.; Sanchez-Vidal, A.; Canals, M.; Paterson, G.L.J.; Coppock, R.; Sleight, V.; Calafat, A.; Rogers, A.D.; Narayanaswamy, B.E.; Thompson, R.C. The deep sea is a major sink for microplastic debris. *R. Soc. Open Sci.* **2014**, *1*. [CrossRef] [PubMed]

98. Besseling, E.; Foekema, E.M.; van den Heuvel-Greve, M.J.; Koelmans, A.A. The Effect of Microplastic on the Uptake of Chemicals by the Lugworm *Arenicola marina* (L.) under Environmentally Relevant Exposure Conditions. *Environ. Sci. Technol.* **2017**, *51*, 8795–8804. [CrossRef]

99. Long, M.; Paul-Pont, I.; Hegaret, H.; Moriceau, B.; Lambert, C.; Huvet, A.; Soudant, P. Interactions between polystyrene microplastics and marine phytoplankton lead to species-specific hetero-aggregation. *Environ. Pollut.* **2017**, *228*, 454–463. [CrossRef] [PubMed]

100. Zhang, W.W.; Zhang, S.F.; Wang, J.Y.; Wang, Y.; Mu, J.L.; Wang, P.; Lin, X.Z.; Ma, D.Y. Microplastic pollution in the surface waters of the Bohai Sea, China. *Environ. Pollut.* **2017**, *231*, 541–548. [CrossRef] [PubMed]

101. Kooi, M.; van Nes, E.H.; Scheffer, M.; Koelmans, A.A. Ups and Downs in the Ocean: Effects of Biofouling on Vertical Transport of Microplastics. *Environ. Sci. Technol.* **2017**, *51*, 7963–7971. [CrossRef]

102. Tata, T.; Belabed, B.E.; Bououdina, M.; Bellucci, S. Occurrence and characterization of surface sediment microplastics and litter from North African coasts of Mediterranean Sea: Preliminary research and first evidence. *Sci. Total Environ.* **2020**, *713*, 136664. [CrossRef]

103. Jiang, Y.; Yang, F.; Zhao, Y.; Wang, J. Greenland Sea Gyre increases microplastic pollution in the surface waters of the Nordic Seas. *Sci. Total Environ.* **2020**, *712*, 136484. [CrossRef]

104. Sfriso, A.A.; Tomio, Y.; Rosso, B.; Gambaro, A.; Sfriso, A.; Corami, F.; Rastelli, E.; Corinaldesi, C.; Mistri, M.; Munari, C. Microplastic accumulation in benthic invertebrates in Terra Nova Bay (Ross Sea, Antarctica). *Environ. Int.* **2020**, *137*, 105587. [CrossRef]

105. Mehdinia, A.; Dehbandi, R.; Hamzehpour, A.; Rahnama, R. Identification of microplastics in the sediments of southern coasts of the Caspian Sea, north of Iran. *Environ. Pollut.* **2020**, *258*, 113738. [CrossRef]

106. Falahudin, D.; Cordova, M.R.; Sun, X.; Yogaswara, D.; Wulandari, I.; Hindarti, D.; Arifin, Z. The first occurrence, spatial distribution and characteristics of microplastic particles in sediments from Banten Bay, Indonesia. *Sci. Total Environ.* **2020**, *705*, 135304. [CrossRef]

107. Mataji, A.; Taleshi, M.S.; Balimoghaddas, E. Distribution and Characterization of Microplastics in Surface Waters and the Southern Caspian Sea Coasts Sediments. *Arch. Environ. Contam. Toxicol.* **2020**, *78*, 86–93. [CrossRef] [PubMed]

108. Gomiero, A.; Oysaed, K.B.; Agustsson, T.; van Hoytema, N.; van Thiel, T.; Grati, F. First record of characterization, concentration and distribution of microplastics in coastal sediments of an urban fjord in south west Norway using a thermal degradation method. *Chemosphere* **2019**, *227*, 705–714. [CrossRef] [PubMed]

109. Rose, D.; Webber, M. Characterization of microplastics in the surface waters of Kingston Harbour. *Sci. Total Environ.* **2019**, *664*, 753–760. [CrossRef]

110. Pan, Z.; Guo, H.G.; Chen, H.Z.; Wang, S.M.; Sun, X.W.; Zou, Q.P.; Zhang, Y.B.; Lin, H.; Cai, S.Z.; Huang, J. Microplastics in the Northwestern Pacific: Abundance, distribution, and characteristics. *Sci. Total Environ.* **2019**, *650*, 1913–1922. [CrossRef]

111. Li, J.; Zhang, H.; Zhang, K.N.; Yang, R.J.; Li, R.Z.; Li, Y.F. Characterization, source, and retention of microplastic in sandy beaches and mangrove wetlands of the Qinzhou Bay, China. *Mar. Pollut. Bull.* **2018**, *136*, 401–406. [CrossRef]

112. Lusher, A.L.; Tirelli, V.; O'Connor, I.; Officer, R. Microplastics in Arctic polar waters: The first reported values of particles in surface and sub-surface samples. *Sci. Rep.* **2015**, *5*. [CrossRef]

113. Desforges, J.P.W.; Galbraith, M.; Dangerfield, N.; Ross, P.S. Widespread distribution of microplastics in subsurface seawater in the NE Pacific Ocean. *Mar. Pollut. Bull.* **2014**, *79*, 94–99. [CrossRef]

114. Alimba, C.G.; Faggio, C. Microplastics in the marine environment: Current trends in environmental pollution and mechanisms of toxicological profile. *Environ. Toxicol. Pharmacol.* **2019**, *68*, 61–74. [CrossRef] [PubMed]

115. Guzzetti, E.; Sureda, A.; Tejada, S.; Faggio, C. Microplastic in marine organism: Environmental and toxicological effects. *Environ. Toxicol. Pharmacol.* **2018**, *64*, 164–171. [CrossRef]

116. Savoca, S.; Capillo, G.; Mancuso, M.; Bottari, T.; Crupi, R.; Branca, C.; Romano, V.; Faggio, C.; D'Angelo, G.; Spano, N. Microplastics occurrence in the Tyrrhenian waters and in the gastrointestinal tract of two congener species of seabreams. *Environ. Toxicol. Pharmacol.* **2019**, *67*, 35–41. [CrossRef]

117. Savoca, S.; Capillo, G.; Mancuso, M.; Faggio, C.; Panarello, G.; Crupi, R.; Bonsignore, M.; D'Urso, L.; Compagnini, G.; Neri, F.; et al. Detection of artificial cellulose microfibers in Boops boops from the northern coasts of Sicily (Central Mediterranean). *Sci. Total Environ.* **2019**, *691*, 455–465. [CrossRef] [PubMed]

118. Mao, Y.; Ai, H.; Chen, Y.; Zhang, Z.; Zeng, P.; Kang, L.; Li, W.; Gu, W.; He, Q.; Li, H. Phytoplankton response to polystyrene microplastics: Perspective from an entire growth period. *Chemosphere* **2018**, *208*, 59–68. [CrossRef] [PubMed]

119. Sjollema, S.B.; Redondo-Hasselerharm, P.; Leslie, H.A.; Kraak, M.H.S.; Vethaak, A.D. Do plastic particles affect microalgal photosynthesis and growth? *Aquat. Toxicol.* **2016**, *170*, 259–261. [CrossRef] [PubMed]

120. Besseling, E.; Wegner, A.; Foekema, E.M.; van den Heuvel-Greve, M.J.; Koelmans, A.A. Effects of Microplastic on Fitness and PCB Bioaccumulation by the Lugworm *Arenicola marina* (L.). *Environ. Sci. Technol.* **2013**, *47*, 593–600. [CrossRef]

121. Lee, K.W.; Shim, W.J.; Kwon, O.Y.; Kang, J.H. Size-dependent effects of micro polystyrene particles in the marine copepod Tigriopus japonicus. *Environ. Sci. Technol.* **2013**, *47*, 11278–11283. [CrossRef]

122. Messinetti, S.; Mercurio, S.; Parolini, M.; Sugni, M.; Pennati, R. Effects of polystyrene microplastics on early stages of two marine invertebrates with different feeding strategies. *Environ. Pollut.* **2018**, *237*, 1080–1087. [CrossRef]

123. Nobre, C.R.; Santana, M.F.M.; Maluf, A.; Cortez, F.S.; Cesar, A.; Pereira, C.D.S.; Turra, A. Assessment of microplastic toxicity to embryonic development of the sea urchin *Lytechinus variegatus* (Echinodermata: Echinoidea). *Mar. Pollut. Bull.* **2015**, *92*, 99–104. [CrossRef]

124. Balbi, T.; Camisassi, G.; Montagna, M.; Fabbri, R.; Franzellitti, S.; Carbone, C.; Dawson, K.; Canesi, L. Impact of cationic polystyrene nanoparticles (PS-NH(2)) on early embryo development of Mytilus galloprovincialis: Effects on shell formation. *Chemosphere* **2017**, *186*, 1–9. [CrossRef]

125. Martinez-Gomez, C.; Leon, V.M.; Calles, S.; Gomariz-Olcina, M.; Vethaak, A.D. The adverse effects of virgin microplastics on the fertilization and larval development of sea urchins. *Mar. Environ. Res.* **2017**, *130*, 69–76. [CrossRef] [PubMed]

126. Gandara, E.S.P.P.; Nobre, C.R.; Resaffe, P.; Pereira, C.D.S.; Gusmao, F. Leachate from microplastics impairs larval development in brown mussels. *Water Res.* **2016**, *106*, 364–370. [CrossRef]

127. Sussarellu, R.; Suquet, M.; Thomas, Y.; Lambert, C.; Fabioux, C.; Pernet, M.E.J.; Le Goic, N.; Quillien, V.; Mingant, C.; Epelboin, Y.; et al. Oyster reproduction is affected by exposure to polystyrene microplastics. *Proc. Natl. Acad. Sci. USA* **2016**, *113*, 2430–2435. [CrossRef]

128. Lu, Y.; Zhang, Y.; Deng, Y.; Jiang, W.; Zhao, Y.; Geng, J.; Ding, L.; Ren, H. Uptake and Accumulation of Polystyrene Microplastics in Zebrafish (Danio rerio) and Toxic Effects in Liver. *Environ. Sci. Technol.* **2016**, *50*, 4054–4060. [CrossRef]

129. Yu, P.; Liu, Z.; Wu, D.; Chen, M.; Lv, W.; Zhao, Y. Accumulation of polystyrene microplastics in juvenile Eriocheir sinensis and oxidative stress effects in the liver. *Aquat. Toxicol.* **2018**, *200*, 28–36. [CrossRef] [PubMed]

130. Heindler, F.M.; Alajmi, F.; Huerlimann, R.; Zeng, C.; Newman, S.J.; Vamvounis, G.; van Herwerden, L. Toxic effects of polyethylene terephthalate microparticles and Di(2-ethylhexyl)phthalate on the calanoid copepod, Parvocalanus crassirostris. *Ecotoxicol. Environ. Saf.* **2017**, *141*, 298–305. [CrossRef]

131. Snell, T.W.; Hicks, D.G. Assessing Toxicity of Nanoparticles Using *Brachionus manjavacas* (Rotifera). *Environ. Toxicol.* **2011**, *26*, 146–152. [CrossRef] [PubMed]

132. Della Torre, C.; Bergami, E.; Salvati, A.; Faleri, C.; Cirino, P.; Dawson, K.A.; Corsi, I. Accumulation and Embryotoxicity of Polystyrene Nanoparticles at Early Stage of Development of Sea Urchin Embryos Paracentrotus lividus. *Environ. Sci. Technol.* **2014**, *48*, 12302–12311. [CrossRef]

133. Paul-Pont, I.; Lacroix, C.; Gonzalez Fernandez, C.; Hegaret, H.; Lambert, C.; Le Goic, N.; Frere, L.; Cassone, A.L.; Sussarellu, R.; Fabioux, C.; et al. Exposure of marine mussels *Mytilus* spp. to polystyrene microplastics: Toxicity and influence on fluoranthene bioaccumulation. *Environ. Pollut.* **2016**, *216*, 724–737. [CrossRef]

134. Gomiero, A.; Strafella, P.; Pellini, G.; Salvalaggio, V.; Fabi, G. Comparative Effects of Ingested PVC Micro Particles With and Without Adsorbed Benzo(a)pyrene vs. Spiked Sediments on the Cellular and Sub Cellular Processes of the Benthic Organism Hediste diversicolor. *Front. Mar. Sci.* **2018**, *5*. [CrossRef]

135. Sleight, V.A.; Bakir, A.; Thompson, R.C.; Henry, T.B. Assessment of microplastic-sorbed contaminant bioavailability through analysis of biomarker gene expression in larval zebrafish. *Mar. Pollut. Bull.* **2017**, *116*, 291–297. [CrossRef]

136. Oliveira, M.; Ribeiro, A.; Hylland, K.; Guilhermino, L. Single and combined effects of microplastics and pyrene on juveniles (0+group) of the common goby *Pomatoschistus microps* (Teleostei, Gobiidae). *Ecol. Indic.* **2013**, *34*, 641–647. [CrossRef]

137. Murray, F.; Cowie, P.R. Plastic contamination in the decapod crustacean *Nephrops norvegicus* (Linnaeus, 1758). *Mar. Pollut. Bull.* **2011**, *62*, 1207–1217. [CrossRef]

138. Scherer, C.; Brennholt, N.; Reifferscheid, G.; Wagner, M. Feeding type and development drive the ingestion of microplastics by freshwater invertebrates. *Sci. Rep.* **2017**, *7*. [CrossRef] [PubMed]

139. Setala, O.; Norkko, J.; Lehtiniemi, M. Feeding type affects microplastic ingestion in a coastal invertebrate community. *Mar. Pollut. Bull.* **2016**, *102*, 95–101. [CrossRef]

140. Setala, O.; Fleming-Lehtinen, V.; Lehtiniemi, M. Ingestion and transfer of microplastics in the planktonic food web. *Environ. Pollut.* **2014**, *185*, 77–83. [CrossRef]

141. Watts, A.J.R.; Lewis, C.; Goodhead, R.M.; Beckett, S.J.; Moger, J.; Tyler, C.R.; Galloway, T.S. Uptake and Retention of Microplastics by the Shore Crab Carcinus maenas. *Environ. Sci. Technol.* **2014**, *48*, 8823–8830. [CrossRef]

142. Romeo, T.; Pietro, B.; Peda, C.; Consoli, P.; Andaloro, F.; Fossi, M.C. First evidence of presence of plastic debris in stomach of large pelagic fish in the Mediterranean Sea. *Mar. Pollut. Bull.* **2015**, *95*, 358–361. [CrossRef]

143. Lunov, O.; Syrovets, T.; Loos, C.; Beil, J.; Delacher, M.; Tron, K.; Nienhaus, G.U.; Musyanovych, A.; Mailander, V.; Landfester, K.; et al. Differential uptake of functionalized polystyrene nanoparticles by human macrophages and a monocytic cell line. *ACS Nano* **2011**, *5*, 1657–1669. [CrossRef]

144. Mazurais, D.; Ernande, B.; Quazuguel, P.; Severe, A.; Huelvan, C.; Madec, L.; Mouchel, O.; Soudant, P.; Robbens, J.; Huvet, A.; et al. Evaluation of the impact of polyethylene microbeads ingestion in European sea bass (*Dicentrarchus labrax*) larvae. *Mar. Environ. Res.* **2015**, *112*, 78–85. [CrossRef] [PubMed]

145. Rolton, A.; Vignier, J.; Volety, A.K.; Pierce, R.H.; Henry, M.; Shumway, S.E.; Bricelj, V.M.; Hegaret, H.; Soudant, P. Effects of field and laboratory exposure to the toxic dinoflagellate Karenia brevis on the reproduction of the eastern oyster, Crassostrea virginica, and subsequent development of offspring. *Harmful Algae* **2016**, *57*, 13–26. [CrossRef]

146. Collard, F.; Gilbert, B.; Compere, P.; Eppe, G.; Das, K.; Jauniaux, T.; Parmentier, E. Microplastics in livers of European anchovies (*Engraulis encrasicolus*, L.). *Environ. Pollut.* **2017**, *229*, 1000–1005. [CrossRef]

147. Kolandhasamy, P.; Su, L.; Li, J.; Qu, X.; Jabeen, K.; Shi, H. Adherence of microplastics to soft tissue of mussels: A novel way to uptake microplastics beyond ingestion. *Sci. Total Environ.* **2018**, *610–611*, 635–640. [CrossRef]

148. Collignon, A.; Hecq, J.H.; Glagani, F.; Voisin, P.; Collard, F.; Goffart, A. Neustonic microplastic and zooplankton in the North Western Mediterranean Sea. *Mar. Pollut. Bull.* **2012**, *64*, 861–864. [CrossRef]

149. Welden, N.A.; Abylkhani, B.; Howarth, L.M. The effects of trophic transfer and environmental factors on microplastic uptake by plaice, Pleuronectes plastessa, and spider crab, Maja squinado. *Environ. Pollut.* **2018**, *239*, 351–358. [CrossRef] [PubMed]

150. Wardrop, P.; Shimeta, J.; Nugegoda, D.; Morrison, P.D.; Miranda, A.; Tang, M.; Clarke, B.O. Chemical Pollutants Sorbed to Ingested Microbeads from Personal Care Products Accumulate in Fish. *Environ. Sci. Technol.* **2016**, *50*, 4037–4044. [CrossRef]

151. Bottari, T.; Savoca, S.; Mancuso, M.; Capillo, G.; GiuseppePanarello, G.; MartinaBonsignore, M.; Crupi, R.; Sanfilippo, M.; D'Urso, L.; Compagnini, G.; et al. Plastics occurrence in the gastrointestinal tract of *Zeus faber* and *Lepidopus caudatus* from the Tyrrhenian Sea. *Mar. Pollut. Bull.* **2019**, *146*, 408–416. [CrossRef]

152. Capillo, G.; Savoca, S.; Panarello, G.; Mancuso, M.; Branca, C.; Romano, V.; D'Angelo, G.; Bottari, T.; Spano, N. Quali-quantitative analysis of plastics and synthetic microfibers found in demersal species from Southern Tyrrhenian Sea (Central Mediterranean). *Mar. Pollut. Bull.* **2020**, *150*, 110596. [CrossRef]

153. Barboza, L.G.A.; Vieira, L.R.; Branco, V.; Carvalho, C.; Guilhermino, L. Microplastics increase mercury bioconcentration in gills and bioaccumulation in the liver, and cause oxidative stress and damage in Dicentrarchus labrax juveniles. *Sci. Rep.* **2018**, *8*, 15655. [CrossRef]

154. Ogonowski, M.; Schur, C.; Jarsen, A.; Gorokhova, E. The Effects of Natural and Anthropogenic Microparticles on Individual Fitness in Daphnia magna. *PLoS ONE* **2016**, *11*. [CrossRef]

155. Ward, J.E.; Kach, D.J. Marine aggregates facilitate ingestion of nanoparticles by suspension-feeding bivalves. *Mar. Environ. Res* **2009**, *68*, 137–142. [CrossRef]

156. Rochman, C.M.; Hoh, E.; Kurobe, T.; Teh, S.J. Ingested plastic transfers hazardous chemicals to fish and induces hepatic stress. *Sci. Rep.* **2013**, *3*, 3263. [CrossRef]

157. Teuten, E.L.; Saquing, J.M.; Knappe, D.R.; Barlaz, M.A.; Jonsson, S.; Bjorn, A.; Rowland, S.J.; Thompson, R.C.; Galloway, T.S.; Yamashita, R.; et al. Transport and release of chemicals from plastics to the environment and to wildlife. *Philos. Trans. R. Soc. Lond. B Biol. Sci.* **2009**, *364*, 2027–2045. [CrossRef]

158. Wen, B.; Jin, S.R.; Chen, Z.Z.; Gao, J.Z.; Liu, Y.N.; Liu, J.H.; Feng, X.S. Single and combined effects of microplastics and cadmium on the cadmium accumulation, antioxidant defence and innate immunity of the discus fish (*Symphysodon aequifasciatus*). *Environ. Pollut.* **2018**, *243*, 462–471. [CrossRef]

159. Ma, Y.; Huang, A.; Cao, S.; Sun, F.; Wang, L.; Guo, H.; Ji, R. Effects of nanoplastics and microplastics on toxicity, bioaccumulation, and environmental fate of phenanthrene in fresh water. *Environ. Pollut.* **2016**, *219*, 166–173. [CrossRef]

160. Diepens, N.J.; Koelmans, A.A. Accumulation of Plastic Debris and Associated Contaminants in Aquatic Food Webs. *Environ. Sci. Technol.* **2018**, *52*, 8510–8520. [CrossRef]

161. Eerkes-Medrano, D.; Thompson, R.C.; Aldridge, D.C. Microplastics in freshwater systems: A review of the emerging threats, identification of knowledge gaps and prioritisation of research needs. *Water Res.* **2015**, *75*, 63–82. [CrossRef]

162. Lusher, A.L.; Burke, A.; O'Connor, I.; Officer, R. Microplastic pollution in the Northeast Atlantic Ocean: Validated and opportunistic sampling. *Mar. Pollut. Bull.* **2014**, *88*, 325–333. [CrossRef]

163. Gonzalez-Fernandez, C.; Toullec, J.; Lambert, C.; Le Goic, N.; Seoane, M.; Moriceau, B.; Huvet, A.; Berchel, M.; Vincent, D.; Courcot, L.; et al. Do transparent exopolymeric particles (TEP) affect the toxicity of nanoplastics on Chaetoceros neogracile? *Environ. Pollut.* **2019**, *250*, 873–882. [CrossRef]

164. Jeong, C.B.; Kang, H.M.; Lee, Y.H.; Kim, M.S.; Lee, J.S.; Seo, J.S.; Wang, M.; Lee, J.S. Nanoplastic Ingestion Enhances Toxicity of Persistent Organic Pollutants (POPs) in the Monogonont Rotifer Brachionus koreanus via Multixenobiotic Resistance (MXR) Disruption. *Environ. Sci. Technol.* **2018**, *52*, 11411–11418. [CrossRef]

165. Almeida, M.; Martins, M.A.; Soares, A.M.V.; Cuesta, A.; Oliveira, M. Polystyrene nanoplastics alter the cytotoxicity of human pharmaceuticals on marine fish cell lines. *Environ. Toxicol. Pharmacol.* **2019**, *69*, 57–65. [CrossRef]

166. Oberbeckmann, S.; Kreikemeyer, B.; Labrenz, M. Environmental Factors Support the Formation of Specific Bacterial Assemblages on Microplastics. *Front. Microbiol.* **2018**, *8*. [CrossRef]

167. Debroas, D.; Mone, A.; Ter Halle, A. Plastics in the North Atlantic garbage patch: A boat-microbe for hitchhikers and plastic degraders. *Sci. Total Environ.* **2017**, *599*, 1222–1232. [CrossRef]

168. Dussud, C.; Hudec, C.; George, M.; Fabre, P.; Higgs, P.; Bruzaud, S.; Delort, A.M.; Eyheraguibel, B.; Meistertzheim, A.L.; Jacquin, J.; et al. Colonization of Non-biodegradable and Biodegradable Plastics by Marine Microorganisms. *Front. Microbiol.* **2018**, *9*, 1571. [CrossRef]

169. Zettler, E.R.; Mincer, T.J.; Amaral-Zettler, L.A. Life in the "Plastisphere": Microbial Communities on Plastic Marine Debris. *Environ. Sci. Technol.* **2013**, *47*, 7137–7146. [CrossRef]

170. McCormick, A.; Hoellein, T.J.; Mason, S.A.; Schluep, J.; Kelly, J.J. Microplastic is an Abundant and Distinct Microbial Habitat in an Urban River. *Environ. Sci. Technol.* **2014**, *48*, 11863–11871. [CrossRef]

171. Oberbeckmann, S.; Loeder, M.G.J.; Gerdts, G.; Osborn, A.M. Spatial and seasonal variation in diversity and structure of microbial biofilms on marine plastics in Northern European waters. *FEMS Microbiol. Ecol.* **2014**, *90*, 478–492. [CrossRef]

172. Reisser, J.; Shaw, J.; Hallegraeff, G.; Proietti, M.; Barnes, D.K.A.; Thums, M.; Wilcox, C.; Hardesty, B.D.; Pattiaratchi, C. Millimeter-Sized Marine Plastics: A New Pelagic Habitat for Microorganisms and Invertebrates. *PLoS ONE* **2014**, *9*. [CrossRef]

173. Amaral-Zettler, L.A.; Zettler, E.R.; Slikas, B.; Boyd, G.D.; Melvin, D.W.; Morrall, C.E.; Proskurowski, G.; Mincer, T.J. The biogeography of the Plastisphere: Implications for policy. *Front. Ecol. Environ.* **2015**, *13*, 541–546. [CrossRef]

174. Webb, H.K.; Crawford, R.J.; Sawabe, T.; Ivanova, E.P. Poly(ethylene terephthalate) Polymer Surfaces as a Substrate for Bacterial Attachment and Biofilm Formation. *Microbes Environ.* **2009**, *24*, 39–42. [CrossRef]

175. Muthukumar, T.; Aravinthan, A.; Lakshmi, K.; Venkatesan, R.; Vedaprakash, L.; Doble, M. Fouling and stability of polymers and composites in marine environment. *Int. Biodeter. Biodegr.* **2011**, *65*, 276–284. [CrossRef]

176. Shah, A.A.; Hasan, F.; Hameed, A.; Ahmed, S. Biological degradation of plastics: A comprehensive review. *Biotechnol. Adv.* **2008**, *26*, 246–265. [CrossRef]

177. Negoro, S. Biodegradation of nylon oligomers. *Appl. Microbiol. Biotechnol.* **2000**, *54*, 461–466. [CrossRef] [PubMed]

178. Yang, Y.; Yang, J.; Wu, W.-M.; Zhao, J.; Song, Y.; Gao, L.; Yang, R.; Jiang, L. Biodegradation and Mineralization of Polystyrene by Plastic-Eating Mealworms: Part 1. Chemical and Physical Characterization and Isotopic Tests. *Environ. Sci. Technol.* **2015**, *49*, 12080–12086. [CrossRef] [PubMed]

179. Yang, Y.; Yang, J.; Wu, W.M.; Zhao, J.; Song, Y.; Gao, L.; Yang, R.; Jiang, L. Biodegradation and Mineralization of Polystyrene by Plastic-Eating Mealworms: Part 2. Role of Gut Microorganisms. *Environ. Sci. Technol.* **2015**, *49*, 12087–12093. [CrossRef] [PubMed]

180. Yang, J.; Yang, Y.; Wu, W.M.; Zhao, J.; Jiang, L. Evidence of polyethylene biodegradation by bacterial strains from the guts of plastic-eating waxworms. *Environ. Sci. Technol.* **2014**, *48*, 13776–13784. [CrossRef]

181. Balasubramanian, V.; Natarajan, K.; Hemambika, B.; Ramesh, N.; Sumathi, C.S.; Kottaimuthu, R.; Rajesh Kannan, V. High-density polyethylene (HDPE)-degrading potential bacteria from marine ecosystem of Gulf of Mannar, India. *Lett. Appl. Microbiol.* **2010**, *51*, 205–211. [CrossRef]

182. Nakajima-Kambe, T.; Onuma, F.; Kimpara, N.; Nakahara, T. Isolation and characterization of a bacterium which utilizes polyester polyurethane as a sole carbon and nitrogen source. *FEMS Microbiol. Lett.* **1995**, *129*, 39–42. [CrossRef] [PubMed]

183. Nakajima-Kambe, T.; Onuma, F.; Kimpara, N.; Nakahara, T. Determination of the polyester polyurethane breakdown products and distribution of the polyurethane degrading enzyme of Comamonas acidovorans, strain TB-35. *J. Food Sci. Technol.* **1997**, *83*, 456–460. [CrossRef]

184. Peng, Y.H.; Shih, Y.H.; Lai, Y.C.; Liu, Y.Z.; Liu, Y.T.; Lin, N.C. Degradation of polyurethane by bacterium isolated from soil and assessment of polyurethanolytic activity of a Pseudomonas putida strain. *Environ. Sci. Pollut. Res. Int.* **2014**, *21*, 9529–9537. [CrossRef]
185. Peng, R.T.; Qin, L.F.; Yang, Y. Biodegradation of polyurethane by the spacecraft-inhabiting bacterium [C]. *China Soc. Biotechnol. Young Sci. F* **2018**, *9*. [CrossRef]
186. De Tender, C.A.; Devriese, L.I.; Haegeman, A.; Maes, S.; Ruttink, T.; Dawyndt, P. Bacterial Community Profiling of Plastic Litter in the Belgian Part of the North Sea. *Environ. Sci. Technol.* **2015**, *49*, 9629–9638. [CrossRef] [PubMed]
187. Klaeger, F.; Tagg, A.S.; Otto, S.; Biemuller, M.; Sartorius, I.; Labrenz, M. Residual Monomer Content Affects the Interpretation of Plastic Degradation. *Sci. Rep.* **2019**, *9*. [CrossRef] [PubMed]
188. Harrison, J.P.; Boardman, C.; O'Callaghan, K.; Delort, A.M.; Song, J. Biodegradability standards for carrier bags and plastic films in aquatic environments: A critical review. *R. Soc. Open Sci.* **2018**, *5*, 171792. [CrossRef] [PubMed]

MDPI

St. Alban-Anlage 66

4052 Basel

Switzerland

Tel. +41 61 683 77 34

Fax +41 61 302 89 18

www.mdpi.com

Toxics Editorial Office

E-mail: toxics@mdpi.com

www.mdpi.com/journal/toxics